DEEP-LEVEL GEODYNAMICS

DEEP-LEVEL GEODYNAMICS

N.L. DOBRETSOV
A.G. KIRDYASHKIN
United Institute of Geology,
Geophysics and Mineralogy
Russian Academy of Sciences
Siberian Branch
Novosibirsk
Russia

A.A. BALKEMA/ROTTERDAM/BROOKFIELD/1998

Translation of: *Glubinnaya geodinamika*, Russian Academy of Sciences, Siberian Branch, Novosibirsk, 1994. Revised and updated for English edition in 1997.

A.A. Balkema, P.O. Box 1675, 3000 BR Rotterdam, Netherlands
Fax: +31.10.4135947; E-mail: balkema@balkema.nl
Internet site: http://www.balkema.nl

Distributed in USA and Canada by
A.A. Balkema Publishers, Old Post Road, Brookfield, VT 05036-9704, USA
Fax: 802.276.3837; E-mail: Info@ashgate.com

ISBN 90 5410 734 0

Preface to the English Edition

This book, translated from Russian for the benefit of English-speaking readers, represents the English version of the book published in Russian by Novosibirsk in late 1994. In this English edition, Chapters 1, 4 and 5 have been thoroughly revised and Chapters 2 and 3 slightly modified on the basis of papers and books published during 1994 to 1997 and, therefore, not covered in the Russian edition. The Literature Cited has been supplemented with a hundred more new titles, including papers by Dobretsov and Kirdyashkin on unstable convection in the mantle, modelling of subduction processes, ultrahigh-pressure metamorphic complexes (coauthored with N.V. Sobolev and V.S. Shatsky) and problems of tectonic exhumation of rocks.

Chapter 1 has been supplemented with Table 1 and Fig. 1 showing the composition of the Earth and its envelopes. In Chapter 4, sections dealing with mantle convection, geological examples of 'plumes', unstable convection of the lower mantle and interaction of the lower mantle and core have been modified and broadened with new illustrations. In Chapter 5, sections on melting and metamorphism in subduction zones and models for collisional processes have been rewritten, with invaluable assistance from Prof. Sh. Maruyama.

In general, the book deals with original models and hypotheses developed by the authors and partly investigated experimentally, on the problems of deep-level geodynamics. The authors have also reviewed recent publications covering these aspects, including many works by Russian geoscientists which are not adequately covered in the English literature (Chapters 1, 3, 4 and 5). The principal concepts of thermal physics used in the book are presented in Chapters 2 and 3. The main feature of the book is a combination of geological-petrological data with thermophysical calculations and experimental data.

Contents

Introduction

This book represents the first attempt at generalising the recent information on geodynamics in general and deep-level geodynamics, studied in particular by the authors both theoretically and experimentally and published in recent years in Russia and elsewhere. The course material developed for students of the geology-geophysics faculty of Novosibirsk State University during 1992 to 1994 has also been used in this work.

The very title of this publication, Deep-level Geodynamics, suggests that attention will be mainly devoted to the subcrustal processes occurring at the lithosphere-asthenosphere boundary in the Earth's active zones and at the boundaries of upper-lower mantle and lower mantle-core. Two books published in Russia have dealt with geodynamics in detail: *Introduction to Geodynamics* (Zonenshain and Savostin, 1979) and *Palaeodynamics* (Zonenshain and Kuz'min, 1993a); only the latter has partly examined the aspects discussed in the present publication. It also recommends the definition of geodynamics as a science dealing with deep-level forces and processes arising as a result of the evolution of the Earth as a planet and causing the movement of material masses and energy in the interior of the Earth and its upper envelopes. The term geodynamics is used in a broad sense and includes general or deep-level geodynamics as well as special geodynamics developed on the basis of lithospheric plate tectonics (Zonenshain and Kuz'min, 1993a, p. 3).

Based on an integrated analysis of geological and geochemical data, Zonenshain and Kuz'min (1983, 1993a, 1993b) and Zonenshain et al. (1991) proposed a two-level model of tectonic structure and development of the Earth: a) lithospheric plate tectonics reflects plate movements responsible for the sealing and opening up of oceans and the formation of folded belts and mountains and is associated with the activity of the upper mantle (primarily of the asthenosphere) and b) tectonics of hot spots aid in the investigation of the evolution of deeper envelopes (primarily uplift of the plumes from the core-mantle boundary) and explains the general trend and cyclic nature of the Earth's development.

Independent of the above studies, several authors (Dobretsov and Kirdyashkin, 1991, 1992, 1993a, b, 1994; and Dobretsov et al., 1993), using petrological data and experimental and numerical modelling, have developed the model of two-layer (or two-level) mantle convection with simultaneous active influence of mantle

2

plumes of three levels rising from the boundaries of core-mantle, lower-upper mantle and plates subducted into the upper mantle.

Some American and Japanese scientists have also recently developed similar models. In a special issue of the *Journal of the Geological Society of Japan*, Maruyama (1994) and Fukao et al. (1994) have substantiated a 'new paradigm' on the structure, evolution and geodynamics of the Earth and planets of the terrestrial group, combining plate and plume tectonics. These studies show that 'the hot spot tectonics' developed by L.P. Zonenshain and N.I. Kuz'min since 1983 have been replaced (with no reference to the aforesaid authors whatsoever) by plume tectonics, following several articles of foreign authors on mantle plumes and superplumes (Morgan, 1971; O'Nions et al., 1979; Larson, 1991; Cazenave and Thoraval, 1994; and others).

A comparison of the results of our investigations with the published data of L.P. Zonenshain and M.I. Kuz'min, the latter developed on an altogether different factual base, will confirm that the main conclusions coincide or supplement each other. These were formulated as hypotheses and presented by N.L. Dobretsov and M.I. Kuz'min at the First International Symposium organised in Honour of L.P. Zonenshain, and held at Zvenigorod in October, 1993.

1) Geological, geochemical and experimental data confirm the two-layer mantle convection model together with two-stage mantle plumes.

2) Plate tectonics is restricted to the upper envelopes of the Earth and is related to convection flows in the asthenosphere and to the intense influence of subduction processes which are correlated with downward flows in the lower mantle.

3) Plume tectonics (hot spots) reflects the processes in the lower mantle and is associated with rising flows and (or) mantle plumes generated at the core-mantle boundary and transformed later at the upper-lower mantle boundary.

4) Experimental and numerical modelling indicates the possible thermogravitational nature of mantle plumes with short (less than five million years) duration of their rising to the surface and substantiates that conditions of their stability are correlatable with geological observations.

The material presented in this book is not, however, confined exclusively to substantiating the above hypotheses. Pursuing an academic objective, we have attempted to incorporate the principles of physical modelling and its utilisation for solving key problems of geodynamics such as structure and formation mechanism of mid-oceanic ridges, subduction zone dynamics, two-layer convection in the upper and lower mantle and possible models and parameters of mantle plumes. The first chapter, introductory, briefly systematises the fundamental information available on the structure and development of the Earth.

To emphasise the importance of the aspects discussed, the priorities in the field of Solid Earth Sciences assigned by the US Academy of Sciences are listed in the Table below (Wyllie et al., 1993).

It can be seen from this Table that in the category of fundamental investigations (A-1) to (A-5), geodynamic investigations (A-4) and (A-5) are assigned a priority of 24% out of a total of 32% for this group. Among them, the investigation of active crustal formations, study of crustal evolution, study of mantle convection and nature of core-mantle boundary form the subject matter of discussions in the present book too. Other fundamental investigations (fluid pressure and composition

Priorities of investigations in solid Earth sciences (Wyllie et al., 1993)

	Topmost priority		High priority
	A-1. Global changes in palaeoenvironment and biological evolution (2%)		
a)	Preceding 2.5 Ma	b) c)	Preceding 150 Ma Earlier to the preceding 150 Ma
	A-2. Global geochemical and biogeochemical cycles (4%)		
a)	Biochemistry and evolution of rock cycles	b) c)	Model of interactions between cycles Effect of geochemical cycles on contemporary world
	A-3. Fluids in and on the Earth (2%)		
a)	Fluid pressure and composition of fluids in the crust	b) c)	Fluid flow in sedimentary basins Microbiological effect on the chemistry of fluids
	A-4. Crustal dynamics in oceans and on continents (19%)		
a)	Active crustal deformations	b) c)	Study of crustal evolution Dependence of topography on climate, tectonics and hydrology
	A-5. Core and mantle dynamics (5%)		
a)	Mantle convection	b) c)	Nature of core-mantle boundary Origin and variation of magnetic field
	B. Identifying essential natural resources (53%)		
a)	National water resources of required quality	b) c) d)	Study of sedimentary basins Thermodynamics and kinematics of water-rock interactions Strategy of developing energy and mineral resources
	C. Predicting geological disasters (4%)		
a)	Defining and characterising regions of seismic disasters	b) c)	Defining and characterising regions of major landslides Defining and characterising potential volcanic disasters
	D. Minimising the effects of global environmental variations (11%)		
a)	Purification of contaminated groundwaters by microbiochemical methods	b) c)	Burying toxic and radioactive wastes Geochemistry and health of man

A to D—Titles of general directions and themes of topmost (a) and high (b and c) priorities. The percentages given in parentheses indicate the approximate quantum of researches assigned by the US federal agencies during 1990 to 1993.

4

of fluids in the crust, models of interaction of geo- and biogeochemical cycles, etc.) are related to the study of geodynamic problems. No less important is the fact that, without solving these problems, it is impossible to solve the applied priority problems, such as origin of sedimentary (including oil and gas) basins, defining the strategy for developing energy and mineral resources or predicting the regions of seismic and other natural disasters.

Similar assessments undertaken in Japan, Germany and other industrial countries mostly coincide with the priorities indicated in the Table above. Unfortunately, no such assessments have been made in Russia thus far although our survey reports for the institutes of the Russian Academy of Sciences do not contradict the above-stated priorities.

The Introduction and Chapters 1 and 5 (Sections 5.1, 5.3 to 5.7, 5.10 and 5.11) were authored by N.L. Dobretsov, Chapters 2 and 3 by A.G. Kirdyashkin and Chapter 4 and the rest of Chapter 5 jointly by the two authors. I.N. Gladkov and A.A. Kirdyashkin took part in the experimental investigations while Ch.B. Borukaev and M.I. Kuz'min, Corresponding Members of the Russian Academy of Sciences, as well as I.V. Aschepkov, N.A. Berzin and M.M. Buslov were involved in collecting material for the book and its discussion.

The authors are sincerely grateful to all their colleagues who helped in the preparation, discussion and publication of this book. They thank V.N. Sharapov, S. Sherman and A.N. Cherepanov, in particular for constructive comments.

Both the Russian and English editions of this book were supported, in part, by the Russian Foundation of Fundamental Research (Grant nos. 93-05-14021 and 96-05–66049).

1

General Information about the Structure and Dynamics of the Earth

This introductory chapter provides fundamental information about the structure and dynamics of the Earth compared to the other terrestrial planets, basic characteristics of the structure and composition of its crust and deeper envelopes, preliminary information on the nature of geodynamic processes and their intensity, as well as a brief description of plate tectonics and its relation to convective movements in the upper mantle. The fundamental physical parameters used by the authors for designing the models presented later in the book are also described here.

1.1. Origin and Internal Structure of the Earth and Terrestrial Planets

In the solar system, the Earth falls among the terrestrial group of planets between Venus and Mars and has a large satellite, the Moon, whose origin continues to be debated (Galimov, 1995). The fundamental parameters of terrestrial planets are shown in Table 1.1.

These parameters vary systematically with increase in distance of the planet from the Sun and correspond to Thesius-Bode's empirical law of planetary distances (Bott, 1971). It is thought that planets were formed by the accretion of cosmic dust near the Sun, probably in several stages. There are other hypotheses too, in particular the comet hypothesis of Marakushev (1992).

The first stage of accretion may have been a short-duration (about 10^5 years) gravitational collapse of protoplanetary condensation (Cameron, 1978; Safronov, 1995) with the formation of primary planetesimal bodies with diameters of up to hundreds of kilometres. In the second stage, extending for a much longer period, of the order of 10^8 years, collisions and aggregations of planetesimals resulted in the formation of two groups of planets: the inner group including the Earth described in Table 1.1 and the outer group (Jupiter, Saturn and others) with a belt of asteroids between them (Vityazev et al., 1990). The Schmidt hypothesis of slow cold accretion prevailed for a long time. The most widely accepted concept at present is the significant heating of the Earth right up to the melting of the outer envelope, even in the stage of accretion, as a result of heat released during the collisions of planetesimals, especially those of large diameter (Hayashi et al., 1979).

Table 1.1. Terrestrial group of planets*

Planet	Average distance from Sun, million km	Period of revolution around the Sun, days	Equatorial radius, km	Period of rotation around axis	Mass	Average density	Mass of core, % of total planet mass	Fe (+ Ni) in core, %	Major elements in general (in descending order of abundance)	Composition of atmosphere
				relative to Earth						
Mercury	57.91	87.97	2439	59	0.054	5.45	91	68	Fe, Ni	Not present
Venus	108.21	224.70	6052	(243.3)	0.815	5.03	35	35	Fe, O, Si, Mg, Al	CO_2 (SO_3, O)
Earth**	149.60	365.26	6378	1.0	1.0	5.517	32.5	26.79	Fe, O, Si, Mg, Al	$N_2 + O_2$
(Moon)***	(0.34)	(27.322)	1738	27.32	0.012	3.3437	< 5%	(?)	Si, Mg, Fe, O, Al	Not present
Earth + Moon	about 150	—	—	—	1.012	4.97	30	27.5	O, Si, Fe, Mg, Al	—
Mars	227.94	686.98	3370	1.026	0.107	3.95	About 15	28	O, Si, Mg, Fe	CO_2 (O, H_2O)

*According to Bott, 1971; and Galimov, 1995.
**For Earth, period of rotation 24 hr and moment of inertia 0.8038×10^{45}; see Table 1.2 for the composition of Earth.
***For Moon, distance from the Earth and revolution around the Earth.

A logical consequence of this concept is the conclusion that the Earth differentiated early into an iron core, solid silicate mantle and molten envelope (to depth 400 to 760 km), resulting in a primary difference in the composition of upper and lower mantles (Hayashi et al., 1979; Kumazawa and Maruyama, 1994). The greenhouse effect generated by the dense primary atmosphere, similar to that of Venus, may have promoted the melting of the outer envelope (Hayashi et al., 1979). Geochemical proofs of early isolation of the core and melting of the outer envelope are primarily found in the isotopic systems (Turner, 1989; Azbel and Tolstikhin, 1993) as well as in the distribution characteristics of siderophile elements (Newsom and Sims, 1991; also see Sec. 1.2). Separation of the core may have occurred even in the early stages of accretion while the continuing accretion of planetesimals brought in very light and oxidised material. This conclusion is supported by a review of recent geochemical data, especially regarding the abundance of siderophile elements in the mantle (Oho, 1992; Guyot, 1995), giving rise to the earlier hypothesis of heterogeneous accretion (Kuskov and Khitarov, 1982; Sobotovich, 1984).

The Moon stands anomalously among the planets of the terrestrial group described in Table 1.1. It evolved almost simultaneous with the Earth. The oldest rock on the Moon is troctolite from the primary anorthosite-norite-troctolite crust aged 4530 ± 40 Ma (Papanastassiou and Wasserburg, 1976), which is close to the age of the oldest meteorites Allende 4567 ± 40 Ma and Andra dos Reis 4500 ± 40 Ma (Anders and Grevasse, 1989). While the composition and density of the Moon are close to the composition and density of the Earth's mantle (Taylor, 1986; Galimov, 1995), the iron core 350 to 500 km in radius constitutes less than 4% of its mass (Nakamura, 1983; Nozette et al., 1994). It has therefore been suggested that the Moon was formed not by simultaneous accretion with the Earth, but as a result of the oblique impact of a large body of the size of Mars against the Earth during its accretion and ejection of the mantle material in a near-terrestrial orbit (Hartmann and Davis, 1975; Galimov, 1995). By 'combining' the Earth with the Moon, the combined parameters fall in the successive planetary series (Table 1.1). Another variant of explanation is the seizure of the Moon by the Earth followed by disintegration of the Protomoon and birth of the Moon from the fragments of the Protomoon and loss of considerable mass (Sorokhtin and Ushakov, 1991). This variant is less acceptable from the geochemical standpoint (Galimov, 1995).

Venus is closest to the Earth in composition and physical parameters (Table 1.1). It is for this reason that several authors compare the early history of the Earth with the contemporary state of Venus. It has been suggested that Venus illustrates the transition from plume tectonics to embryonal plate tectonics characteristic of the early stage of the Earth (Sandwell and Shubert, 1992; Kumazawa and Maruyama, 1994; Khain, 1994). Thus, a comparative study of planets offers an effective method of understanding the geodynamics of the Earth.

Among the planets of its group, Earth possesses the most powerful magnetic field. There is no magnetic field on Venus since its surface temperature is close to the Curie point. A moderate magnetic dipole is noticed on Mercury while it is very weak on Mars. On the Earth, as on Jupiter, the present-day magnetic field has suppressed the preceding magnetic history but meteorites and other planets of the terrestrial group have preserved traces of ancient magnetic fields of the

8

period of formation of chondrules and formation of planetesimal swarms about 4.6 billion years ago (Strangway, 1980). The intense magnetic field of the Earth prevailed even 3.5 billion years ago (McElhinny and Senanatake, 1980). The most important characteristic of the Earth's magnetic field is the changing polarity, sometimes rising and sometimes attenuating. Reversals run as brief migrations of the Pole through North and South America or East Asia and Australia, along the two sides of the Pacific Ocean for a distance of 180° along the latitude (Runcorn, 1992; Aurnou et al., 1996). These reversals are extensively used in discussions of core-mantle couplings of the Earth (see also Chapter 4).

The most important characteristics of the Earth (Table 1.1) are its powerful magnetic field, oxygen-nitrogen atmosphere, hydrosphere and continents with a granitic crust abundant in Si, Al and K (potassic sial). These features are probably related not only to the dimensions and position of the Earth in the solar system, but also to the characteristics of its evolution during which granitoidal (andesite-dacite) magmas rich in water were long liberated (particularly active in the early stages) (Dobretsov, 1980; Bogatikov et al., 1987). The presence of the hydrosphere and atmosphere gave rise to life and its development on the Earth, which in turn enriched the atmosphere with free oxygen and promoted development of oxidation processes close to the surface.

The geological structure of the surface and surface envelopes of the Earth (primarily of the Earth's crust) has been studied in utmost detail through objective geographic and geological investigations for more than two centuries. The largest features on the Earth's surface are the continents (including the shallow shelf) and oceans with continental and oceanic crust respectively. The latter account for two-thirds of the Earth's surface (Fig. 1.1).

Several ancient Precambrian cratons are delineated in the structure of continents and comprise about 70% of the area of the latter. In between these cratons are disposed orogenic (or folded) belts which, as a first approximation, are classified according to their ages as Palaeozoic and Mesozoic-Cenozoic. The extensive distribution of slices of ophiolites, blueschists and island arc volcanites in them reveals that these belts represent a mixture of rocks of ancient oceans, island arcs, microcontinents, rocks of forearc and back-arc basins and passive margins, tectonically displaced during the proximation and collisions of cratons, island arcs, seamounts and microcontinents. These complex processes will be dealt with in detail below. For the present, it is important to emphasise that, apart from the contemporary oceans and island arcs, ancient oceans and island arcs prevailed and folded belts formed at their site.

A weakly deformed cover and crystalline basement exposed in the uplifted blocks or shields are delineated in the structures of cratons. The basement structure reveals Precambrian folded belts of different ages and the oldest blocks of 'primary' crust so that the history of ancient oceans and island arcs can be traced into the older stages of the Earth's evolution, at the least up to the border of 1.5 billion years (according to other assessments, up to 3 to 3.5 billion years). It is, however, more important to point out here that the oldest blocks (older than 3 billion years) constitute more than one-half (about 60%) of the volume of cratons. Considering that the crustal thickness in the cratons (40 to 50 km) exceeds the average thickness of the rest of the continental crust (30 to 35 km) and that part of the sediments

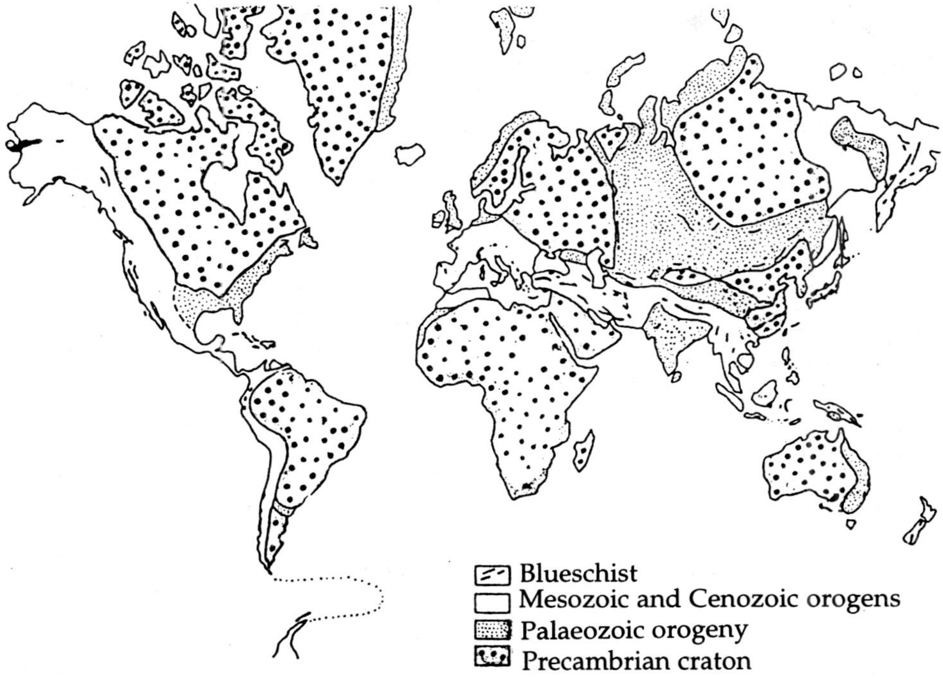

Fig. 1.1. Distribution of Precambrian cratons and Phanerozoic orogenic belts (Maruyama, 1994). Thick dashes and lenses in the folded regions represent blueschists and ophiolites. Palaeozoic belts formed at the site of Palaeozoic oceans are shown as fine dots.

and microcontinents (about 30%) in the cratons and folded regions formed from the very same material as the oldest blocks, it can be concluded that most part of the continental crustal material was formed during the first 1.5 billion years of the Earth's evolution. (A very simple calculation shows that the volume of the cratons is equal to 73.5% of the volume of the crust and that the volume of the oldest rocks is 52% of the crustal volume.)

According to other assessments, 80% or even the entire volume of the continental crust was formed in the Early Precambrian in the first 1.5 billion years and was later redistributed on the Earth's surface (Wise, 1978; Dobretsov, 1980). Some other estimates are also known about the relatively uniform evolution of acidic island (= continental) crust throughout the geological past.

Studies of the structure of the ocean floor commenced only 30 to 40 years ago. They have shown that the oceanic crust is thin (7 to 15 km) and young (younger than 200 million years). Its volume, V_{oc} constitutes 33% of the volume of the entire Earth's crust. Considering the prolonged period of formation of oceanic crust (about 2.5 billion years), the total volume of the newly formed oceanic crust clearly exceeds by 5 to 10 times the volume of the continental crust but almost the whole of the old oceanic crust, according to the subduction mechanism, disappeared into the mantle and was transformed into secondary melts of andesite composition and restites (see below). Continental crust represents the 'preserved' element of the Earth formed for the most part in the early stages of the Earth's development and only slightly supplemented in the much later epochs. Oceanic crust, on the contrary, is a more mobile, regularly renewed element of the Earth.

The most important components of the structure of the ocean floor are the mid-oceanic ridges representing the divergent boundaries of lithospheric plates. Figure 1.2 depicts the most important lithospheric plates and divergent boundaries of the mid-oceanic ridges located in the eastern part of the Pacific Ocean (boundary of the Pacific Ocean plate); in the midportion of the Atlantic Ocean (boundaries of the South American and African plates, North American and Eurasian plates); and in the Indian Ocean (boundary of the Indo-Australian plate). Converent boundaries (consuming plate boundaries) for the most part (apart from the Asian continent) coincide with deepwater trenches serving simultaneously as the outer boundaries of subduction zones in which the oceanic plate is submerged under

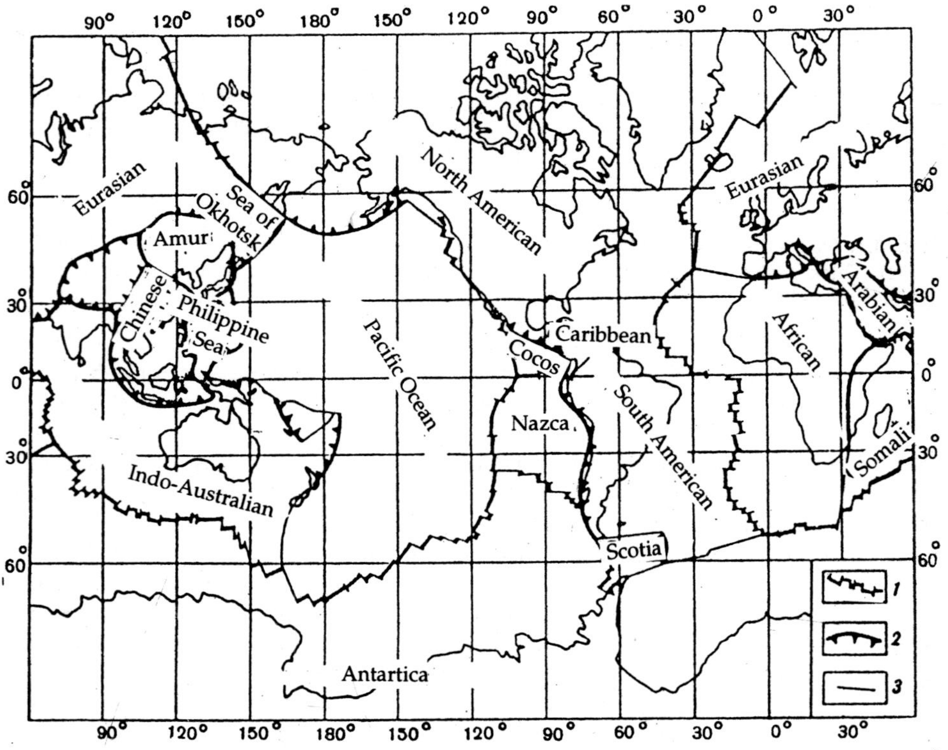

Fig. 1.2. Map showing the lithospheric plates of the Earth (Zonenshain and Kuz'min, 1993a). Plate boundaries: 1—divergent, 2—convergent and 3—transform-fault.

the island arc (as in the western part of the Pacific Ocean) or under the continent (as in the eastern part of the Pacific Ocean). Finally, there are transform-fault boundaries of plates in which one plate is displaced relative to another along a transform fault (strike-slip fault).

A similar scheme is shown in Figure 1.3 in which the vectors of plate movement, the oldest sections of the oceanic lithosphere aged 95 to 200 Ma as well as the elements of other important structures in oceans and under the continents reflecting the presence of mantle plumes, are shown. Among them, the two most important and major plumes (superplumes), Pacific and African, have been delineated from the data of seismic tomography in the lower mantle. The most important 'hot spot' (zones of active deep-level mantle magmatism) fall within or around the periphery of the African superplume and trend less distinctly towards the outer periphery the Pacific Ocean plume.

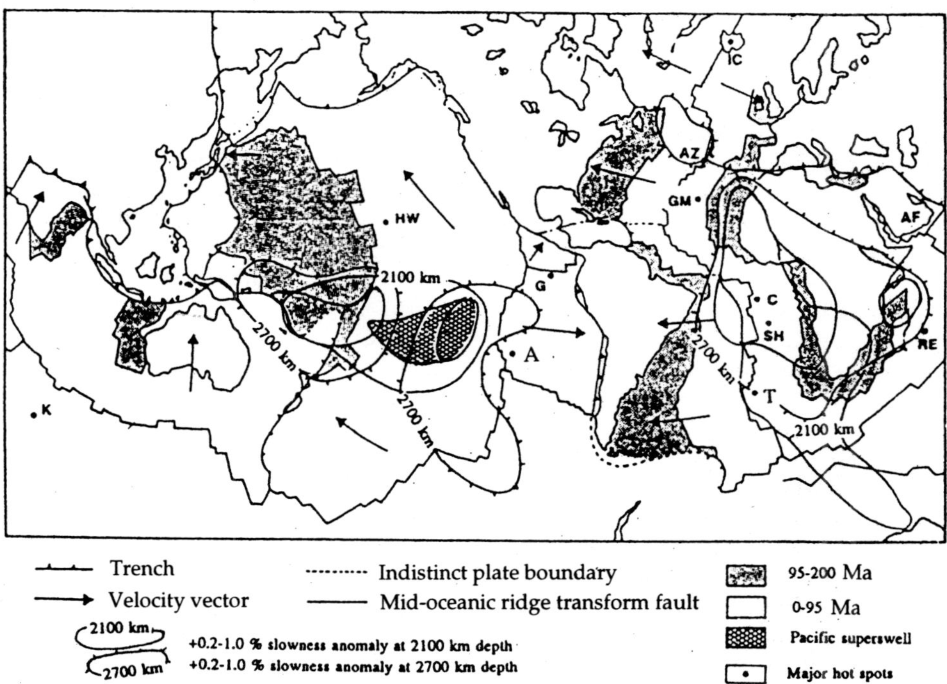

⊢—⊣— Trench	·········· Indistinct plate boundary	95-200 Ma
——▶ Velocity vector	——— Mid-oceanic ridge transform fault	0-95 Ma
⌒2100 km⌒ +0.2-1.0 % slowness anomaly at 2100 km depth		Pacific superswell
⌣2700 km⌣ +0.2-1.0 % slowness anomaly at 2700 km depth		Major hot spots

Fig. 1.3. Lithospheric plates with vectors of their movement, continental part of plates close to the boundary of continents: the older part of the oceanic plates (95 to 200 Ma) is hatched. In the Pacific Ocean and in the region of Africa are shown isolines of 2100 and 2700 km denoting the position of the Pacific and African superplumes at depths of 2100 and 2700 km based on the data of seismic tomography (Fukao et al., 1994, modified).

Names of hot spots: A—Ascension, AF—Afars, AZ—Azores, C—Circe, G—Galapagos, GM—Great Meteorite, HW—Hawaii, IC—Iceland, K—Kerguelen, RE—Reunion, SH—St. Helena, T—Tristan.

The concept of the system of mobile (revolving along a geoid surface) thin and solid lithospheric plates is the most important element of plate tectonics or the new global tectonics. The kinematics and possible mechanisms of movement of lithospheric plates are discussed below. For the present, it is important to point out that the viscosity of lithospheric plates differs from that of the underlying asthenosphere by 2 to 3 orders. Lithospheric plates include the crust and the underlying highly viscous (solid) part of the upper mantle. Their thickness thus varies from the thickness of the crust to values tens of times exceeding it.

Oceanic plates are thinner and uniform; the thickness of the lithosphere in them is minimal (5 to 7 km around the mid-oceanic ridge axis) and rapidly increases to 80 km and remains stable (70 to 80 km) for the most part. The thickness of continental plates varies from 30 to 50 km in the rift zones to 250 to 300 km under the old and cold cratons (Zonenshain and Kuz'min, 1993a). The average thickness of continental plates is assumed to be 150 to 200 km but this figure too is extremely tentative (Turcotte and Schubert, 1985).

1.2. Chemical Composition, Properties and Rheology of Different Envelopes of the Earth

The solid Earth disposed beneath the atmospheric-hydrospheric envelope of the Earth is divided into several shells on the basis of geophysical (primarily seismic) data. As a first approximation, these shells are spherically symmetrical (Fig. 1.4). Following B. Gutenberg, these envelopes are conventionally designated by the English letters A to G although most of them have their own names. The position of the boundaries and their physical nature have been continuously improved during the past five decades.

A — Crust (oceanic, about 10 km; continental, about 40 km)
B_1 — Lithospheric mantle (A + B_1—oceanic lithosphere, about 80 km)
B_2 — Asthenosphere (at depths of 80 km under oceans or 200 km under continents)
C — Transition zone, up to 670 km (B_1 + B_2 + C—upper mantle)
D_1 — Lower mantle (C + D_1, mesosphere)
D_2 — Transition zone (up to 2900 km)
E — Outer core
F — Transition zone
G — Inner core

Some boundaries of the envelopes are sharp while a transition zone occurs in other cases. The most important boundaries are: boundary between surface of the Earth and hydrosphere-atmosphere, varying in height from + 8.9 to – 11.2 km (± 20 km); between lithosphere and asthenosphere, varying in depth from 5 to 350 km (70 to 40 km under oceans and 200 to 100 km under continents); boundary between upper and lower mantles MB (average 670 ± 20 km); and core and mantle boundary CMB (2900 ± 20 km).

Determination of the chemical composition of the Earth and its principal envelopes is an important task of Earth sciences. Assessment of differentiation processes of the Earth and its envelopes depends on it, in particular the growth and evolution of continental crust which, along with the hydrosphere, is an important

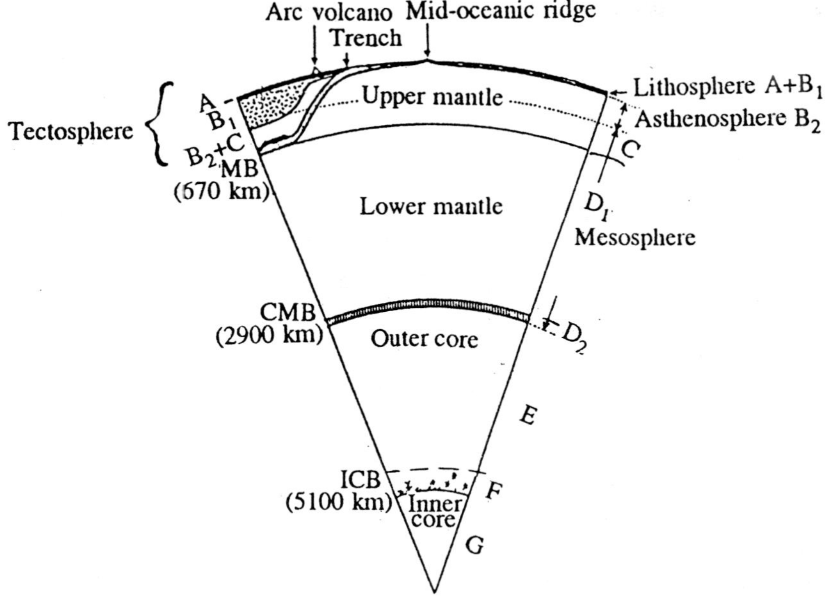

Fig. 1.4. Stratified envelopes of the Earth, A to G, also delineated as crust A, lithosphere A + B₁, asthenosphere B₂, upper mantle B + C and lower mantle D₁ + D₂.

feature of the Earth compared to other planets. Table 1.2 presents some of the recent evaluations (Allegre et al., 1995) based on the meteorite hypothesis, i.e., similarity between composition of the Earth and meteorites relative to major and minor elements. These calculations have undergone improvements over a long time (Ringwood, 1966, 1977, 1979; O'Nions et al., 1979; Jagoutz et al., 1979; Allegre et al., 1986, 1995; Hart and Zindler, 1986; Kuskov et al., 1995).

It has been assumed in Table 1.2 (Allegre et al., 1995) that the ratios of different lithophile elements, concentrated in the mantle and not entering core (Al, Mg, most of Si, Ca, Na and Ti), are identical in the bulk Earth (BE) and in the primitive mantle (PRIMA), and can be evaluated from the extreme compositions of mantle xenoliths in basalts and kimberlites (Jagoutz et al., 1979; Hart and Zindler, 1986; see Al/Mg-Si/Mg in Fig. 1.5). For the bulk Earth the ratio of siderophile and lithophile elements present in the core and mantle (Fe/Al, Fe/Mg, Ni/Al and others) was determined from the correlation diagrams in meteorites (Fig. 1.5) where the value of Al/Mg (or Mg/Al) was determined from xenolith compositions. Thus the bulk composition of the Earth could be assessed (Table 1.2).

The composition of the core determined from the difference between the bulk composition of the Earth and primitive mantle, is independent of assumed core-formation processes. It has also been found (Table 1.2) that the core contains 7.3%

Table 1.2. Composition of bulk Earth and geospheres

Element	Bulk Earth, wt. %	Core, wt. %	Primitive mantle, wt. %	Oxides	Lower mantle, wt. %	Upper mantle, wt. %	Crust, wt. %	Pyrolite I, wt. %	Pyrolite II, wt. %
1	2	3	4	5	6	7	8	9	10
O	32.44	4.10 ± 0.5	44.79						
Si	17.22	7.35 ± 1.0	21.52	SiO_2	46.12	45.8	55.4	45.16	43.1
Al	1.5		2.18	Al_2O_3	4.09	3.58	14.6	3.54	3.3
Mg	15.37		22.78	MgO	37.77	38.8	5.38	37.5	38.8
Fe	28.18	79.39 ± 2	5.82 (6.5)	FeO	7.49 (8.2)	7.45	8.07	8.45	8.0
Ni	1.71	4.87 ± 0.3	0.20	NiO	0.25	0.26	0.003	0.2	0.39
S	0.75	2.3 ± 0.2	0.1	S	0.1	0.11	0.08	—	—
Ca	1.56		2.31	CaO	3.23	3.08	8.01	3.08	3.1
K	0.02 (?)	(?)	0.03 (?)	K_2O	0.035 (?)	– (0.07)	1.63	0.13	0.22
Na	0.18 (?)		0.26 (?)	Na_2O	0.36	0.3 (0.41)	2.42	0.57	0.61
Ti	0.07		0.11	TiO_2	0.18	0.17 (0.2)	0.86	0.71	0.58
Mn	0.26	0.582	0.12	MnO	0.15	0.15	0.145	0.14	0.13
Cr	0.43	0.779	0.27	Cr_2O_3	0.38	0.40	0.03	0.43	0.42
Co	0.083	0.253					1.37 (CO_2)		0.02
P	0.127	0.369	0.01	P_2O_5	0.3	0.025	0.172	0.1	0.08
				H_2O	0.2 (?)	0.18 (?)	1.53		0.21
Total	100	99.993	100.38	Total	100.28	100.20	99.70	100.01	99.96
Mass, %	100	32.5 (2% inner core)	67.50		48.90	18.20 (97.8%)	0.41 (2.2%)		
Mass, kg $\times 10^{24}$	6.057	1.967	4.09		2.96	1.105	0.0245		

Col. 2—Allegre et al., 1995, calculated from cols. 3 and 4; cols. 3, 4 and 6—Allegre et al., 1995, with possible correction for Fe; col. 7—calculated from cols, 6 and 8; col. 8—Ronov and Yaroshevskii, 1978; col. 9—Ringwood, 1966; col. 10—Ringwood, 1979.

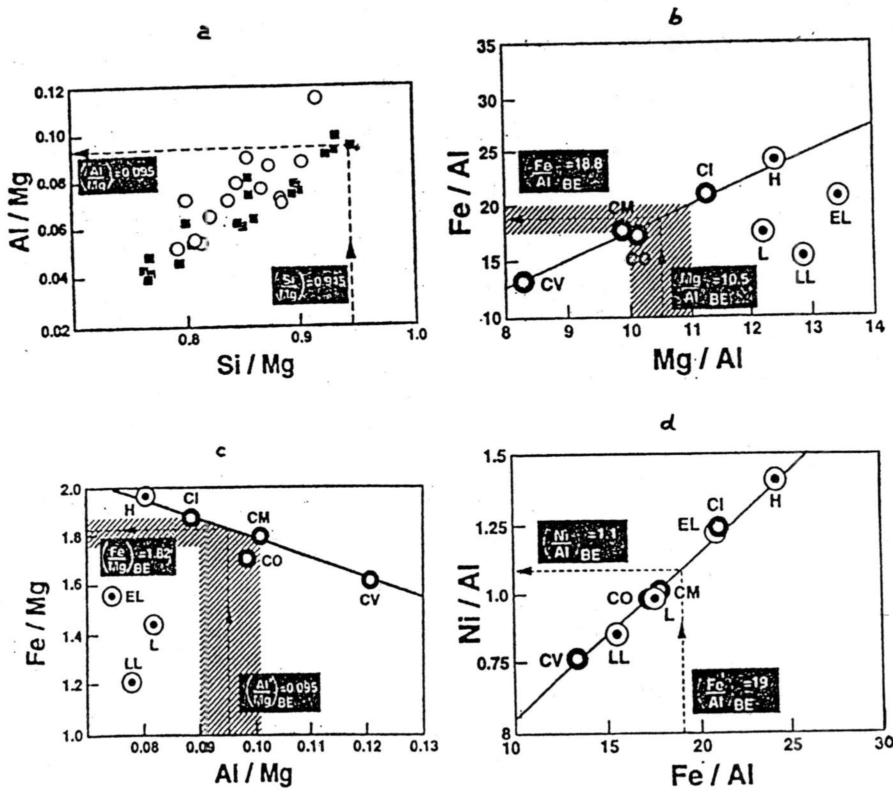

Fig. 1.5. Ratios of elements for a suite of xenoliths and orogenic lherzolites (a) and for a suite of chondrites (b to d) (compiled from Allegre et al., 1995).
a) assuming that PRIMA corresponds to the least differentiated samples, we obtain Al/Mg = 0.095 and Si/Mg = 0.945. Quadrangles represent xenoliths and circles orogenic lherzolites.
b, c and d) black-edged and white-edged symbols represent carbonaceous and normal chondrites respectively. Regression line passes through carbonaceous chondrite points. Using the values $(Al/Mg)_{BE}$ = 0.095 or $(Mg/Al)_{BE}$ = 10.5 derived in (a), we obtain $(Fe/Al)_{BE}$ = 18.80; assuming $(Fe/Al)_{BE}$ = 19 gives Ni/Al = 1.1, where BE is Bulk Earth.

Si and 2.3% S as well as about 4% oxygen. The latter has been added to make good the density 'deficit' as emerging from seismological data. Some uncertainty exists regarding the maximum contents of Si and O in the core; for Si it has been determined as 4.5 to 8.3%. At constant content S = 2.3% and low Si = 4.5%, the oxygen content in the core may go up to 7.5%.

The composition of the bulk Earth (Table 1.2) is similar to other estimates that have been given, falling in particular between carbonaceous chondrites CI

16

and CM[1], but closer to the latter (Fig. 1.5). This composition is also similar to that of 'solar chondrite' assumed as the original material of planets of the terrestrial group (Kuskov et al., 1995). After deducting 22.9% Fe, 1.7% Ni and 0.48% FeS, this chondrite contains 48.3% SiO_2, 3.4% Al_2O_3, 34.7% MgO, 10.7% FeO, 2.9% CaO, 1.3% Na_2O and 0.1% K_2O. This composition is close to that of the primitive mantle recalculated as oxides and adopted as the composition of the lower mantle (Table 1.2). The major difference lies in the much lower ferruginous state (low FeO and high MgO, Table 1.2) and low alkalis (Na_2O = 0.4% and K_2O = 0.03% instead of Na_2O = 1.3% and K_2O = 0.1%). The high ferruginous content of the lower mantle has been suggested by other investigators also (Kuskov and Parfenov, 1992; Wang et al., 1996) but the very high alkali content arises also from the composition of the crust, released from the mantle, and from the composition of mantle xenoliths.

The composition of the upper mantle shown in Table 1.2 was calculated assuming that the entire crust was separated from the upper mantle and that the proportional sum of the composition of the upper mantle (98%) and crust (2%) corresponds to the lower (primitive) mantle. With respect to major constituents, the resultant composition of the upper mantle (Table 1.2) is proximate to that of pyrolite (Ringwood, 1966, 1979), assumed as the substance of the upper mantle and calculated with allowance for the composition of mantle xenoliths. The major difference here too lies in the higher content of K_2O, Na_2O and TiO_2 in pyrolite. The values shown in parentheses in column 7 (Table 1.2) for K_2O, Na_2O and TiO_2 were obtained by adding (and not deducting) 2% content of these constituents in the crust to the contents in the upper mantle; nevertheless these values also, albeit close to those of the assumed pyrolite, are notably low.

Thus the contents of K, Na and Ti and probably hydrogen in the composition of the Earth and mantle are not at all certain. H_2O content was estimated for the crust and pyrolite and shown approximately for the mantle (Table 1.2). But, hydrogen as well as K and Na may be dissolved in the metallic core and serve as the major alternative to the light elements Si and O in the core (Marakushev, 1992; Oho, 1992; Breuer and Spohn, 1993).

It must be emphasised that the composition of the Earth and geospheres was determined from the meteorite hypothesis with almost no regard for geophysical data. Even the proportions of mantle and core can be determined from meteorite data and average density (5.517 g/cm^3) and mass of the Earth (6.057 $\times$ 10^{24} kg). In general, the data on composition given in Table 1.2 accord with geophysical data (Fig. 1.4, Table 1.3 etc.). In many instances this comparison enables ascertaining the required corrections or estimating the separation dynamics of geospheres. For example, for the bulk Earth the ratio $(Fe/Al)_{BE}$ = 18.8 was obtained from the correlation diagrams in meteorites (Fig. 1.5). Using the ratio of core and mantle masses from geophysical data, we obtain (Allegre et al., 1995):

$$(Fe/Al)_{BE} = \frac{(Fe/Al)_{PRIMA}}{Fe_M \; m_M} (Fe_M \; m_M + Fe_C \; m_C) = 20.4,$$

---

[1]Carbonaceous chondrites are classified by Allegre, 1982; Allegre et al., 1995, into four groups: CI, CO, CV and CM.

where m_M and m_C are the masses of the mantle and the core; Fe_M and Fe_C are the Fe concentration in the mantle and core. These values may be associated with an error in determining the Fe content in the mantle. Using FeO = 8.3 (Fe = 6.5) as in pyrolite, we obtain $(Fe/Al)_{BE} = 19.0$ which is equal to the meteorite value. Similar small deviations are recorded for Si and Mg when calculated from meteorite ratios or from mass balance. In the early stage of the Earth's development, with other values of m_M and m_C and composition of the Earth remaining constant, the concentrations of all elements in the crust and mantle would vary systematically.

Knowing the present-day composition of envelopes and the likely mineral composition (from experimental data), the probable density, propagation velocity of seismic waves and other properties of each of the geospheres can be calculated. However, many properties, especially the variation gradients of these properties at the geosphere boundaries and within them, can be determined only from direct geophysical measurements.

The nature of sharp and smooth variations of physical parameters at the boundaries of geospheres as well as the depth variation of the boundaries per se represent the most important indicators of geodynamic processes. The parameters actually measured in all the envelopes of the Earth are the propagation velocities of longitudinal V_p and transverse V_s seismic waves which depend on density ρ, bulk (incompressibility) modulus k and shear modulus G as follows:

$$V_p^2 = [k + (4/3)G]/\rho \tag{1.1}$$

$$V_s^2 = (G/\rho). \tag{1.2}$$

Table 1.3 shows the depth-wise distribution of these parameters based on the data for experimental and theoretical seismic models (Dziewonski and Anderson, 1981; Zharkov, 1983; Montagner and Anderson, 1989) and also the values calculated from the anticipated phase relations in the upper mantle for 'solar chondrite' composition (Kuskov et al., 1995). The calculated values accord well with the experimental results. A similar calculation for pyrolite composition of upper and lower mantles showed (Calderwood, 1996) good agreement between the calculated and observed values of V_p and ρ, with an error of $\pm 1\%$ for the upper mantle and $\pm 0.5\%$ for the lower mantle, for a temperature distribution corresponding to the common adiabatic curve commencing at 1680° K at a depth of 100 km. This is considerably higher than the values given in Table 1.3. Taking into consideration the actual physics of minerals, other investigators (Zerr and Boehler, 1993; Wang et al., 1996) have predicted a perceptible difference between the composition of lower and upper mantles and 'colder' adiabatic curve (250° K lower than the one calculated for the upper mantle).

The reduction of velocities V_p and V_s in the asthenosphere (Table 1.3 and Fig. 1.6), composition remaining constant, is explained by the reduced viscosity and shear modulus G due to rising temperature and partial melting. Sharp variations in density, V_p and V_s at the boundaries of layer C are explained by phase transitions studied experimentally (see Chapter 4). A detailed study of characteristic heterogeneities of V_p and V_s in envelopes B, C and D (seismotomography) enables estimation of characteristic variations in density and/or temperature in the mantle

Table 1.3. Physical parameters of geospheres for various depths (*Sources:* 1—Dziewonski and Anderson, 1981; Zharkov, 1983; 2—Montagner and Anderson, 1983; 3—Kuskov et al., 1995; 4—Fig. 1.6; 5—Figs. 1.8 and 1.9)

Geo-sphere	h, km	P (av) kb	TK (av), (1, 2, 3)	V_p			V_s			ρ			K (1)	G (1)	Q (4)	lg η (5) poises
				1	2	3	1	2	3	1	2	3				
A (conti-nental)	15	3.5	673	6.75	6.8		3.8	3.9		2.85	2.8		0.54–0.65	0.35	450	24 (dry)
B₁	60	14	873	8.0	8.1		4.45	4.5		3.34	3.31		1.10	0.72	60	22
	100	29	1103	8.0	8.3	8.01	4.45	4.52	4.58	3.37	3.31	3.38	1.13	0.72	100	22
B₂	150	46.5	1343	7.95	8.35	8.2	4.35	4.45	4.58	3.37	3.36	3.40	1.13	0.72	>100	18
B₂	200	64.0	1473	7.9	8.4	8.3	4.32	4.36	4.1	3.362	3.35	3.43	1.25	0.70	"	18
B₂	300	99.0	1623	8.6	8.53	8.55	4.63	4.67	4.7	3.48	3.51	3.5	1.74	0.75	"	18.5
C	400	130	1673	8.72	8.7	8.72	4.7	4.9	4.8	3.54	3.58	3.57	1.76	0.82	300	19
C	600	190	1723	9.65			5.2			4.13			2.7	1.3	450	20
D₁	1100	30	1873	11.6			6.0			4.74			3.25	1.7	2000	21
D₁	1600	530		12.1			6.5			5.03			4.3	2.2	2000	22
D₁	2700	1240		13.6			7.25			5.55			6.2	2.9	2000	22
D₂	2870	1340		13.45			7.18			5.68			6.55	2.95	?	20
E	2900	1350		9.35			0			9.89			6.5	0	4000	5
E	4700	3040		10.0			0			12.26			12.4	0	4000	10

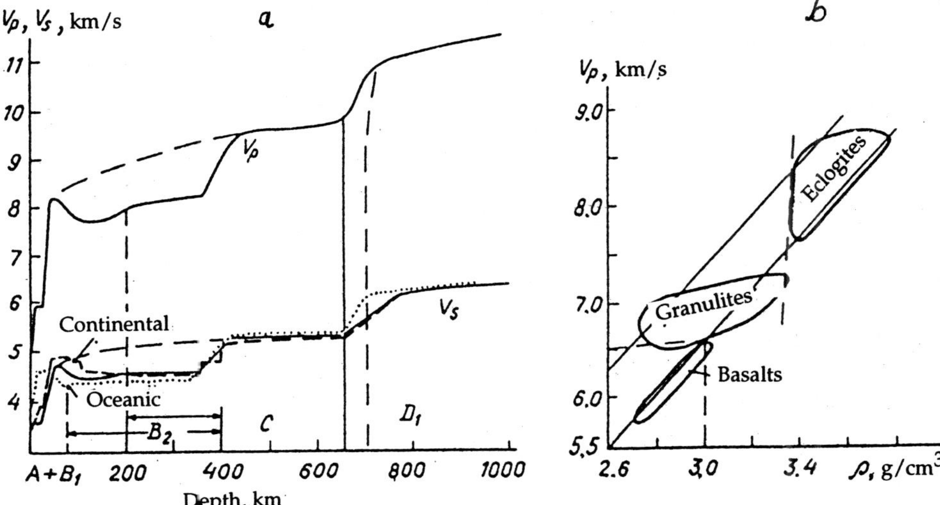

Fig. 1.6. a) Distribution of seismic wave velocities in outer envelopes A to D₁ of solid Earth (Bott, 1971, supplemented). For V_s, the average distribution of velocities under continents (broken line) and under oceans (dotted line) are shown; b) dependence of velocity of longitudinal waves (V_p) on density for basalts, granulites and eclogites (Manghnani et al., 1974).

(Inoue et al., 1990; Spakman, 1990; Spakman et al., 1993; Fukao et al., 1994; Gossler and Kind, 1996). Seismotomographic data convey the most vital information on deep-level processes and are used time and again in this book (Chapters 4 and 5).

The most significant density jump at the core boundary (by 1.8 times) is related to transition from the oxide-silicate mantle to the metallic core. Transition layer D₂ is very important for understanding the general geodynamics of the Earth and is currently under active study (Ringwood et al., 1992; Zerr and Boehler, 1993; Wysession, 1995; Aurnou et al., 1996; and others). The outer core in which $G \rightarrow 0$ and $V_s \rightarrow 0$, i.e., transverse waves do not pass through, represents a fluid Fe melt with admixtures of Ni, Si, O, S, Cr, Mn and Co (see Table 1.2). The inner core is regarded as solid. Investigations of the splitting of seismic waves traversing through the core (Tromp, 1995) showed that core anisotropy is highly intense and significantly exceeds the possible error of measurement. Depending on the degree of anisotropy, the properties of the inner core resemble a single crystal, which is readily explained by the settlement of highly oriented magnetic crystals of iron during formation of the inner core. In turn, this poses the question about the early separation of the core and early origin of the magnetic field of the Earth (see Sec. 1.1).

Table 1.3 and Figure 1.7a also depict such parameters as seismic quality factor Q and electrical conductivity Sc which strongly depend on temperature and

20

presence (or absence) of melt (fluid). The minimum of these values in the crust, especially below Baikal region (Logachev, 1993) is mainly associated with phase transitions and/or the presence of zones saturated with pore fluid in the front of regional metamorphism at $T = 300$ to $500°C$. The asthenospheric minimum is positively associated with the presence of pore melt while the high gradients of Q and Sc in layer C are essentially related to innumerable phase transitions.

Any type of conduction depends on temperature, pressure and phase state:

$$Sc = Sc_0 \cdot e^{-E/kT} \qquad\qquad (1.3)$$

where k is Boltzmann's constant and E activation energy depending, like Sc_0, on the phase state (phase transitions). For olivine (Fig. 1.7b), conduction increases with rise in temperature by three orders, with increase in pressure (at $400°C$) by two orders and by two more orders as a result of olivine-spinel transition (at $T = 650°C$ and $P = 4$ kb). Seismic quality Q also makes it possible to delineate the cold and brittle lithospheric plates in the subduction zones (see below).

The most important properties of the Earth's envelopes required for thermophysical calculations and geodynamic modelling are density, viscosity and thermal conductivities. The distribution of density in the Earth's envelopes has been determined quite reliably (Table 1.3, Fig. 1.6a) using seismic data, seismological experiments for different rocks at different pressures (see Fig. 1.6b), analogy with meteorites and theoretical models (Bott, 1971; Zharkov, 1983; Cazenave et al., 1988; Kuskov et al., 1995). Deviations from axisymmetrical density distribution cause the residual anomalies of gravitational forces (after allowing for the effect of surface and crustal heterogeneity) or, more accurately, the anomalies in the shape of the geoid determined from the satellites. Geoid anomalies together with global anomalies of topography and heat flux represent the most important indicators of thermal and density heterogeneities in the upper and lower mantles and possible flows in them (Hager, 1984; Hager et al., 1985; Richards et al., 1988; Cazenave and Thoraval, 1994).

Dynamic and kinematic viscosities have not been well estimated. Viscosity assessments of the lithosphere and asthenosphere have been done according to various assumptions. In the lithosphere, however, viscosity depends markedly on the rock composition, deformation rate and the presence of fluid—dry or wet rocks (Fig. 1.8). The range of these values is generally 10^{19} to 10^{26} poises (Strehlau and Meissner, 1987). Another characteristic feature is the minimum viscosity at the base of the sedimentary layer, at the base of the granite layer due to the high creep of quartz and/or partial melting of rocks and at the base of the lower crust. This determines the multilayer rheology of the continental lithosphere (Lobkovskii, 1988; Burov et al., 1994). In several studies the lower crust has been considered the most plastic layer of the lithosphere on the continents and the concept of effective elastic thickness (EET or Te) of the lithosphere has been introduced (McNutt et al., 1988; Burov and Diament, 1986):

$$Te^3 = \sum_{i=1}^{n} \Delta h_i^3 = f(T, \xi, R), \qquad\qquad (1.4)$$

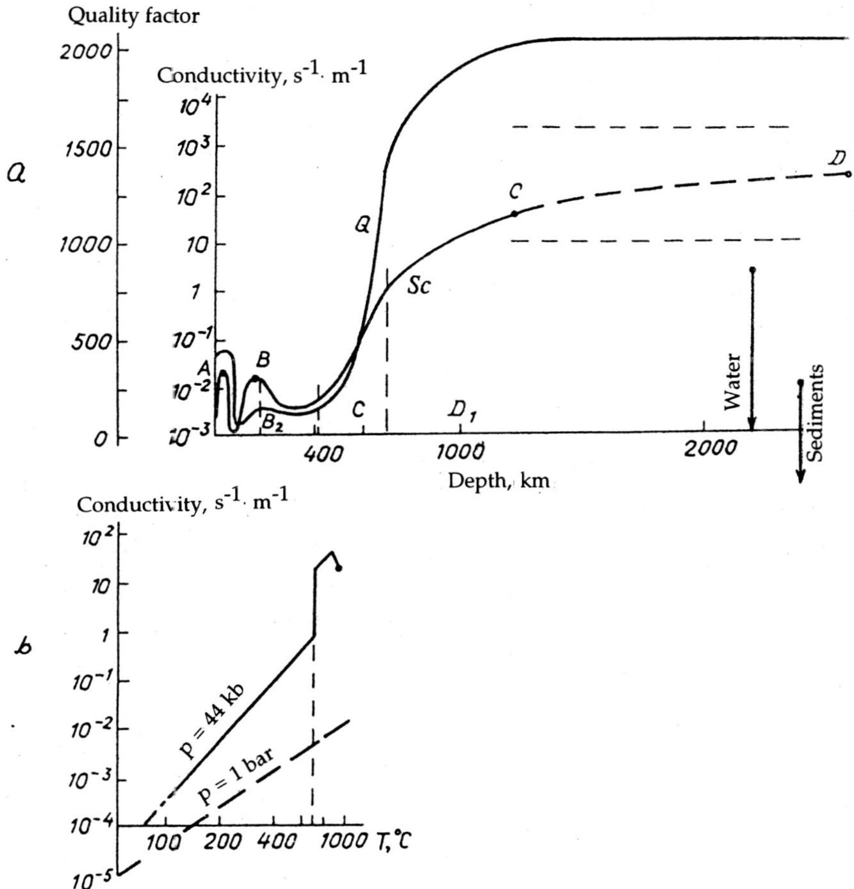

Fig. 1.7. a) Average values of seismic quality factor (*Q*) and electrical conduction (*Sc*) of crust and mantle rocks (Bott, 1971). Interval A-B—magnetotelluric measurements and B-C—diurnal variations and magnetic storms; broken lines show the ambiguous interval in the lower mantle (based on secular variations); b) dependence of the conduction of fayalite on temperature at 1 bar and 44 kb (about 200 km) (Akimoto and Fujisawa, 1965).

where Δh_i is the thickness of the *i*-th layer in the lithosphere (upper crust, lower crust and lithospheric mantle), ξ the deformation rate and R the radius of curvature of the envelope. The thickness of the mechanically rigid lithosphere h_m depends on *Te*, crustal thickness *Tc* and thickness of the more rigid upper crust h_l (Burov and Diament, 1936):

$$h_m = 3 \sqrt{(Te^3 - h_l^3)} + Tc \leq h \ (T = 600 \text{ to } 700°C),$$

22

i.e., usually approaches the depth where T = 600 to 700°C is reached (Burov and Diament, 1986) although for old platforms of the African type Te corresponds to the depth of 400°C isotherm (Hartley et al., 1986).

Viscosity values of about 10^{18} poises have been recorded for the asthenosphere when calculating the isostatic compensation of topography and processes of topographic restoration after the disturbance of isostasy, e.g., melting of the ice sheet in Siberia, Scandinavia and Canada (Artyushkov, 1979, 1993; Fjeldskaar, 1994). Figure 1.9 depicts the possible distributions of viscosities and their correlation with anomalies of topography and geoid (Cazenave and Thoraval, 1994). The viscosity of transition layer C (about 10^{20} poises) has been adopted as one and, relative to it, the viscosity of the lithosphere may be 1 to 2 orders higher (by 10, 50 and 100 times). The viscosity of the asthenosphere is 1 to 2 orders less (0.1, 0.05 and 0.01). Viscosity variations of the lower mantle have been estimated within smaller limits (10, 20 and 50 times higher than viscosity of the transition layer). For comparison, it may be recalled that the seismic quality in the lower mantle is 5 to 10 times better and electrical conductivity three orders more than in layer C (see Fig. 1.7).

Correlation functions shown at the right in Figure 1.9 were derived as correlations of the averaged anomalies of topography (surface topography) and geoid for two different viscosity profiles. It can be seen from this Figure that surface anomalies of topography are determined by the properties of the lithosphere and underlying asthenosphere since correlation becomes insignificant at the boundary of the transition zone (400 km) while remaining insignificant at minimum viscosity of the asthenosphere (0.01 relative to C) and lower mantle (one-fiftieth) even at a depth of 250 to 300 km (curves 3 in Fig. 1.9a, b at right). On the contrary, geoid anomalies correlate to the maximum extent with the transition layer (400 to 700 km), the highest correlation coefficient (0.6) being established at minimum viscosities in the asthenosphere (0.01) and in the lower mantle (≤ 20 relative to C).

Dynamic topography calculated relative only to mantle flows including subduction (Gurnis et al., 1996) provides maximum agreement with the actual topographic variations if the viscosity of the lower mantle is 50 to 100 times higher than the average viscosity of the upper mantle. Thus, viscosity values shown in Table 1.3 are quite consistent.

Thermal conduction coefficients λ (W/m · K) measured for different rocks (Dortman et al., 1984) are as follows: sedimentary layer to depths of 8 to 10 km, λ = 1.7 W/m · K; granite-metamorphic layer to depths of 15 to 18 km, λ = 1.9 W/m · K; and basalt layer to depths of 40 to 50 km, λ = 2.5 W/m · K. With increasing depth, λ rises to values 20 W/m · K at a depth of 200 km according to Lyubimova (1968) with very little variation throughout the mantle thereafter. According to Zharkov (1983) the thermal conduction coefficient decreases to a depth of 100 km as λ = 3.4 × 418/T (W/m · K) and later the value of λ rises. For the Earth's core, λ = 41.8 W/m · K. The coefficient of temperature conduction $a = \lambda/Cp\ \rho$ depends also on heat power Cp and density. Values of $a = 10^{-6}$ to 3×10^{-6} m^2/s have been adopted in different studies (McKenzie et al., 1974; Turcotte and Shubert, 1982; and Zharkov, 1983).

The above-mentioned averaged or varying values of density, viscosity and coefficients of thermal conductivity and temperature conductivity will be used

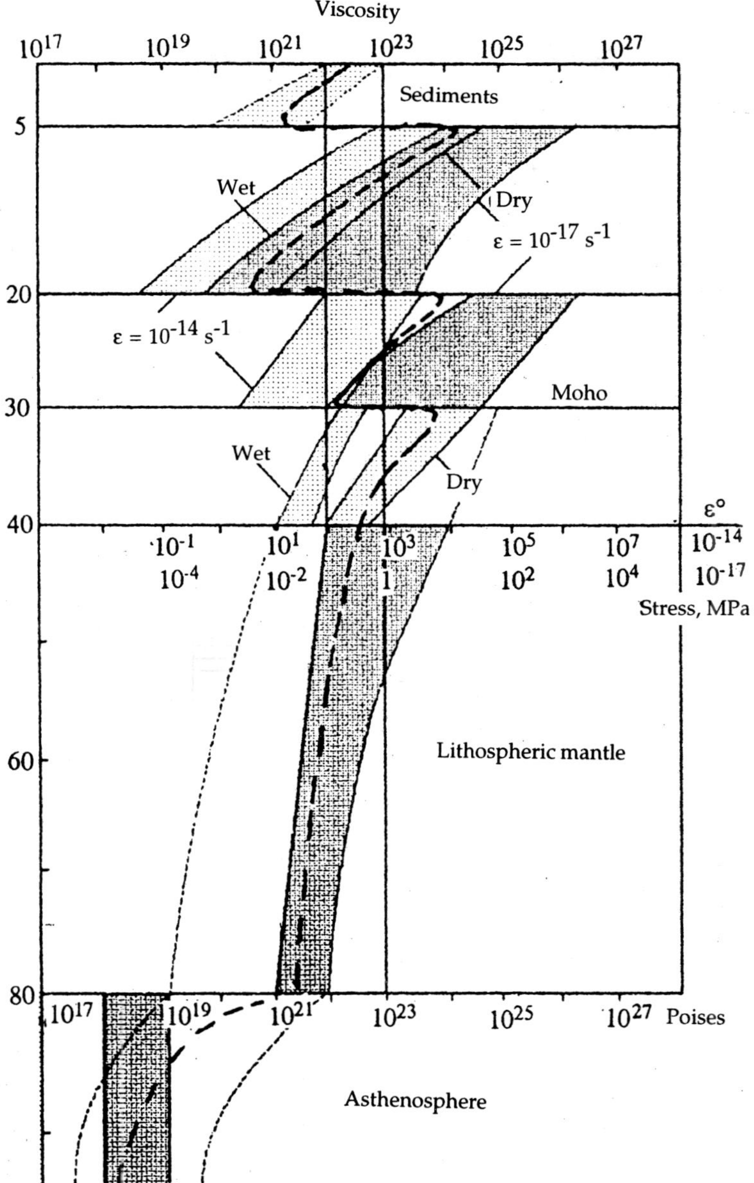

Fig. 1.8. Viscosity distribution with depth in the lithosphere: upper part (to a depth of 40 km), for western Europe with crustal thickness of 30 km, thermal flow 70 W/m^2 and T at the base of the crust 600°C (Strehlau and Meissner, 1987). Regions covered with dots correspond to deformation rate $\varepsilon = 10^{-14}\,s^{-1}$ and hatched regions to $\varepsilon = 10^{-17}\,s^{-1}$. Viscosity to a depth of 20 km is controlled by the creep of quartz, in the interval 20 to 30 km by the creep of pyroxene and plagioclase and below 30 km (for dunite-hartzburgite mantle) by the creep of olivine. The lower part of the Figure (deeper than 40 km) was extrapolated by the authors. Broken line represents the average viscosity of lithosphere and asthenosphere.

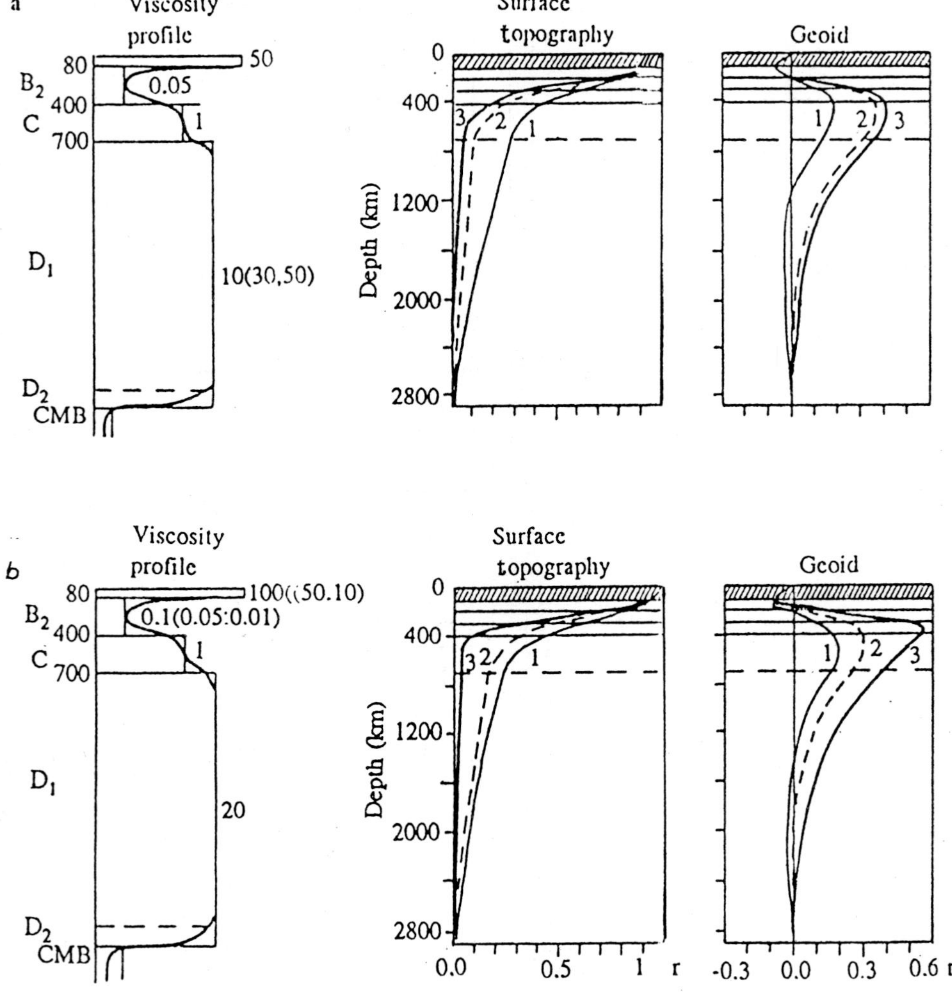

Fig. 1.9. Possible viscosity profiles in the lithosphere, asthenosphere (B₂) and lower mantle (D₁) relative to the viscosity of intermediate layer C adopted as one (a, b at left):
at right (a and b) are shown the calculated correlation functions relating anomalies of viscosity in the mantle and anomalies of averaged surface topography and geoid for different viscosity profiles.
a—at asthenosphere viscosity 0.05 of the viscosity of layer C and varied low mantle viscosity (curves 1, 2 and 3) and b—at lower mantle viscosity 20 times higher than that of layer C and asthenosphere viscosity variations 0.1, 0.05 and 0.01 of the viscosity of layer C (after Cazenave and Thoraval, 1994).

hereafter for thermophysical calculations and for studying the geodynamic processes in the Earth's crust and mantle.

1.3. Localisation and Intensity of Geodynamic Processes

The scheme of geodynamic processes in the Earth's envelopes is depicted in Table 1.4 and the characteristic space-time scales of these processes in Figure 1.10.

Circulation processes in the atmosphere and hydrosphere studied in the physics of the atmosphere, meteorology and hydrology are rapid processes (minutes on local scales and days or a few years on global scales) and cannot be strictly classified as geodynamic processes. However, geodynamic processes are associated with many surface and cosmic phenomena. There are also some analogies between the circulatory process in the hydroatmosphere and mantle convection. Hence the spatial and temporal scales of each circulation are also given in Figure 1.10.

Geological methods actively study the surface processes, such as volcanic degassing, volcanic eruptions, hydrothermal circulation in volcanic systems and deposits, sedimentation, biogeochemical cycles involving the participation of bacteria and plants in material transport, primarily of carbon, hydrocarbons and oxygen. The typical duration of these processes (days, years or tens of years)

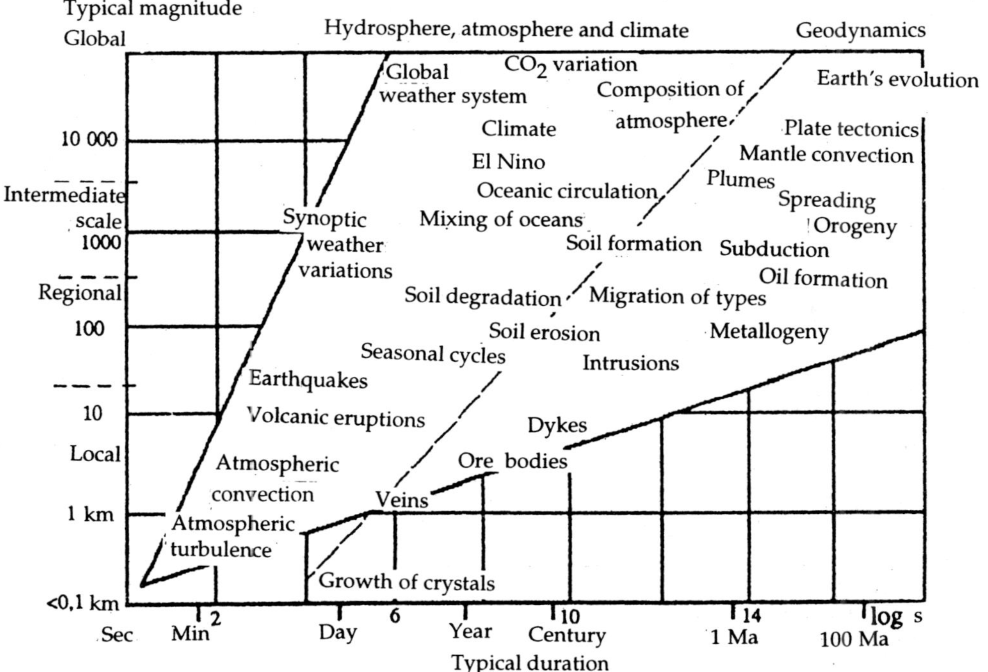

Fig. 1.10. Scale of typical durations and spatial dimension of principal processes in the Earth (Wyllie et al., 1993; modified).

Table 1.4. Scheme of geodynamic processes in the Earth's envelopes (Wyllie et al., 1993, modified)

Type of observation	Long-duration processes			Short-duration manifestations
Meteorology and hydrology	Circulation in oceans and atmosphere			Weather and climatic changes
Geological observations and measurements	↓ *Surface processes* ↑ Volcanic degassing Volcanism Hydrothermal circulation Sedimentation Biogeochemical cycles			Past climatic variations; volcanic eruptions
Geology, geodesy, cosmology (deformations, stress fields and plate movements)	↓ *Lithospheric processes* ↑ Subsidences, deformations, intrusions, dykes, metamorphism, plate thrusting etc.			Earthquakes and landscape changes
Geophysics (gravimetry, magnetometry, heat flux etc.). Geological observations and reconstructions	↓ *Lithosphere-asthenosphere coupling* ↑ Processes causing plate movements and spreading Subduction and collision Geochemical cycles			Formation of ores and energy resources
Geophysics and astronomy (fluctuations of Earth's rotation rate, pole migration and seismology). Geology and geochemistry of magmatism	↓ *Convection and plumes in Earth's mantle* ↑			Pole nutations
Geophysics, meteorites and experiments (variations of geomagnetic field, meteorite composition etc.)	↓ *Core-mantle coupling* ↑ *Convection in outer fluid core*			Secondary geomagnetic field Variations of Earth's rotation

enables a direct study of them and a comparison with products of similar processes in the past. This is the basis of the principle of actualism, which is well known in geology.

Significant fluctuations in topography, temperature and density on the Earth's surface represent an important feature. Relief varies from a depth of –11.6 km (in Mariana Trench) to + 8.9 km (Everest), i.e., $\Delta H = 20.5$ km, while the land temperature in the different climatic zones and elevations varies from –60 to +40°C ($\Delta T = 100°$C). Density variations too are significant from 1 (water) to 3.5 g/cm^3 (dense rocks) which, together with topographic and temperature variations ensure effective weathering and erosion. Conditions on the floors of seas and oceans are 'smoother' where the temperature is fairly constant (3 to 4°C) while a bed of unconsolidated silt (thickness of up to a few metres) provides a transition zone in density (1.5 to 2.5 g/cm^3). Maximum contrasts are produced by conditions associated with surface manifestations of endogenic processes, i.e., volcanism and thermal springs. The temperature of gas discharges of Kudryavyi volcano in the Kuril islands reaches 1000°C and rare minerals such as rhenium sulphide are deposited here. The temperature of the boiling lava lake of Kilauea in the Hawaiian islands exceeds 1200°C. Fluctuations of oxidation and reduction conditions of endogenic lavas and fluids compared to the surface oxidising conditions amount to 6 to 10 orders (log PO_2 from –9 to 0; pH from 2 to 9). Such contrasting conditions make for non-equilibrium rapid processes although final states closer to equilibrium are attained under oxidising conditions: P = 1 to 100 atmospheres and $T = 25$ to 50°C. Sedimentation conditions in seas and oceans represent much slower equilibrium processes although here too the sedimentation rate varies by several orders, from 0.01 (10 m in a million years) to 10 mm/year (10 km in a million years). For example, in Baikal, 3 to 5 km of sediments (average 0.1 to 0.2 mm/yr) accumulated in the first 25 to 30 million years and up to 2 to 3 km of sediments (average 1 mm/yr) in the later 2 to 3 million years.

Lithospheric processes (see Table 1.4) such as the uplift of mountains and plate subsidences and deformations are studied in their present state by airborne geophysical methods (including satellite measurements) and some geological and surface and subsurface geophysical methods (observations in boreholes and seismic methods) and the results of these movements in the past (deformations, metamorphism, intrusions, dykes and ore formation) are reconstructed by the usual geological methods. Most important at present are the direct measurements of plate movements by very long base interferometry (VLBI) and satellite observations (Carter and Robertson, 1987) which provide direct confirmation of plate tectonics. At the same time, a particularly important fact is that the measured current plate-movement velocities are close to values independently determined from geological data for the preceding 1 to 3 million years, thus confirming stable plate movements (Fig. 1.3).

The very movement of lithospheric plates is determined by lithosphere-asthenosphere coupling. This can be regarded as generally recognised, although the explanation of forces and processes causing plate movements continues to be actively debated. Information here is provided predominantly by geophysical methods (gravimetry, magnetometry, study of heat fluxes, geoid forms etc.) but their interpretation is impossible without geological observations and reconstructions. This is particularly relevant to the study of volcanism in the spreading zones, i.e., mid-oceanic ridges and Red Sea type rifts and to the study of subduction and collision zones (Himalayan type) by not only geophysical, but also geological methods (study of volcanism, associated deposits, island arc slope deformations etc.). The synthesis of these geological and geophysical data which help model the basic processes will be dealt with later (Chapter 4–5). Nevertheless, it is clear right now that processes at the lithosphere-asthenosphere boundaries control plate tectonics and geology as a whole. The rate of plate movements and its constituents (e.g., subduction rate) varies in the range 1 to 10 cm/yr (10 to 100 km in a million years).

Mantle convection and mantle plumes represent a deeper level of movements. The rate of plate movements is probably similar to the convection rate in the upper mantle while convection in the lower mantle is slower. These convective motions are global since they encompass a large part of the Earth's mass. They are inaccessible for direct observation and study. Plate movements serve as an indirect manifestation of convection in the asthenosphere while magmatism of hot spots and fields is an indication of the manifestation of mantle plumes (Hawaii, Iceland, Azores etc., see Fig. 1.3). A study and comparison of the geology and geochemistry of the magmatism of hot spots and spreading zones not only constitutes a source of knowledge about mantle plumes, but also serves as a proof of two-layer (two-level) convection in the mantle.

Nevertheless, geophysical and cosmophysical methods represent the major and direct sources of information about mantle heterogeneities and movements: study of variations of Earth's rotation, polar migration, seismic tomography, comparative planetology etc.

Finally, convective processes in the liquid metallic (outer) core of the Earth and processes in the transition layer of lower mantle D_2, caused by the core-mantle coupling, represent extremely deep-level geodynamic movements. It is gradually becoming clear that these are the most important movements from the energy point of view which possibly determine all other geodynamic movements and the main properties of the 'heat engine' of the Earth. Here can be seen the maximum jump of density ($\Delta\rho$ = 4.3 to 4.5 g/cm^3) and probably of viscosity. The viscosity in the lower mantle which is of the order of 10^{21} to 10^{22} poises decreases in the fluid core by 15 orders or more. The typical duration of processes also changes correspondingly: tens and hundreds of millions of years in the lower mantle and very rapid processes in the fluid core which, according to many workers, are reflected in the annual and secular variations of the magnetic field and rotation of the Earth. Processes in layer D_2 responsible for the origin of lower-mantle plumes are perhaps of an intermediate scale.

This represents the most difficult level of processes for investigation. Direct measurements are possible only for seismic tomography (structure of D_2, topographic variation of core-mantle boundary (CMB) and fluid nature of core) and variations of the geomagnetic field whose origin is positively associated with the core although the theory of geomagnetic dynamo has not yet been perfectly understood. High-pressure experiments (especially, explosive adiabatic curves, solid-phase transitions and investigations of intensely compressed plasma) as well as comparisons with ferruginous (Fe-Ni) meteorites and study of lower-mantle magmatism are of indirect importance. However, investigations in this direction have commenced and substantial progress can be anticipated in the near future.

Thus, only geological investigations and reconstructions together with geophysical and physical (thermophysical) modelling enable comparison of geodynamic movements at different levels and correlate with divergent, often highly specialised information. The surface zone, lithosphere-asthenosphere coupling zone and CMB represent the most important levels of localisation of geodynamic processes.

The intensities of different processes (Fig. 1.10) enable distinguishing the slower and often global geodynamic processes in the high-viscosity Earth from the more rapid processes occurring in the hydroatmosphere responsible for the climate and partly for surface processes. According to L.P. Zonenshain (Zonenshain and Savostin, 1979; Zonenshain and Kuz'min, 1993a), geodynamic processes can be of a global level (formation of the Earth, mantle convection and plumes), or regional level in the actual spreading, subduction, collision, orogenic and metallogenic zones and intrusions in magmatic zones (with a linear scale of 100,000 to a billion km) or local level (0.1 to 100 km). Microlevel (1 mm to 1 m) processes including the growth of crystals, mineral aggregates and microstructural variations (deformations) complete this series. This last level can hardly be placed among geodynamic processes.

In this book attention is devoted mainly to the global and regional (intermediate scale) levels measurable on scales of hundreds and thousands to millions of years (see Fig. 1.10).

1.4. Plate Tectonics

Lithospheric plate tectonics represents the principal paradigm of contemporary theoretical geology. Of late, a second paradigm, i.e., tectonics of hot spots (or plume tectonics), has been added. The concept of plate tectonics arose in the early 60s based on the hypothesis of Hess (1962) and its variant postulated by Dietz (1961) about ocean-floor spreading and was conclusively formulated at the end of the 60s (Wilson, 1965, 1973; Morgan, 1968, 1971; Le Pichon, 1968; McKenzie and Morgan, 1969; Isacks et al., 1974; Le Pichon et al., 1977).

Many premises of plate tectonics have been recounted above. We shall now attempt to present them in a systematic, albeit extremely brief manner based on the works of Dobretsov (1980), Zonenshain and Savostin (1979) and Zonenshain and Kuz'min (1993a).

Kinematics of lithospheric plate movements. The principal geodynamic movements on the Earth's surface and close to it can be regarded as the result of the rotation of relatively rigid and comparatively thin lithospheric plates. According to Euler's theorem, this rotation is described by a turning of the plate on a sphere at a definite angular velocity relative to the pole of rotation (Fig. 1.11). This causes a relative displacement of plates and results in three types of boundaries: divergent (deviation or crystallisation of plates), convergent (collision or absorption of plates, also called subduction) and transform (or sliding line). Sliding of plates relative to each other usually occurs along transform faults parallel to Euler's parallels (see Fig. 1.11).

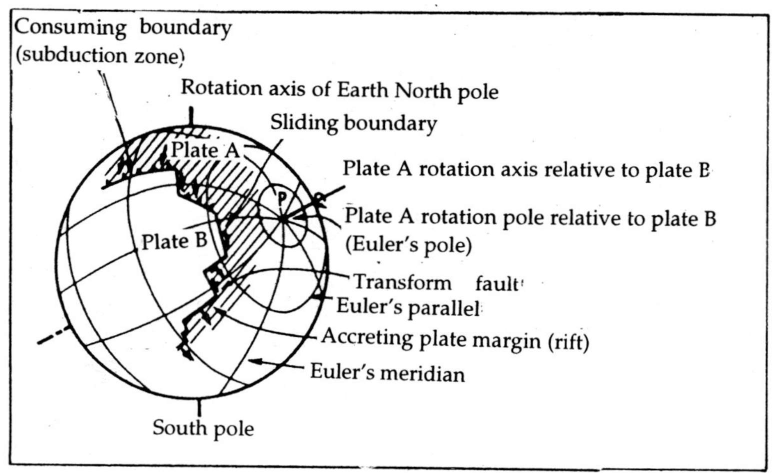

Fig. 1.11. Scheme of movements of plates A and B on the Earth's sphere around the axis with pole P.

When studying plate movements, a distinction should be made between present-day differential and final movements. Present-day plate movements occurring over a few tens of years are currently estimated on the basis of VLBI and satellite geodesy (Carter and Robertson, 1987). It has been found that these estimates are close to the measurements of vectors of plate displacements covering a very prolonged time interval (1 to 2.5 million years) along faults on the continent or along the distance between the youngest magnetic anomalies in the oceans.

The directions of present-day displacements can also be determined from stress orientations at the earthquake foci and from the strike of active segments of transform faults in the oceans and on the continents. Final displacements over a longer time interval (tens and hundreds of millions of years) can be estimated on the basis of reconstructions of the positions of continents. For example, Figure 1.12 depicts the final positions of Eurasia and Africa with North America which prevailed before the opening up of the Atlantic some 190 Ma ago (Zonenshain and Gorodnitskii, 1978; Zonenshain and Kuz'min, 1993a). With such an arrangement, opening of the Tethys ocean between Eurasia and Africa seemed inevitable. A comparison with the present position of these continents enables an estimation of the final displacements (represented here as rotations relative to the supposedly immobile North America). Maximum rotation was established for Africa. Nevertheless, these final movements also include relative differential displacements which may not coincide with the final displacement curves.

A more detailed theory and methods describing the kinematics of lithospheric plate movements have been incorporated in monographs of Zonenshain and Savostin (1979), Cox and Hart (1989) and Zonenshain and Kuz'min (1993a).

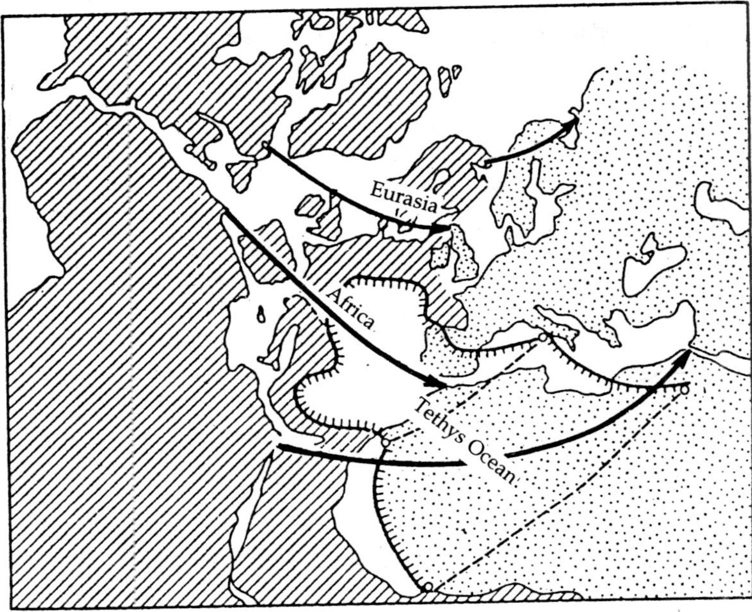

Fig. 1.12. Final positions of Eurasia and Africa with North America 190 Ma ago (hatched) and reconstruction of the Tethys ocean: dots reveal the present position of the characteristic points of Eurasian and African continents; arrows point to their final displacements while the toothed and broken lines show the contour and extent of the opening up of the Tethys ocean (Zonenshain and Kuz'min, 1993a).

Theoretical principles of lithospheric plate tectonics. Historically and logically, the theory of lithospheric plate tectonics emerged from many outstanding studies and hypotheses, among which the following can be distinguished:

a) The hypothesis regarding ocean-floor spreading (flow or widening) relative to mid-oceanic ridges (Hess, 1962; Dietz, 1961). It was later confirmed by the dependence of mountain topography and heat flux on age and other parameters (Parsons and Sclater, 1972; Sclater and Francheteau, 1970; Sclater et al., 1977). This hypothesis can now be regarded as proven. The fundamental mechanisms of this spreading-flow have been studied (see Chapter 5).

b) Magnetic anomaly belts established in the 60s in the oceans were interpreted as the result of spreading of the ocean floor under conditions of a variable magnetic field that has experienced reversals (Vine and Matthews, 1963, 1974). The magnetic anomalies were compared with the age of magnetic reversals established in the sediments on the continents. As a result the age of magnetised ocean floor (under the sediments) could be assessed and thus the velocity and direction of displacement of oceanic plates could be quantitatively estimated. In recent years, the age of the ocean floor under sediments has been determined at more than four hundred

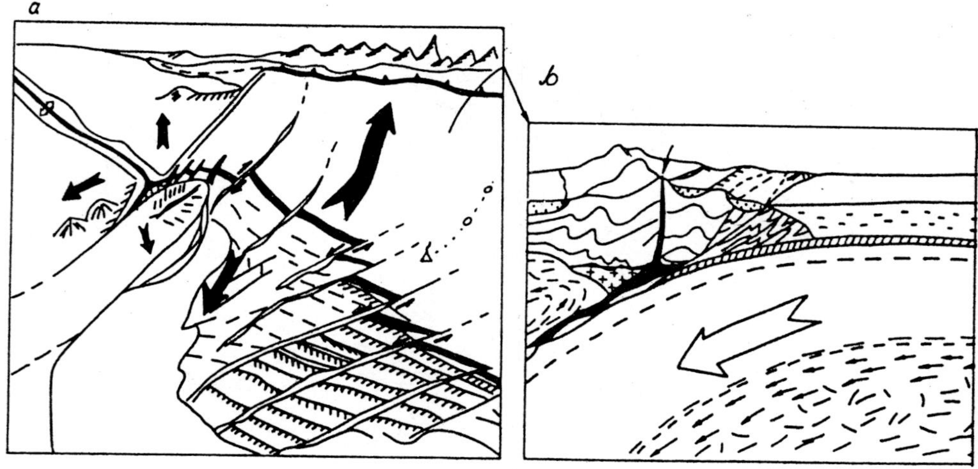

Fig. 1.13. Idealised sketches illustrating plate tectonics:
a) Divergent plate boundaries in rifts and ocean (spreading zones) with transform faults;
b) section of subduction zone; oceanic crust hatched; above it, the deformed rocks of accretionary prism and island arc; black portion represents melt in the subduction zone and volcanic systems.

points by deepwater drilling and, in 95% of them, the results coincided with the ages determined from the magnetic anomalies (*Oceanology...*, 1979; Scotese et al., 1988). Thus, this hypothesis may also be regarded as proven.

c) A new class of faults—transform faults—has been suggested in which the apparent displacement of segments of a mid-oceanic ridge is opposite to the true displacement of the moving plates (Wilson, 1965, 1974) (see Fig. 1.13). Examples of such displacements brilliantly confirmed by seismic and geological data are also discussed in Chapter 5.

d) Another hypothesis postulated that oceanic plates moving away from mid-oceanic ridges are absorbed under island arcs and active margins of continents. The absorption (subduction) zones, also called Benioff-Zavaritskii-Wadatti zones, were delineated primarily on the basis of seismic data (Isacks et al., 1974; Ueda, 1982). The subduction hypothesis was later confirmed by seismic profiling data, which clearly indicated the subsidence of an oceanic plate under an island arc (Zonenshain and Savostin, 1979; Magee and Zoback, 1993), by studies on geology and geochemistry of volcanism in these zones, structure of trench and island-arc slope, and by experimental modelling. Nevertheless, the mechanism and processes in subduction zones are extremely complex and debatable (see Chapter 5).

e) A very important hypothesis facilitating the use of plate rotation kinematics, is the concept of rigidity and integrity of plates relative to the plastic asthenosphere and 'plastic' (plastic-deformable) plate boundaries. This hypothesis was based on the identified pattern of seismic activity concentration only at the plate boundaries

(Isacks et al., 1974) and observations on the non-deformability of oceanic sediments and floor outside the axial part of the mid-oceanic ridge and transform faults. These features are far better confirmed for oceanic plates for which plate tectonics are applicable to the maximum extent. Insofar as parts of continents and active zones are concerned, the situation is more complex. Seismicity is widespread even inside the continents (e.g., in Eurasia) and in broad belts in many active zones (e.g., Alpine-Himalayan). This led to the need to delineate increasingly smaller plates and to complications in kinematics of their movement.

f) Hot spots (of the Hawaiian type) have been suggested as the result of the activity of independent mantle plumes rising from the lower mantle. Traces of plate movements above the hot spot remain in the form of a chain of volcanic islands of gradually increasing age (e.g., Hawaiian-Emperor Seamount Chain in the north-western part of the Pacific Ocean; see Fig. 4.18). This feature was also used as a direct proof of plate movement (Wilson, 1973; Molnar and Atwater, 1973; Berk and Wilson, 1977) and as a possible model of convective flows in the mantle (Morgan, 1971). The latter aspect continues to be studied (Irvine, 1988, 1991).

The kinematics of lithospheric plate movements can be described as an ensemble of three groups (see Fig. 1.3):

1) eight extremely large plates (Pacific Eurasian, Indo-Australian, Antarctic, South American, North American, Nazca and African) have been recognised right from the very postulation of the plate tectonics theory (Morgan, 1968; Le Pichon, 1968; Münster and Jordan, 1978).

2) 'small' plates 1000 to 3000 km in size, roughly double in number (Sea of Okhotsk, Amur, Chinese, Tibetan and Philippine in south-east Asia; Scotia, Coco and Caribbean together with Nazca and other smaller plates around the South American plate; Somalian, Arabian, Asia Minor and other smaller plates around the African plate); and

3) microplates within broad active zones (not indicated in Fig. 1.3) on the western fringe of the Pacific Ocean and in the Alpine-Himalayan belt (Lut, Menderes, Moesian plate, Sardinia etc.). The extent of these plates (300 to 1000 km) is commensurate with the thickness of the lithosphere, i.e., they are in the form of blocks and not plates; their presence confirms the more complex nature of deformations in the active zones.

Figure 1.13 depicts the overall qualitative plate tectonics based on the above-listed theoretical and empirical premises. To the left is shown the divergence of some plates whose boundaries are represented by rifts or mid-oceanic ridges. A triple junction of rifts is similar to that of African, Arabian and Somalian plates (see also Fig. 5.1). In the oceanic plates, transform faults, 'banded' topography of plates coinciding with the magnetic anomaly bands and traces of plate movements above a hot spot are shown schematically. In the upper part of Figure 1.13a is shown a subduction zone a section of which is depicted in Figure 1.13b. The section shows the subducting oceanic lithosphere with a thin oceanic crust, island arc above the subduction zone with intensely deformed accretionary wedge in the arc front and local sedimentary basins—forearc and back-arc. Convection flows in the asthenosphere below the subducting plate and in the back-arc region are shown schematically.

It must be emphasised that, notwithstanding the empirical and 'mechanistic' character of many propositions of plate tectonics, most of the observed geological and geophysical phenomena can be satisfactorily explained within its framework (Zonenshain et al., 1976, 1990; Zonenshain and Kuz'min, 1993a).

1.5. Theoretical Models and Mechanisms of Lithospheric Plate Movements

These models and mechanisms are still under debate and are hypothetical. The general proposition about the relation between lithospheric plate movements and convective flows in the mantle has been reported before (Morgan, 1971; Wilson, 1973, 1990). However, the actual mechanisms ensuring the relation of asthenospheric flows with plate tectonics continue to be discussed. Various mechanisms have been proposed: viscous coupling between asthenospheric convective flows and lithospheric plates; gravitational sliding of plates from axial uplifts of mid-oceanic ridges; sinking of lithospheric plates into the mantle as a result of negative buoyancy of slabs generated by lower temperature and (or) eclogitisation of subducting plates.

Repeated attempts were made to quantitatively assess the contribution of each mechanism to plate movement (Artyushkov, 1979, 1993; Ueda, 1982; Turcotte and Schubert, 1982). Recent evaluations (Lithgow-Bertelloni and Richards, 1995; Lithgow-Bertelloni et al., 1996) have led to the conclusion that, at least during the Cenozoic, the major cause of plate movements was the negative buoyancy of subducting slabs and other effects in the subduction zones, which provide 90% of the driving forces for plate movements; uplift of mid-oceanic ridges and zones of lithospheric thickening provide the remaining 10%. This conclusion is inapplicable, however, to the southern part of the Atlantic, African, South American and Nazca plates (see Fig. 1.2). Here, large-scale upwellings in the lower and upper mantle enjoy greater importance. They cause the movement of the South American plate, mainly under the action of viscous friction with intense mantle upwelling later flowing parallel to the Peruvian trench along the subducting Nazca plate (Lithgow-Bertelloni et al., 1996; Russo and Silver, 1996).

Several problems of plate tectonics continue to be debated. These are: the large dimensions of plates, the sizes of anticipated convection flows in the asthenosphere remaining small; initial subduction stages; causes for the break-up of supercontinents etc. To explain these and other debatable aspects, models that are more complex than whole-mantle convection are used: two-layer convection, convection with independent mantle jets (plumes) etc.

Most of the models discussed above still contain several arbitrary assumptions and are not developed on the basis of physical modelling that can reduce the ambiguities. These aspects will be discussed in Chapters 4 and 5 after presentation of the fundamentals of physical modelling and the possibility of experimental testing of the models. General models of mantle convection with various plume types in the Earth as well as models of the three basic processes responsible for the origin and stability of lithospheric plate movements will be viewed. These are:

a) spreading processs in mid-oceanic ridges,

b) subduction processes in island arcs and active margins and

c) collision processes of Himalayan and other types.

To conclude this section, Table 1.5 is given in which the basic principles of plate tectonics in the classic form, as presented in 1968 (Le Pichon, 1968) and in a more recent version (Khain, 1994), have been formulated with the improvements and modifications discussed in part above.

Palaeogeodynamic reconstructions. One of the most important outcomes of plate tectonics is the possibility of palaeogeodynamic reconstructions which facilitate determination of the past positions of continents, oceans and island arcs, as well as analysis of the evolution of folded belts that arose after the sealing of ancient oceans.

These reconstructions were made most reliably for the Cenozoic and Mesozoic epochs (Zonenshain et al., 1987; Scotese et al., 1988; Zonenshain and Kuz'min, 1993a) based on knowledge of the age of the ocean floor and palaeomagnetic data. By sequentially 'eliminating' the older sections (bands) of the ocean floor commencing from the youngest (up to 5 million years), a picture of the relative position of continents at different periods can be ascertained. Palaeomagnetic and geological data help verify these reconstructions.

The structure of the Permian supercontinent Pangaea that existed 200 to 250 Ma ago was obtained from Mesozoic reconstructions. Figures 1.14 (1) and (2) show

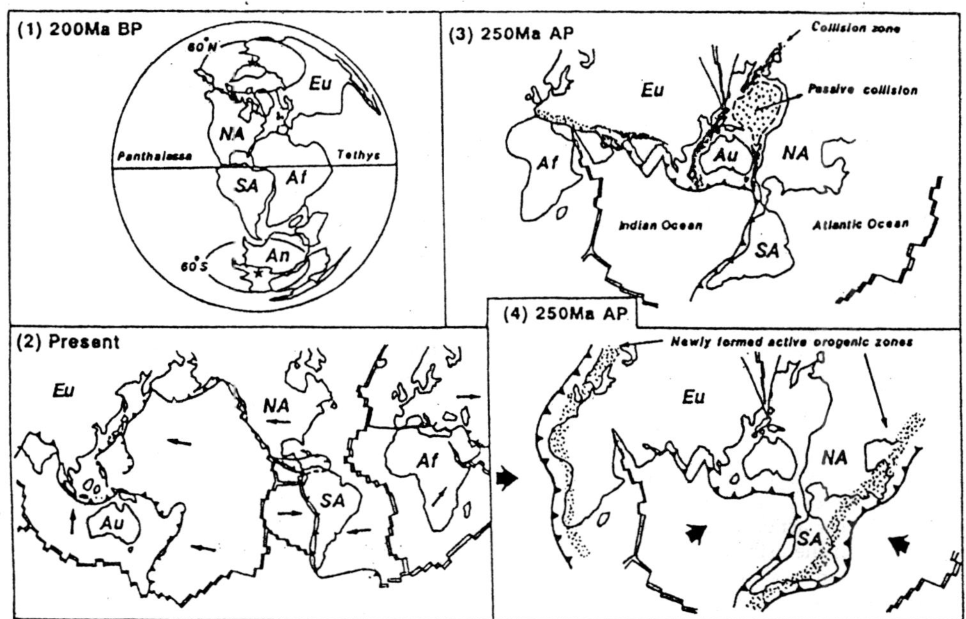

Fig. 1.14. Reconstruction of Pangaea 200 to 250 Ma ago (1) compared to the present position (2) of continents, oceans and subduction zones (toothed lines); compare also with Fig. 1.12. (3) and (4): possible position of continents 250 Ma hence (Maruyama, 1994); 3—collision zones in the new supercontinent and 4—transformation of Atlantic-type passive margins into Andean-type active margins.

Table 1.5. Comparison of basic principles of classic and modern plate tectonics (Khain, 1994, modified)

Classic plate tectonics, 1968	Modern plate tectonics, 1993
1) Upper part of solid Earth is divided into brittle lithosphere and plastic asthenosphere	1) Upper part of solid Earth is divided into lithosphere and asthenosphere but *the former has undergone stratification and deformations while thickness and viscosity in asthenosphere change significantly in lateral directions*
2) Lithosphere is divided into a limited number of large- and medium-size plates; major tectonic, seismic and magmatic activity is concentrated at the plate boundaries, when plates are moving relative to each other along the top of the asthenosphere	2) Lithosphere is divided into large-, medium- and *small-size plates. Between the large-size plates are belts involving a mosaic of small-size plates* while the *large plates per se are heterogeneous* along the vertical and lateral directions; major activity is concentrated at plate boundaries but is *manifest to a small extent even within the inner parts of plates*
3) Horizontal movements of lithospheric plates obey Euler's theorem	3) Horizontal movements of large- and medium-size plates obey Euler's theorem but *small-size plates may experience more complex movements*
4) Three principal types of relative plate movements are noticed: i) divergence manifested as rifting and spreading, (ii) convergence manifested as subduction and collision and (iii) shear displacements along transform faults	4) Three principal types of plate movements are noticed: i) divergence manifested as rifting and spreading, ii) convergence manifested as subduction, *obduction*, collision, *squeezing of mass in lateral direction also along shears or as deep-level injection* and (iii) shear movements along transform faults, *often together with compression (transpression) or strain*
5) Spreading in oceans is compensated by subduction and collision along their periphery, the radius and volume of Earth remaining constant	5) Spreading in oceans is compensated *not only by subduction and collision, but also by other processes* (obduction, shear and squeezing); the *question* of constant radius and volume of the Earth *is debatable.*
6) Displacement of lithospheric plates is caused by their drag under the action of convective flows in the asthenosphere; convection is thermal and encompasses the entire mantle	6) Displacement of lithospheric plates is caused not only by drag of convective flows but also *by tightening in subduction zone due to increase in weight (eclogitisation) and moving away from the axis of mid-oceanic ridges; convection is multilayered and complex*
7) Several very important geo-dynamic processes have not been covered in plate tectonics or have been unduly simplified	7) *Periodicity of endogenic processes, intraplate deformations, magmatism and process complexities at the plate boundaries should be considered*

the reconstruction of Pangaea 200 Ma ago compared to the present disposition of the continents. An attempt was also made to predict the possible position of continents with the formation of a new supercontinent 250 Ma hence (Maruyama, 1994). This reconstruction is based on present-day plate movements using which the position of continents in the near future can be judged quite rationally (Duncan and Turcotte, 1994). The cycle 'from Pangaea to Pangaea' spanning about 450 Ma (see below) and some additional assumptions were also used (Maruyama, 1994). Justification for these additional assumptions is quite problematic and hence depiction of the future supercontinent shown in Fig. 1.14 (3) and (4) is only one of many possible alternatives.

Palaeogeodynamic reconstructions for the Palaeozoic in the time interval 220 to 550 Ma ago are fairly well substantiated (Zonenshain and Gorodnitskii, 1978; Zonenshain et al., 1985, 1987) although many assumptions had to be made here also, besides extensive use of palaeomagnetic data. The largest transform faults (strike-slip faults) and traces of the movement of assumed hot spots have been used as the direction of movement of palaeoplates. Euler's pole of plate rotation can be determined from these directions.

One of the most widely used methods of plate tectonic reconstructions is the method of final superpositions. For this purpose, it is important to know the points suitable for superposition. These points are connected by the arc of a great circle and the tangent to this arc between the points gives the general azimuth of final displacement (see Fig. 1.10). If some arcs and azimuths corresponding to them are known, the position of Euler's pole of rotation can be calculated (Zonenshain and Kuz'min, 1993). Further, the relative positions of continents and microcontinents in different periods are inferred from the palaeomagnetic data.

For palaeotectonic reconstructions, it is extremely important to identify geological complexes that constitute indicators of the past geodynamic environments, using the principle of actualism. Such an approach represents a further development of formation analysis in geology. Especially important among such indicators are certain geological complexes at ancient margins, such as ophiolites with a suite of parallel dykes representing the oldest spreading zones: boninite series as indicators of 'primitive' island arcs; volcanites and granitoids of calc-alkali series indicating 'normal' island arcs; eclogites, glaucophane schists and melange zones as evidence of ancient fragments of subducted rocks (more precisely, a definite stage in the evolution of these zones leading to their 'jumping over' or cessation of subduction). Examples of these and other complexes and proof of the characteristics used are discussed in Chapter 5.

One result of plate-tectonic reconstructions of the Early Palaeozoic period is the formation of one or two more supercontinents in the Late Precambrian 750 to 800 or 950 to 1000 Ma ago (Hoffman, 1991; Dalziel et al., 1994; Young, 1995). Their break-up led to the birth of the Palaeo-Asiatic and Palaeo-Pacific oceans (Maruyama, 1994; Dobretsov et al., 1995a). Palaeoreconstructions of the ancient epochs encounter increasingly greater difficulties and it is hardly possible to reconstruct the precise position of the palaeo-oceans and palaeocontinents for these epochs.

Nevertheless, it has been established with a high degree of probability that Pangaean supercontinents existed many times in the history of the Earth, breaking up and accreting again at an average periodicity of 500 Ma, called the Wilson cycle (Wilson, 1990; Khain, 1994). Trubitsyn (Trubitsyn and Nikolaichik, 1991; Rykov and Trubitsyn, 1995) discussed the reasons for the break-up and formation of supercontinents within the framework of convection models. Further, there are also shorter Bertrand cycles of a duration of about 200 Ma (Khain, 1994) and a main geological periodicity of a duration of 30 or 60 Ma (Rampino and Caldeira, 1993; Dobretsov, 1988, 1994a, c). The reasons for these periodicities and the overall evolution of the Earth will be discussed at the end of Chapter 5.

To construct satisfactory geodynamic models, a sound physical basis is very important along with optimal consideration of geological facts. The theoretical principles of thermophysical modelling are discussed in Chapters 2 and 3.

2

Theoretical Principles of Heat and Mass Transfer Modelling

2.1. Earth as Heat Engine

The preceding chapter showed that transfer phenomena (primarily heat and mass transfer) of different spatial scales constitute the basis of geodynamic processes. These spatial scales vary from levels comparable to the Earth's radius to those of growing crystals although attention in future will be mainly devoted to large-scale processes. The largest geodynamic processes correspond to the relative movements of continents which give rise to global variations of environment and climate of the Earth and their occurrence involves maximum periods.

Processes of crystal growth or ore formation (depending on the size of vein or ore-bearing stratum) fall among small-scale phenomena which depend on large scales of movements in the crust and upper mantle of the Earth. Tracing their interrelations is quite a complex problem and confusion often occurs between cause and effect. Gradation of the spatial scales of geodynamic processes is tentative but nevertheless global, intermediate, regional and local scales can be distinguished, although qualitatively, as done by Zonenshain and Savostin (1979) (see also Fig. 1.10).

A definite interrelationship or interaction of movements of different scales is seen. Global flows caused by thermogravitational forces encompass the lower and upper mantle and give rise to displacements of intermediate spatial scale, which are manifest in the relative movements of lithospheric plates. Such displacements in turn influence global motions. During movements, plates accrete in some zones and are absorbed in others. Accretion occurs in the zone of ascending asthenospheric flows and convergent boundaries, while the plates are submerged in the region of descending movement (subduction). Subduction phenomena also influence global movements since they provide the boundary conditions for the latter. The interaction of these movements reveals in inverse relation: as the plate movement in the lithosphere changes, the structures of mantle flows also changes.

The lithospheric plate movements manifest themselves in the regions of interaction at their boundaries as orogenic, volcanic, metamorphic, seismic and relief-forming processes. For geologists, these processes serve as indicators of plate boundaries.

Thermogravitational forces, i.e., forces that arise in the gravitational field during density changes due to temperature variations play a decisive role in all these movements. For example, the fluid above the heated surface becomes heated and ascends while a colder fluid descends. Atmospheric and ocean flows as well as geodynamic displacements occur in the 'solid' Earth under the action of thermogravitational forces.

As a first approximation, the Earth can be regarded as a heat engine: part of the internal (generated) heat of the Earth is converted into the work of moving masses. In this sense, the primary form of energy is thermal while thermogravitational flows represent a consequence of heat transfer. Various processes can provide the source of heat energy: radioactive decay, exothermic reactions, gravitational contraction and friction. Thermogravitational flows have governed the evolution of the planet Earth and its structure at various periods. By studying the structure of flows in the Earth's interior, the history of its development can be retrospectively assessed. Thus processes of heat and mass transfer play a decisive role in the study of geodynamics and evolution of the Earth.

Heat transfer may occur by heat conduction when heat is transferred by way of molecular interactions. Heat transfer may also occur by convection when a moving mass of material functions as the carrier of heat energy. Convective heat transfer is observed in a viscous fluid when displacement of fluid layers relative to each other is possible.

2.2. Elastic and Viscous Properties of the Earth

Under varying conditions, the bulk Earth behaves as an elastic or viscous body (e.g., as a result of earthquakes or powerful tides). The elastic and viscous states of a substance are primarily distinguished by their mechanical properties. Let us study this on the example of the dependence of shear stress on deformation.

The elastic state is characterised by the linear relation of shear stress on strain (Fig. 2.1) known as Hook's law:

$$\tau_{xy} = \frac{G\partial\xi}{\partial y} = Ge, \tag{2.1}$$

where G (N/m^2) is the shear modulus and $e = \partial\xi/\partial y$ the strain. If stress exceeds the ultimate strength, the body breaks up. Under multidirectional compression

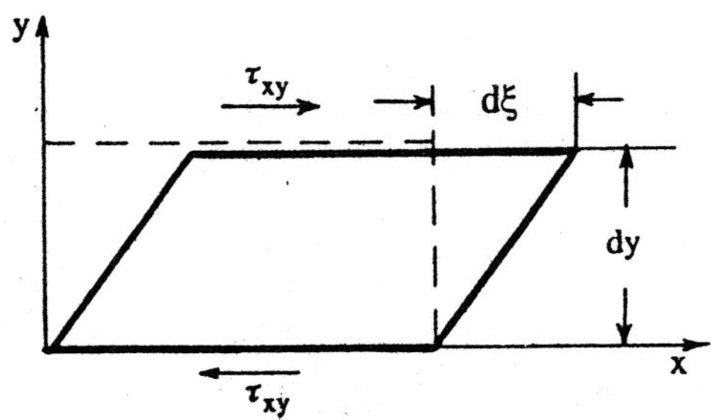

Fig. 2.1. Deformation caused by shear stress τ_{xy}.

greater than the level of ultimate strength, the body enters the state of creep, i.e., the viscous state.

For a viscous material it is experimentally established that the shear stress (stress of friction) depends not on the strain, but on the rate of its variation:

$$\tau_{xy} = \frac{\eta \partial e}{\partial t} = \eta \left(\frac{\partial}{\partial t}\right)\left(\frac{\partial \xi}{\partial y}\right)$$

$$= \eta \left(\frac{\partial}{\partial y}\right)\left(\frac{\partial \xi}{\partial t}\right). \tag{2.2}$$

Since deformation rate $u = \dfrac{\partial \xi}{\partial t}$,

$$\tau_{xy} = \frac{\eta \partial u}{\partial y}. \tag{2.3}$$

Equation (2.3) shows that the shear stress is proportional to the velocity gradient. Such fluids are called viscous or Newtonian fluids and eqn (2.3) Newton's law. Proportionality coefficient η represents the coefficient of dynamic viscosity and is expressed as $N \cdot s/m^2$ or $Pa \cdot s$. For example, at 20°C, $\eta = 10^{-3} N \cdot s/m^2$ for water and $\eta = 1.48 \, N \cdot s/m^2$ for glycerine. At temperatures of about 1500°C, $\eta = 10^{19} \, N \cdot s/m^2$ for the asthenosphere.

To illustrate the properties of a viscous fluid, let us study its flow between two parallel plates, of which the lower one is fixed and the upper one moves in its own plane at velocity U; the distance between the two plates is l (Fig. 2.2). The experiment shows that the fluid adheres to both the plates and that the rate depends linearly on y:

$$u\,(y) = U\left(\frac{y}{l}\right). \tag{2.4}$$

For such a motion to occur, tangential force A in the direction of flow should be applied to the fluid from the upper plate. This force counterbalances the friction of the fluid. The experiment shows (Fig. 2.2) that this force acting on area S of the

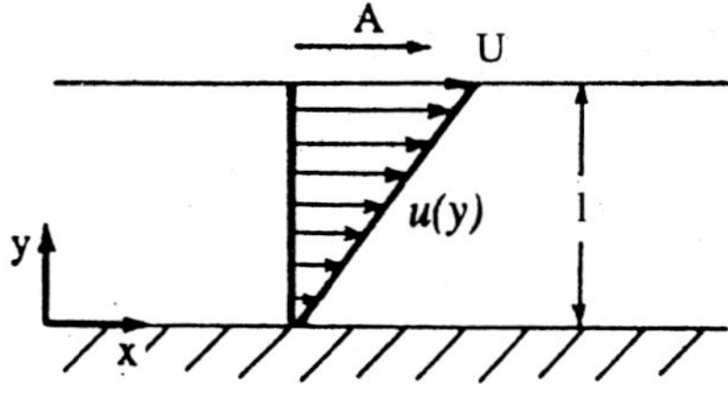

Fig. 2.2. Velocity distribution in a viscous fluid flow between two parallel flat walls (Couette flow).

plate is directly proportional to velocity U and inversely proportional to the distance between the plates. It follows from eqns (2.3) and (2.4) that

$$\tau = \eta \partial/\partial y \, (Uy/l) = \eta \, U/l. \tag{2.5}$$

What then is our Earth—viscous or elastic? This depends on the time scales of the forces operation. For seismic waves in which the periods of oscillation are 0.1 to 10 s, the bulk Earth behaves as an elastic body. The velocities of seismic longitudinal and transverse waves for such a body are determined using relations (1.1) and (1.2).

Even for these periods, the outer core is fluid with no transverse seismic waves in it ($V_s = 0$). Elastogravitational waves (intrinsic oscillations of the Earth) have an oscillation period of 3 to 59 min. At these periods, the Earth represents an elastic body. The Earth, however, responds to the pull of the Moon as a viscous fluid.

Let us attempt to estimate for what oscillation periods or, in other words, for what characteristic periods of the process, the Earth stratified along the radius can be regarded as elastic or viscous. For this purpose, let us examine the characteristic period of the process in a viscoelastic fluid. These fluids exhibit elastic properties over short time-intervals but viscous properties over a long period.

Let us design the model of a viscoelastic medium for which the elastic and viscous properties act additively on the deformation rate. For such a medium, the linear deformation rate is the sum of linear velocities of elastic and viscous deformations:

$$\partial e/\partial t = \partial e_y/\partial t + \partial e_b/\partial t, \tag{2.6}$$

where $e = \partial \xi/\partial y$ is the relative deformation (e.g., elongation). Such a viscoelastic medium is called the Maxwell medium. Let us assume that a unidirectional shear stress acts on this medium (see Fig. 2.1). It can be seen from the Figure that elastic deformation $e_y = \tau/G$. In this case, the rate of elastic deformation is determined from the equation

$$\partial e_y/\partial t = (1/G) \, d\tau/dt. \tag{2.7}$$

According to Newton's law, the rate of variation of strain for viscous flow will be:

$$\partial e_b/\partial t = (\partial/\partial t) \, (\partial \xi/\partial y) = (\partial/dy) \, (\partial \xi/\partial t)$$

$$= \partial u/\partial y. \tag{2.8}$$

Since, according to eqn (2.3), $\tau = \eta \, \partial u/\partial y$,

$$\partial e_b/\partial t = \tau/\eta. \tag{2.9}$$

The overall rate of viscoelastic deformations, according to eqns (2.6), (2.7) and (2.9) will be

$$de/dt = (1/G) \, d\tau/dt + \tau/\eta. \tag{2.10}$$

Equation (2.10) represents a relation between the deformation rate, tangential stress and rate of stress variation. Let the viscoelastic medium experience sudden deformation e_0 at $t = 0$ which, at $t > 0$, remains constant. At the moment of rapid deformation, the main term of eqn (2.10) is the first one, which is the derivative with respect to time, $d\tau/dt$. In other words, during small time-intervals of deformation, the medium behaves as an elastic body. Therefore, initially, $\tau_0 = Ge_0$. Later, deformation remains constant, i.e., $de/dt = 0$. In this case, it follows from eqn (2.10) that

$$(1/G)\, d\tau/dt + \tau/\eta = 0;\ d\tau/\tau = -(G/\eta)\, dt. \tag{2.11}$$

Let us integrate eqn (2.11) for the initial condition $\tau = \tau_0$ at $t = 0$:

$$\ln \tau = -Gt/\eta + c;\ \tau = c \exp(-Gt/\eta). \tag{2.12}$$

Considering the boundary condition, we obtain:

$$\tau = \tau_0 \exp(-Gt/\eta). \tag{2.13}$$

Let us determine the time over which initial stress τ_0 changes by e times:

$$Gt_0/\eta = 1;\ t_0 = \eta/G. \tag{2.14}$$

This interval (t_c) is called the time of viscoelastic relaxation of the process.

Let us estimate the relaxation time of the asthenosphere by considering it a viscoelastic medium. According to the above estimates, we adopt the probable value of the dynamic viscosity coefficient for asthenosphere $\eta = 10^{19}\ \mathrm{N \cdot s/m^2}$. The value of $G \approx 7 \times 10^{10}\ \mathrm{N \cdot s/m^2}$. In this case, $t_0 = 1.43 \times 10^8$ s or 4.55 years; for the lower mantle $\eta = 10^{21}\ \mathrm{N \cdot s/m^2}$ and $t_0 = 455$ years; for different sections of the lithosphere $\eta = 10^{23}$ to $10^{27}\ \mathrm{N \cdot s/m^2}$. (see Chapter 1, Fig. 1.8) and $t_0 = 45{,}000$ years to 455 million years.

A medium can be regarded as viscous if the duration of the application of forces or their variation $t \gg t_0$. Thus, the asthenosphere can be regarded as viscous if $t \gg 4.55$ years. For $t \gg 455$ years, the lower mantle is a viscous liquid and for $t \gg 45{,}500$ years or $t > 455$ million years, the lithosphere is a viscous liquid. Although approximate, these time durations are extremely long relative to the periods of seismic waves but concomitantly extremely short compared to the duration of the geological processes. Thus, seismic tomography helps determine quite accurately the structure of the Earth as an elastic body at a given geological moment of time.

A reverse problem has to be solved to understand the geological history of the Earth. One of the methods of such a study is to develop a physical model and, on its basis, a mathematical model, and solve the direct or the inverse problem for the given boundary conditions. The estimates given above point out that for global geophysical processes, the Earth (except for an upper part of the lithosphere) can be regarded as a viscous liquid, i.e., the mathematical model of a viscous fluid can be used. In this case also we have no knowledge of the physical properties of the complex stratified Earth. Most ambiguous are the initial conditions of the evolution of the Earth (e.g., whether the Earth was gaseous, molten or solid at commencement of accretion). Even if these conditions have been clarified, there is still the question about an algorithm for solving this complicated problem. Laboratory modelling

44

based on a simplified physical and mathematical model using the theory of similarities is also possible. Therefore, the principles of this theory and methods of its application will be discussed below. For this purpose, however, equations of energy, momentum and mass conservation have to be derived first.

2.3. Heat Transfer

Depending on pressure and temperature, matter changes from solid (crystalline) into liquid and later into gaseous states. A knowledge of these parameters is therefore essential for understanding the aggregate state of different materials in the Earth's interior. The rheology of rocks also changes with temperature and pressure. Temperature can be ascertained by studying the heat transfer processes in the interior of the Earth; it mainly depends on the mechanism and rate of heat transfer from the interior of the Earth to the surface.

Molecular heat transfer (thermal conduction) occurs when energy is transmitted through collisions between molecules, i.e., as diffusive transfer of kinetic energy from one molecule to another. This kind of heat transfer occurs in the solid as well as the liquid and gaseous states. The influence of this process of heat transfer is most significant in the Earth's upper envelopes.

Convection is associated with the flow of a medium (liquid or gas). For example, under conditions of thermogravitational convection, a hot fluid ascends, transporting accumulated heat. Contrarily, a cold fluid descends and cools the surrounding medium by mixing with it. Thermal convection, as will be shown below, is of decisive importance for the mantle and the outer core of the Earth.

Radial transfer involves energy transport by electromagnetic radiations. Examples are the energy transport in certain solar zones and from the Sun to the Earth. Radial heat transfer deep within the Earth is insignificant and its effect can be taken into account, when required, by introducing a correction to the coefficient of heat conductivity.

2.4. Heat Conduction and Temperature Gradients

In a general case, temperature T at any point in space is a function of co-ordinates x, y and z and time t, i.e., $T = f(x, y, z, t)$. The totality of temperature values for all points in space forms the temperature field for a given moment of time. The heat transfer regime is regarded as an unsteady state if the temperature field depends on time. If, however, the temperature at each point is constant independent of time, the heat transfer regime is in a steady state. The steady-state temperature field can be a function of one co-ordinate $T(x)$, two $T(x, y)$ or three $T(x, y, z)$ co-ordinates. Correspondingly, the temperature field is one-, two- or three-dimensional.

Isothermal surfaces represent geometric sites of points having the same temperature. Such surfaces for different temperatures do not intersect each other but are closed upon themselves or terminate at the boundary of the body. Temperature does not change along an isothermal surface but varies at the intersection of isotherms. The most rapid temperature changes occur at intersections along the normal (n) to the isothermal surface (Fig. 2.3). In this direction (n), the maximum *temperature gradient* which represents the limit of the ratio of temperature variation ΔT to the distance between the isotherms along the normal Δn is:

$$\lim_{\Delta n \to 0} \Delta T/\Delta n = \partial T/\partial z = \operatorname{grad} T \ (^{\circ}\mathrm{C/m}), \tag{2.15}$$

where z is an arbitrary co-ordinate (in particular, along the radius of the Earth) coinciding with the normal.

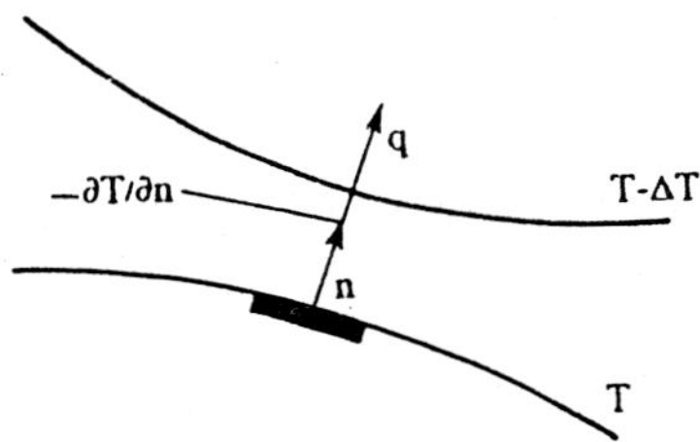

Fig. 2.3. Isothermal surfaces and temperature gradient.

Temperature gradient is a vector oriented along the normal to isothermal surface at a point corresponding to it.

By studying thermal conduction in solid bodies, Fourier established that the quantity of heat transferred is proportional to the temperature gradient, time and cross-sectional area perpendicular to the vector of temperature gradient, i.e., to the direction of heat transfer.

Let us consider the specific quantity of heat per unit area transferred in unit time. This is called the specific heat flux q (W/m^2).

In this case the dependence established experimentally can be expressed by the following equation:

$$q = -\lambda \, \partial T/\partial n = -\lambda \operatorname{grad} T. \tag{2.16}$$

The specific heat flux is a vector coinciding in its direction with the propagation of heat and oriented opposite to the temperature gradient (see Fig. 2.3). Therefore, a minus sign enters into the main equation (2.16), reflecting the objective law of nature: heat is transmitted from a hot to a cold body. Equation (2.16) expressing the main law of heat transfer by thermal conduction is called the *Fourier law*. The proportionality coefficient in eqn (2.16) is called the *coefficient of thermal conductivity and characterises the capacity of a substance to conduct heat. It follows from eqn (2.26) that*

$$\lambda = -q/\operatorname{grad} T \ (\mathrm{W/m} \cdot {}^{\circ}\mathrm{C}). \tag{2.17}$$

The value of the thermal conductivity coefficient can be described as the quantity of heat transmitted in unit time as temperature changes by 1°C per unit length. The thermal conductivity of different substances differs and depends on their struc-

46

ture, density, moisture (fluid saturation), pressure and temperature. For example, for crystalline quartz, $\lambda = 6.5$ to 11.3 W/m · °C depending on the direction parallel to crystallographic crystal axes; for peat, $\lambda = 0.07$ W/m · °C; for shale, $\lambda = 1.7$ W/m · °C at depth of 400 m; and, for chalk, $\lambda = 1.5$ W/m · °C etc.

Let us examine **heat transfer through a homogeneous plate** having constant wall temperatures T_1 and T_2 (Fig. 2.4a). Temperature varies only in the direction of the y-axis. In this case, the isotherms are in the form of planes normal to the y-axis. Equation (2.26) is applicable for an isolated elementary layer dy and represents a differential equation of steady-state heat transfer by thermal conduction is a one-dimensional case. Let us rewrite eqn (2.16) as $dT = -q\,dy/\lambda$ and integrate it for the above case:

$$T = -qy/\lambda + c \tag{2.18}$$

Integration constant c is determined from the boundary conditions at $y = 0$, $T = T_1$; and $c = T_1$. In this case,

$$T = \frac{-qy}{\lambda} + T_1 \tag{2.19}$$

Let us determine the unknown value q from eqn (2.19) and the second boundary condition at $y = l$ and $T = T_2$:

$$q = \frac{-\lambda(T_2 - T_1)}{l} = \frac{-\lambda\Delta T}{l} \tag{2.20}$$

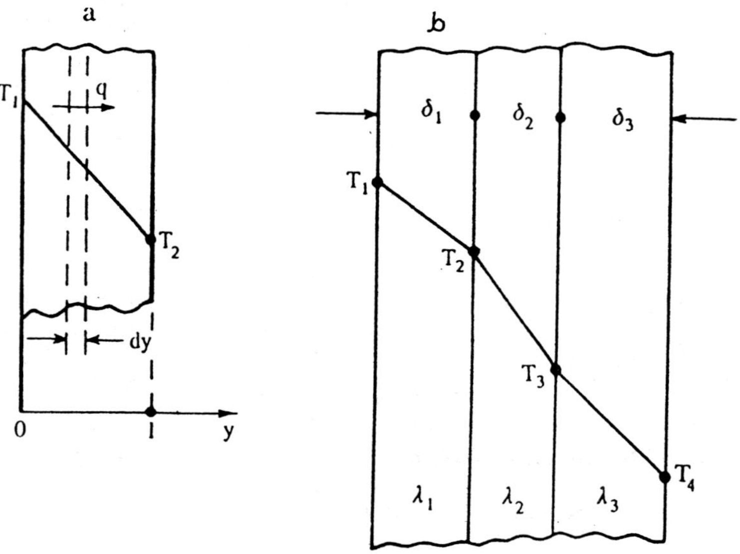

Fig. 2.4. Uniform (a) and multilayer (b) flat walls.

By knowing the value of heat flux q, the quantity of heat transferring through a flat wall of area S (m^2) in time t can be determined using eqn (2.20):

$$Q = qSt = \left\{\frac{\lambda}{l}\right\} \Delta T S t.$$

From eqns (2.19) and (2.20) we obtain the equation for temperature distribution across the wall, which is represented as a linear function (see Fig. 2.4):

$$T(y) = T_1 - (T_1 - T_2)\frac{y}{l}. \tag{2.21}$$

To determine heat flux along a direction for which grad T is known, it is necessary to determine the coefficient of thermal conductivity (λ) of the rock. The *method of multilayer wall* is one of the means of determining λ. A package of three plates immediately adjacent to one another comprises two lateral plates of known thermal conduction λ_1 and λ_3 (see Fig. 2.4b) and the specimen for which λ_2 is to be determined placed between them. Constant temperatures T_1 and T_4 are maintained by thermostats. In this case, q is constant in all the three layers. Based on eqn (2.20), the following relation can be expressed for each layer:

$$q = \frac{\lambda_1(T_1 - T_2)}{\delta_1} = \frac{\lambda_2(T_2 - T_3)}{\delta_2}$$

$$= \frac{\lambda_3(T_3 - T_4)}{\delta_3} \tag{2.22}$$

The temperature variation in each layer is determined from eqn (2.22) as:

$$T_1 - T_2 = \frac{q\delta_1}{\lambda_1}; \; T_2 - T_3 = \frac{q\delta_2}{\lambda_2};$$

$$T_3 - T_4 = \frac{q\delta_3}{\lambda_3}. \tag{2.23}$$

By summing up the right and left sides of eqns (2.23), we obtain the following relation:

$$T_1 - T_4 = q\,(\delta_1/\lambda_1 + \delta_2/\lambda_2 + \delta_3/\lambda_3); \tag{2.24}$$

the value of q for known δ/λ can be determined from the above equation. The value of λ_2 of the specimen at known λ_1/δ_1 and λ_3/δ_3 can be obtained from eqn (2.22):

$$\lambda_1\,(T_1 - T_2)/\delta_1 = \lambda_2\,(T_2 - T_3)/\delta_2;$$

from this the following relation is found:

$$(T_2 - T_3)/(T_1 - T_2) = \lambda_1\delta_2/\lambda_2\delta_1 + c. \tag{2.25}$$

The value of c is determined as the quantity caused by temperature difference at the contact between layers. Usually, the temperature ratio for different specimen thicknesses is found and the dependence of $(T_2 - T_3)/(T_1 - T_2)$ on δ_2 derived. Later, the gradient of straight line $d[(T_2 - T_3)/(T_1 - T_2)]/d\delta_2 = \mathrm{tg}\,\alpha = \lambda_1/\lambda_2\,\delta_1$ is found and finally $\lambda_2 = (\lambda_1/\delta_1)\,\mathrm{tg}\,\alpha$ determined. This excludes the unknown value c associated with temperature drop at the contact.

2.5. Heat Flux at the Earth's Surface

It was noted long ago that the temperature in mines and boreholes increases with increase in depth, the temperature gradient usually being 20 to 30°C/km. Since the thermal conductivity coefficient of rocks varies in the range of 2 to 3 W/m · °C, heat flux at the surface, according to eqn (2.16), will be equal to $q_0 = 40$ to $90\ \mathrm{mW/m^2}$. The heat flux at the Earth's surface is expressed in heat flow units (HFU) as:

$$1\ \mathrm{HFU} = 10^{-6}\ \mathrm{cal/cm^2}\ \mathrm{s} = 41.84\ \mathrm{W/m^2}.$$

Heat flux at the continents can be accurately measured in deep boreholes. This is because heat flux at the Earth's surface is influenced by diurnal, seasonal and annual temperature fluctuations and also probably by groundwaters close to the surface. The initial temperature gradient in the borehole is disturbed by the drilling process itself and by the circulation of drilling mud. Therefore, measurements are made 2 to 3 years after a stable (initial) temperature gradient is established. The latter is usually measured using thermistors of high temperature resolution. Thermistors are lowered into the borehole and temperature difference determined between them.

The average heat flux measured in all the continents is $q_0 = 56.5\ \mathrm{mW/m^2}$ (1.35 HFU) (Zharkov, 1983). A high heat flux of 63.6 $\mathrm{mW/m^2}$ (1.52 HFU) has been reported in the stable cratons in Australia, its least value or 49.8 $\mathrm{mW/m^2}$ (1.19 HFU) being recorded in Africa. In stable regions there is a correlation between the heat flux and concentration of radioactive isotopes, the heat flux (q_0) increasing with an increase in content of radioactive isotopes. Decrease in heat flux is also related to increase in age of the rocks. The distribution of heat flux has a complex character (Fig. 2.5). High heat fluxes are noticed in the zones of active volcanism (in Kamchatka) or in the zones under tension and circulation of deep-seated fluids (in Western Siberia).

The temperature of the ocean floor is fairly constant with little variation (3–4°C) and hence the surface heat flux in the oceanic crust is almost constant over time. However, hydrothermal circulation in the basalt rocks of the floor can influence the value of heat flux. Much of the ocean floor is covered by silts or a bed of soft sediments. Therefore, *heat flux measurement at the ocean floor* is carried out by needle-shaped probes about 3 m in length. The probe lowered from a ship penetrates by gravity into the bed of sediments. Temperatures gradient is determined by means of thermistors located at different depths. The value of λ for the sediments is determined using a heating element set up on the probe. After it is switched on, the soil is heated and temperatures variations with time are recorded by the probe. The coefficient of thermal conductivity of the medium is calculated from these variations.

The average heat flux for all the oceans is $q_0 = 78.2\,\text{mW/m}^2$ (1.87 HFU). The highest heat flux is measured in the northern part of the Pacific Ocean at $q_0 = 95.4\,\text{mW/m}^2$ (2.28 HFU) and the lowest in the South Atlantic at $q_0 = 59\,\text{mW/m}^2$ (1.41 HFU). The concentration of radioactive isotopes in the oceanic crust is two orders less than in the continental crust while the oceanic crust is one-fifth as thick as the continental crust, i.e., the radioactivity of oceanic crust contributes not more than 2% of the total oceanic heat flux.

On the ocean floor, the specific heat flux is related to the age of the rocks: with increase in distance from the axis of a ridge, the age of the rocks increases while the heat flux decreases. Since the contribution of radiogenic heat in the oceans to the heat flux is very small, the average temperature gradient in the oceanic crust and lithosphere, according to Fourier's law, can be expressed as:

$$(\partial T/\partial y) = q_0/\lambda = 78.2 \times 10^{-3}/(2 \text{ to } 4) = (39.1 \text{ to } 19.5) \times 10^{-3}\,°\text{C/m},$$

where $\lambda = (2 \text{ to } 4)\,\text{W/m}°\text{C}$. In this case, the temperature variation with increasing depth can be determined from the equation

$$T = T_0 - (19.5 \text{ to } 39.1)\,y \times 10^{-3}, \tag{2.26}$$

where T_0 is the temperature at the surface (of the ocean floor) and y is the depth co-ordinate (at the surface, $y = 0$). It follows from eqn (2.26) that at depth $y = 5 \times 10^4$ m, the temperature will be 975 to 1950°C without taking into account the rise of temperature with depth due to adiabatic compression (i.e., without considering the adiabatic gradient of temperature). It follows from the averaged estimates that the melting temperature of basalt (1100 to 1200°C) may be reached on the ocean floor at these depths, i.e., basalt melt will be stable and will impregnate the rock or will be concentrated at this depth. Oceanic crust, as noted above, is formed as a result of basaltic melt crystallisation.

If, however, heat transfer were due to thermal conduction alone, a rapid rise in temperature would have been seen and, according to eqn (2.26), even at $y = 100$ km, the temperature would have reached the level of olivine melting. In this case, even mantle residues (restites) enriched with olivine would have been melted. This, however, has not occurred, as evident from the propagation of transverse seismic waves in the mantle under oceans. This suggests that the temperature gradient under constant heat flux decreases with depth. This is possible only when there is a more intense heat transfer, i.e., convection. Thus, heat transfer processes in the mantle under oceans occur predominantly by convection. Our problem therefore is to derive an equation of energy conservation in a general form when heat transfer occurs in a fluid flow capable of absorbing or generating heat (i.e., in the presence of internal heat sources).

2.6. Differential Equation of Heat Transfer (Energy)

The main problem when studying a physical processs is to establish the relation between various parameters characterising this process. Usually, complex phenomena have several characteristic parameters which vary in space and time and hence relations between them are difficult to establish. In such a case, the dependence between variables characterising this process (i.e., between the co-ordinates, time and physical parameters) is established for an elementary differential volume on the basis of the general laws of physics. As a result, a differential equation is derived for the process under study. By integrating it, the analytical dependence between the values for the entire field of integration and for the time interval under review is obtained. Equations are similarly derived for thermal processes. To find the temperature distribution in the flowing fluid, it is necessary to also know the distribution of velocity in the volume and in time. Therefore, the heat transfer is described by a system of equations consisting of differential equations of continuity, motion (hydrodynamics) and heat transfer (energy). First, we write the heat transfer equation for a flowing fluid and then the equations of motion and continuity.

Let us formulate the heat balance for a moving fluid elementary volume. In an incompressible fluid, the heat balance of an elementary volume is determined by this internal energy, thermal conduction, heat convection and heat generation by internal friction. Let us consider an elementary parallelepiped with faces dx, dy and dz in the fluid flow (Fig. 2.6).

Consider the physical parameters, viz., thermal conduction λ, heat capacity c_p and density ρ as constant. Ignoring the heat generated by internal friction and change of pressure, the amount of heat entering the elementary volume $dx\,dy\,dz$, according to the first law of thermodynamics would be equal to the change of its enthalpy. Let us determine the heat inflow through the faces of the element by thermal conduction. The quantity of heat transferring in time dt in the direction

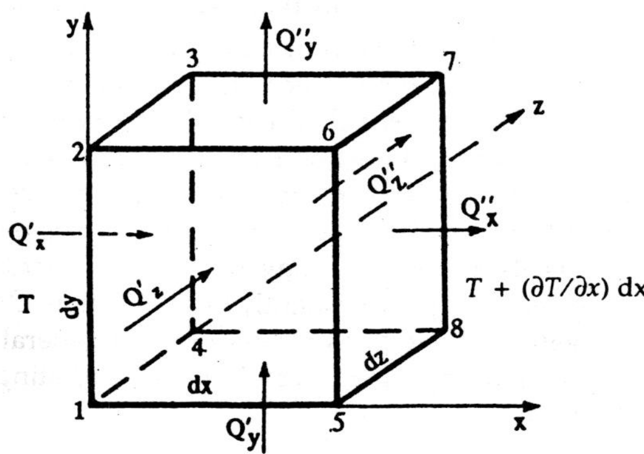

Fig. 2.6. Deriving the energy conservation equation.

of the x-axis through face 1234 (see Fig. 2.6) according to eqn (2.16) will be equal to:

$$Q'_x = -\lambda \left(\frac{\partial T}{\partial x}\right) dy\, dz\, dt. \tag{2.27}$$

The quantity of heat transferring in time dt through face 5678 with temperature $T + (\partial T/\partial x)\, dx$ will be equal to

$$Q''_x = -\lambda \left(\frac{\partial}{\partial x}\right)\left[T + \left(\frac{\partial T}{\partial x}\right) dx\right] dy\, dz\, dt. \tag{2.28}$$

The change of heat content in the elementary volume due to heat transfer in the direction of the x-axis is found by subtracting eqn (2.28) from (2.27):

$$\partial Q_x = Q'_x = Q''_x = \lambda \left(\frac{\partial^2 T}{\partial x^2}\right) dx\, dy\, dz\, dt. \tag{2.29}$$

We find the change in quantity of heat transferring in the direction of y- and z-axes in a similar manner:

$$\partial Q_y = \lambda \left(\frac{\partial^2 T}{\partial y^2}\right) dx\, dy\, dz\, dt, \text{ and} \tag{2.30}$$

$$\partial Q_z = \lambda \left(\frac{\partial^2 T}{\partial z^2}\right) dx\, dy\, dz\, dt. \tag{2.31}$$

Internal heat sources ∂Q_i with specific heat release or absorption Q_1 (W/m^3) are also possible in the elementary volume. The overall change in heat content in body $dx\, dy\, dz$ can be determined by combining eqns (2.29), (2.30) and (2.31) and adding to the result the heat of internal sources:

$$\partial Q = \partial Q_x + \partial Q_y + \partial Q_z + \partial Q_i =$$

$$= \lambda(\partial^2 T/\partial x^2 + \partial^2 T/\partial y^2 + \partial^2 T/\partial z^2)\, dx\, dy\, dz\, dt + Q_1\, dx\, dy\, dz\, dt. \tag{2.32}$$

Due to heat supply in time dt, the total change in temperature of the volume under consideration changes by DT/Dt and enthalpy by value:

$$\partial Q = c_p \rho\, (DT/Dt)\, dx\, dy\, dz\, dt. \tag{2.33}$$

By equating the right sides of eqns (2.32) and (2.33), we find that

$$c_p \rho\, (DT/Dt)\, dx\, dy\, dz\, dt = \lambda(\partial^2 T/\partial x^2 + \partial^2 T/\partial y^2 +$$

$$+ \partial^2 T/\partial z^2)\, dx\, dy\, dz\, dt + Q_1\, dx\, dy\, dz\, dt. \tag{2.34}$$

52

On rewriting eqn (2.34), we obtain the Fourier-Kirchhoff differential equation of heat transfer:

$$DT/Dt = (\lambda/c_p\rho)\,(\partial^2 T/\partial x^2 + \partial^2 T/\partial y^2 + \partial^2 T/\partial z^2) + Q_1/c_p\,\rho. \tag{2.35}$$

The left side of eqn (2.35) represents the total time derivative. It is known from mathematical analysis that the total derivative of a complex function is equal to the sum of products of partial derivatives of this function by an appropriate derivative of an intermediate argument with respect to an independent variable. In our case, the total time derivative is as follows:

$$DT/Dt = (\partial T/\partial x)\,(dx/dt) + (\partial T/\partial y)\,(dy/dt) + (\partial T/\partial z)\,(dz/dt) + \partial T/\partial t, \tag{2.36}$$

where derivatives dx/dt, dy/dt and dz/dt represent the velocity components along the x-, y-, and z-axes: u, v and w. Such a derivative associated with a fluid flow is called a substantial derivative and is expressed as

$$DT/Dt = \partial T/\partial t + u\,(\partial T/\partial x) + v(\partial T/\partial y) + w(\partial T/\partial z). \tag{2.37}$$

Here, $\partial T/\partial t$ is the local change of temperature with time and $[u(\partial T/\partial x) + v(\partial T/\partial y) + w(\partial T/\partial z)]$ is the variation of T due to convection. By substituting eqn (2.37) in (2.35), we obtain the differential equation of energy transport:

$$\partial T/\partial t + u(\partial T/\partial x) + v(\partial T/\partial y) + w(\partial T/\partial z) = a(\partial^2 T/\partial x^2 + \partial^2 T/\partial y^2 +$$

$$+ \partial^2 T/\partial z^2) + Q_1/c_p\rho, \tag{2.38}$$

where $a = \lambda/c_p\rho$ (m^2/s) is the coefficient of thermal diffusion. The polynomial on the right side of eqn (2.38) describes the conductive heat transfer and $\partial T/\partial t$ on the left side the time-dependent heat transfer; the three subsequent terms represent the convective heat transfer.

In the absence of convection and for solid bodies, eqn (2.38) will have the following form:

$$\frac{\partial T}{\partial t} = a\left(\frac{\partial^2 T}{\partial x^2} + \frac{\partial^2 T}{\partial y^2} + \frac{\partial^2 T}{\partial z^2}\right) + \frac{Q_1}{c_p\rho} \tag{2.39}$$

This equation is called the Fourier differential equation.

2.7. Heat Transfer through Continental Lithosphere

If it is assumed that heat is transferred from the lower parts of the continental lithosphere to the surface by conduction according to the Fourier law, melting should occur at continental lithosphere thickness of over 50 km. Neither seismic investigations nor geological reconstructions for the past epochs reveal such melting (excluding specific cases). The reason is that not all the heat is transferred from the top of the asthenosphere and much of the heat is released in the upper part, i.e., continental crust. This heat is generated in the rocks due to radioactive decay of their radioactive elements.

Estimates of content of radioactive elements and generation of radiogenic heat in the upper layer of Aldan and other shields are given in Table 2.1.

Table 2.1. Content of radioactive elements and generation of radiogenic heat in the upper layer Q_n

Shield, region, block	U, g/ton	Th, g/ton	K, %	Q_n, μW/m^3
Aldan shield				
Granite-greenstone region				
Chara-Olekma Block	1.4	9.8	2.37	1.27
Batomga block	1.7	6.7	2.06	1.15
Granulite-gneiss region				
I				
Kurul'ta block	0.8	2.8	1.76	0.58
II				
Zverevo block	1.0	4.7	1.3	0.74
Tyrkanda block	0.9	4.8	2.1	0.81
Seima block	1.1	6.2	1.64	0.9
Tyrkan block	1.2	6.5	1.9	0.98
III				
Melemken block	1.0	9.6	2.42	1.15
Kholbolokh block	1.5	8.6	1.7	1.16
Uchur-Gonam block	1.2	9.0	2.28	1.19
Sunnaga block	1.1	9.8	2.12	1.23
IV				
Sutam block	1.8	11.0	2.04	1.48
Nimnyr block	1.7	14.3	2.96	1.8
Chuga block	2.4	19.7	3.05	2.34
Anabar shield				0.76
Baltic shield				
Karelian south-western block				1.17
Kola megablock				$\dfrac{0.3{-}1.2}{1.0}$

The distribution pattern of radioactive elements with depth is depicted in Figure 2.7. The distribution curves of radiogenic heat release can be approximated by the exponential law of decreasing radiogenic heat release with depth:

$$Q_1 = Q_n \exp\frac{-y}{y_0}, \qquad (2.40)$$

where Q_n (W/m^3) is the specific heat release near the surface. At depth y_0, $Q_1 = Q_n/e$, i.e., y_0 is the characteristic scale of reduction of heat release with depth. Considering the extremely great age of the continental lithosphere, the process of heat transfer by thermal conduction can be regarded as a steady state. In this case, eqn (2.39) can be expressed as:

$$\frac{\lambda \partial^2 T}{\partial y^2} + Q_n e^{-y/y_0} = 0. \tag{2.41}$$

By integrating eqn (2.41)

$$\frac{\lambda \partial T}{\partial y} - Q_n y_0 e^{-y/y_0} = c_l. \tag{2.42}$$

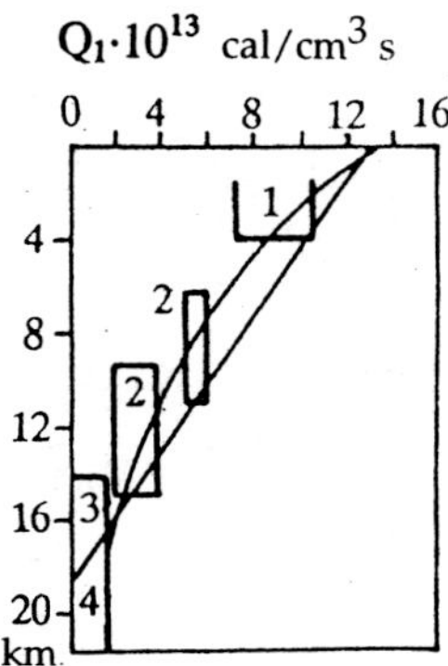

Fig. 2.7. Variation in the energy of radiogenic heat in different parts of the Idaho batholith, USA (Swanberg, 1972). 1—granites; 2—quartz monzonites; 3—quartz diorites; 4—gabbro; curves show the approximation of the data to linear and exponential models at $Q_n = 5.53\,\mu W/m^3$ (1×10^{-13} cal/cm$^3 = 0.419\,\mu W/m^3$) and $y_0 = 9.4$ km.

Since $q = -\lambda \partial T/\partial y$, eqn (2.42) can be represented as:

$$-q - Q_n y_0 e^{-y/y_0} = c_l.$$

The value of constant c_l is found from the following condition: at $y \to \infty$, $q = q_m$, where q_m is the specific heat flux in the mantle. Thus it follows that $c_l = -q_m$. The heat flux at an arbitrary depth will be determined by the equation:

$$q = q_m - Q_n y_0 e^{-y/y_0}. \tag{2.43}$$

On the Earth's surface (at $y = 0$):

$$q_n = q_m - Q_n y_0 \quad \text{and} \quad Q_n y_0 = q_m - q_n. \tag{2.44}$$

It follows from eqn (2.44) that, under exponential reduction of heat release with increasing depth, the surface heat flux (q_n) is determined by linear dependence on specific heat release at the surface (Q_n).

Figure 2.8 depicts the relation between heat flux at the surface and specific heat release at the surface due to radiogenic heat. The latter was determined in the eastern montane regions of the USA.

The intercept of the approximating straight line with axis q_n gives the value of heat flux to the mantle passing from the upper mantle to the continent. This value,

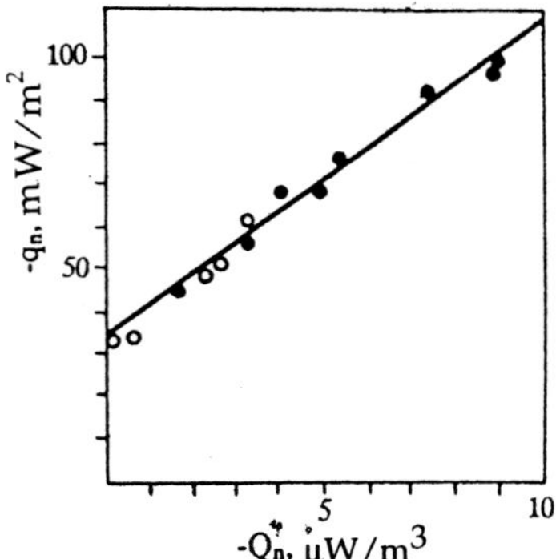

Fig. 2.8. Relation between heat fluxes and rate of radiogenic heat generation for plutons of New England regions (black dots) and the central parts of stable regions (circles) (Roy et al., 1968).

according to Figure 2.8, is $q_m = -34 \times 10^{-3}$ W/m^2. The minus sign points out that the Earth loses heat. At average heat flow $q_n = -60 \times 10^{-3}$ W/m^2 and internal heat release $Q_n = 3.5 \times 10^{-6}$ W/m^3 corresponding to q_n, the characteristic depth can be determined from eqn (2.44) as:

$$y_0 = (q_m - q_n)/Q_n = 7429 \text{ m}.$$

The value of y_0 varies for different regions but ranges usually from 16 to 25 km. The value of $Q_n y_0$ can also be estimated from the following considerations: the thickest and oldest Precambrian shields are exemplified by stable, low heat flows since the heat flux from the mantle is insignificant. Therefore, as a first approximation, it can be suggested that $Q_n y_0 \leq q_n \leq 39$ mW/m^2 as determined for the Precambrian Canadian, Australian and Baltic shields. For granites, $Q_n \approx 2.4 \times 10^{-6}$ W/m^3; then, $y_0 \leq 39 \times 10^{-3}/2.40 \times 10^{-6} \leq 16$ km.

Table 2.2 gives the results of Puzankov's investigations (1989) which show that the value of y_0 varies for different regions in the range 4 to 34 km (average 19 km).

Heat flux (q_n) at the Earth's surface can be expressed as the sum of radiogenic heat flow generated in the upper crust (q_0) and mantle flow incoming from the lithosphere-asthenosphere boundary:

$$q_n = q_m + q_0.$$

Table 2.3 gives the values of q_m, q_0 and q_n for various regions (Duchkov and Sokolova, 1974) showing that in the Baikal rift zone, mantle heat flux is decisive and radiogenic heat comprises 28% of the heat flux at the surface. In the tectonically quiescent regions, the share of mantle heat flux is 20%.

The vertical temperature profiles depicted in Figure 2.9 are after Duchkov and Sokolova (1974).

Table 2.2. Results of statistical analysis of relations between q_n and Q_n values in Altai-Sayan region

Tectonic zone	No. of sections	q_n, mW/m^2	Q_n, μW/m^3	y_0, km
Western and central regions (without Tom'-Kolyvan zone)	33	44	1.25	19.2
Ore Altai	15	47	1.37	19.2
Altai Mountains and Kuznetsk-Priteletsk region	6	40	1.32	16
Salair	4	30	0.9	16
Kuznetsk basin	14	55	1.3	16
Khakassa-Western Tuva	12	42	1.0	22.8
Southern Minusa basin	8	55	1.3	22.8
Eastern Sayan-Sangilen	5	62	1.07	34.5
Granite batholiths of Novosibirsk Pri-Ob'	4	51	6.03	4.3
Altai-Sayan region as a whole	38	48	1.19	23

Note: q_n is the average value of heat flux within the tectonic zone; Q_n the average value of radiogenic heat generated in the surface rocks of the region; and y_0 the scale of reduction of Q_n with depth (average 19 km).

Table 2.3. Heat flux components for various regions

Region	Heat flux, mW/m^2			
	q_n	q_0	q_m	q_0/q_n, %
West Siberian platform				
1. Southern and central regions	41.8	33.5	8.34	80
Baikal region				
2. South of northern platform	41.8	33.5	8.34	80
3. Baikal basin (rift)	117.2	33.5	83.7	28
4. South-western Transbaikal				
a) Uda basin	56.5	36.8	19.7	65
b) Dzhida region	75.3	48	27.3	64
Kamchatka				
5. Kronotsky peninsula	37.7	16.7	21	44
6. Koryak-Avacha and Paratunka depression	62.8	16.7	46.1	27

If it is assumed that heat is transferred only by thermal conduction, there should be total melting of basalt even at a depth of 40 km under the Baikal rift; however, seismic investigations have not supported this phenomenon. This being so, there is more intense convective heat transfer at depth $y > 20$ km; hence the temperature gradient decreases in this region. A similar pattern is noticed for Eastern Transbaikal. According to seismic data, there is no melting at depth exceeding 35 km. In order to understand the temperature distribution in the convective zone, it is necessary to first know the velocity field in it, as can be seen from eqn (2.38).

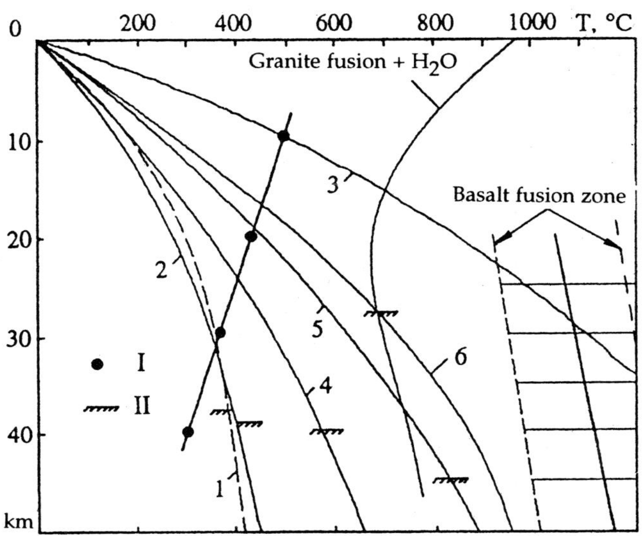

Fig. 2.9. Distribution of temperature in depth for Siberia and Kamchatka. See Table 2.3 for index. I—Curie point for titanomagnetites relative to the depth of their formation; II—position of 'M' boundary.

Therefore, heat transfer can be described by a system of equations including equations of energy, mass and momentum conservation (or equation of motion).

The energy equation has already been derived. Let us now derive the equations of mass and momentum conservation.

2.8. Continuity Equation

The continuity equation is derived on the basis of the law of mass conservation. In a fluid flowing at velocity u, let us fix an elementary parallelepiped with sides dx, dy and dz and determine the fluid mass flowing through it in time dt (Fig. 2.10)

Let us determine the fluid flow in the direction of the x-axis. Fluid mass G'_x flowing through face 1234 will be equal to

$$G'_x = \rho\, u\, dy\, dz\, dt. \tag{2.45}$$

Mass G''_x exiting through the opposite face 5678 will have specific flow $[\rho u + (\partial(\rho u)/\partial x)\, dx]$:

$$G''_x = \left[\rho u + \left(\frac{\partial\,(\rho u)}{\partial x}\right) dx\right] dx\, dy\, dz\, dt \tag{2.46}$$

The resultant flow is found form eqns (2.45) and (2.46):

58

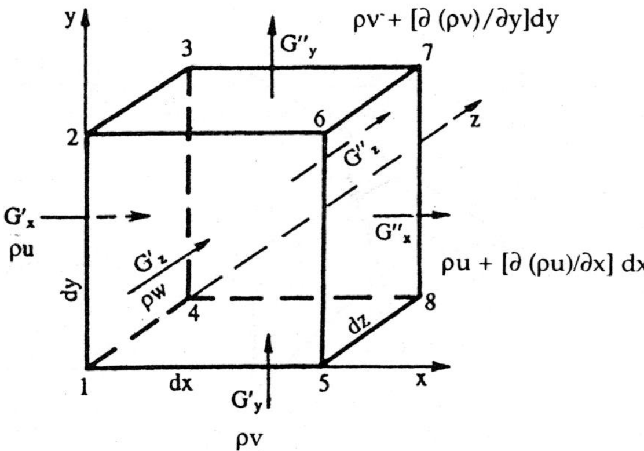

Fig. 2.10. Deriving the differential equation of continuity.

$$\partial G_x = G''_x - G'_x = \left[\frac{\partial (\rho u)}{\partial x}\right] dx\, dy\, dz\, dt. \tag{2.47}$$

The resultant flows in the direction of y and z are similarly found

$$\partial G_y = \left[\frac{\partial (\rho v)}{\partial y}\right] dx\, dy\, dz\, dt \tag{2.48}$$

and

$$\partial G_z = \left[\frac{\partial (\rho w)}{\partial z}\right] dx\, dy\, dz\, dt. \tag{2.49}$$

The total mass balance of the fluid inflowing and outflowing from the elementary volume will be equal to the sum of eqns (2.47) to (2.49):

$$\partial G = \left[\frac{\partial (\rho u)}{\partial x} + \frac{\partial (\rho v)}{\partial y} + \frac{\partial (\rho w)}{\partial z}\right] dx\, dy\, dz\, dt. \tag{2.50}$$

This change of mass of volume $dx\, dy\, dz\, dt$ is associated with the variation in fluid density in it with time $(\partial \rho / \partial t)\, dx\, dy\, dz\, dt$:

$$\left[\frac{\partial (\rho u)}{\partial x} + \frac{\partial (\rho v)}{\partial v} + \frac{\partial (\rho w)}{\partial z}\right] dx\, dy\, dz\, dt = \frac{-\partial \rho}{\partial t}\, dx\, dy\, dz\, dt.$$

This equation can be transferred into a differential equation of continuity

$$\frac{\partial \rho}{\partial t} + \frac{\partial (\rho u)}{dx} + \frac{\partial (\rho v)}{\partial y} + \frac{\partial (\rho w)}{\partial z} = 0. \tag{2.51}$$

For an incompressible fluid (ρ = const) the continuity equation is simplified as:

$$\text{div } \vec{u} = \frac{\partial u}{\partial x} + \frac{\partial v}{\partial y} + \frac{\partial w}{\partial z} = 0. \tag{2.52}$$

Evaluation of motion (momentum conservation) Derivation of the equation of motion is based on the second Newton law, according to which mass multiplied by acceleration is equal to the sum of all the forces acting on the mass under consideration. The following forces act on the isolated mass of flowing fluid: mass force or force of gravity (F) and surface force (P) or force of pressure and friction. The force is a vector quantity and hence this equation in a general form can be expressed for projections on the three axes x, y and z.

Let us denote the mass force as $\vec{F} = \rho\,\vec{g}$ per unit volume, where $\vec{g}$ is the vector of gravity acceleration and the surface force, as well as per unit volume, as $\vec{P}$. Then the equation in the vector form is as follows:

$$\rho\,\frac{D\vec{u}}{Dt} = \vec{F} + \vec{P}, \tag{2.53}$$

where $D\vec{u}/Dt$ is the total or substantial acceleration. We were already concerned with the substantial derivative when deriving the energy eqn (2.37).

By analogy with eqn (2.37), the substantial derivative can be expressed for x-, y- and z-axes respectively as:

$$\frac{Du}{Dt} = \frac{\partial u}{\partial t} + u\,\frac{\partial u}{\partial x} + v\,\frac{\partial u}{\partial y} + w\,\frac{\partial u}{\partial z},$$

$$\frac{Dv}{Dt} = \frac{\partial v}{\partial t} + u\,\frac{\partial v}{\partial x} + v\,\frac{\partial v}{\partial y} + w\,\frac{\partial v}{\partial z}, \tag{2.54}$$

$$\frac{Dw}{Dt} = \frac{\partial w}{\partial t} + u\,\frac{\partial w}{\partial x} + v\,\frac{\partial w}{\partial y} + w\,\frac{\partial w}{\partial z}.$$

The substantial acceleration represents the sum of local components of acceleration $\partial u/\partial t$, $\partial v/\partial t$ and $\partial w/\partial t$ which account for the non-steady state of flux and the convective component of acceleration, which accounts for the motion of fluid particles. The convective component of projection on the x-axis is a polynomial:

$$u\,\frac{\partial u}{\partial x} + v\,\frac{\partial u}{\partial y} + w\,\frac{\partial u}{\partial z}. \tag{2.55}$$

In a vector form, the total acceleration can also be represented as the sum of local acceleration $\partial\vec{u}/\partial t$ and convective acceleration $(\vec{u}\,\mathrm{grad})\,\vec{u}$.

Let us study the volume $\vec{F}$ and surface $\vec{P}$ forces.

Mass forces should be regarded as given external forces. Foremost among them is gravity. Surface forces, however, depend on the state of deformation (state of motion) of the fluid and are represented by pressure and friction.

Consider an elementary volume in a flowing fluid in the form of a parallelepiped with sixes dx, dy and dz (Fig. 2.11).

Gravity and surface forces, i.e., pressure and friction, act on the elementary volume. Let us determine the projections of these forces on the x-axis. Gravity acts at the centre of the elementary volume and is equal to the product of the projection of the gravity acceleration (g_x) by the mass of this volume, $m = \rho\,dV$:

$$\rho\,dVg_x = g_x\,\rho\,dx\,dy\,dz. \tag{2.57}$$

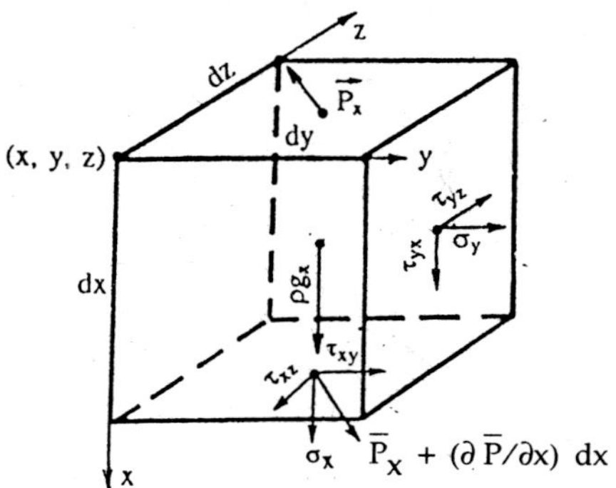

Fig. 2.11. Deriving the differential equation of fluid motion.

Let us divide expression (2.57) by volume dV. We obtain the mass force per unit volume:

$$\vec{F_x} = \vec{g_x}\,\rho\,. \tag{2.58}$$

Surface forces (pressure and friction forces) depend on the state of deformation (state of motion) of the medium (fluid). The sum of surface forces determines the stress state. Therefore, it is also necessary to know the relation between the states of stress and deformation. This relation can be established only empirically and is known as Hooke's law and Stokes' law of friction.

The latter represents a generalised case of Newton's law of friction. The difference between Hooke's law and the Newton-Stokes' laws, as pointed out before (see Sec. 2.3), is as follows. Hooke's law expresses empirically the elastic deformation according to which the forces arising under the deformation of elastic bodies are proportional to the value of strain (2.1). According to the law of viscous friction (Newton's law), however, forces arising under the deformation of the medium (viscous fluid) are proportional to the rate of deformation (2.3).

Let us study the overall stress state for elastic bodies and fluids. For this purpose, we consider the forces acting on the faces of the parallelepiped with sides dx, dy and dz (see Fig. 2.11) and with the co-ordinate on one of its vertices $(x, y$ or $z)$. Resultant stresses, $\vec{P_x}$ and $\vec{P_x}(\partial \vec{P_x}/\partial x)\, dx$, are applied to both faces of the parallelepiped of area $dz\, dy$ perpendicular to the x-axis. These stresses represent vectors. Correspondingly, the forces acting on these faces will be: $\vec{P_x}\, dz\, dy$ and $[\vec{P_x} + (\partial \vec{P_x}/\partial x)\, dx]\, dz\, dy$. The resultant of these forces will be equal to their difference:

$$[\vec{P_x} + (\partial \vec{P_x}/\partial x)\, dx]\, dz\, dy - \vec{P_x}\, dz\, dy = (\partial \vec{P_x}/\partial x)\, dx\, dy\, dz. \tag{2.59}$$

This force is also a vector. Index x points out that the vectors considered above act on an elementary area perpendicular to the x-axis.

The resultant components of surface force for all the three co-ordinate directions are obtained in a similar manner:

for direction x: $(\partial \vec{P_x}/\partial x)\, dx\, dy\, dz$,

for direction y: $(\partial \vec{P_y}/\partial y)\, dx\, dy\, dz$ and

for direction z: $(\partial \vec{P_z}/\partial z)\, dx\, dy\, dz$.

Let us add these components and divide by volume $dx\, dy\, dz$ to obtain the resultant surface force $\vec{P}$ per unit volume caused by the stress state of the elementary volume:

$$\vec{P} = \partial \vec{P_x}/\partial x + \partial \vec{P_y}/\partial y + \partial \vec{P_z}/\partial z. \tag{2.60}$$

In eqn (2.60), vectors $\vec{P_x}$, $\vec{P_y}$ and $\vec{P_z}$ can be resolved into components along the axes. For this purpose, let us introduce the following notation. Components normal to the elementary areas, i.e., normal stresses, are designated as σ with index showing the axis parallel to which the normal stress is oriented. Tangential stresses laying in the plane of elementary areas are shown as τ with two indexes, the first showing to which axis the area is perpendicular and the second the axis parallel to which the shear stress is oriented. In this case, resultant stresses $\vec{P_x}$, $\vec{P_y}$ and $\vec{P_z}$ can be represented in the form of vectorial sum of normal and tangential stresses (see Fig. 2.11):

$$\begin{aligned}
\vec{P_x} &= \vec{i}\,\sigma_x + \vec{j}\,\tau_{xy} + \vec{k}\,\tau_{xz}, \\
\vec{P_y} &= \vec{i}\,\tau_{yx} + \vec{j}\,\sigma_y + \vec{k}\,\tau_{yz} \quad \text{and} \\
\vec{P_z} &= \vec{i}\,\tau_{zx} + \vec{j}\,\tau_{xy} + \vec{k}\,\sigma_z,
\end{aligned} \tag{2.61}$$

where $\vec{i}$, $\vec{j}$ and $\vec{k}$ are unit vectors parallel respectively to x-, y-, and z-axes. It follows from eqn (2.61) that the stress state can be determined by nine scalar values but only six of them are independent. In fact, shear stresses with the same indexes but positioned in a reverse order are equal to one another. This may be seen from the equilibrium of moments relative to an arbitrary axis when the elastic body is in equilibrium. For example, by comparing the moments relative to the z-axis (see Fig. 2.11), we find:

$$\tau_{xy}\, dy\, dz\, dx = \tau_{yx}\, dx\, dz\, dy,$$

hence, it follows that $\tau_{xy} = \tau_{yx}$. In an exactly similar manner, we find:

$$\tau_{xz} = \tau_{zx} \quad \text{and} \quad \tau_{zy} = \tau_{yz}. \tag{2.62}$$

Thus, stress state is determined by six independent scalar values τ_{xy}, τ_{xz}, τ_{yz}, σ_x, σ_y and σ_z. By substituting in eqn (2.60) values $\vec{P_x}$, $\vec{P_y}$ and $\vec{P_z}$ from eqn (2.61) and taking into account the stress equilibrium according to eqn (2.62), we obtain the following equation for surface force $\vec{P}$ per unit volume:

$$\left. \begin{aligned}
\vec{P} = &\ \vec{i}\,(\partial\sigma_x/\partial x + \partial\tau_{xy}/\partial y + \partial\tau_{xz}/\partial z) + && \text{component along } z\text{-axis} \\
&\ \vec{j}\,(\partial\tau_{xy}/\partial x + \partial\sigma_y/\partial y + \partial\tau_{yz}/\partial z) + && \text{component along } y\text{-axis} \\
&\ \vec{k}\,(\partial\tau_{xz}/\partial x + \partial\tau_{yz}/\partial y + \partial\sigma_z/\partial z). && \text{component along } z\text{-axis}
\end{aligned} \right\} \tag{2.63}$$

The equations derived for inertia resolved along the axes (2.54), mass forces $g_x\rho$, $g_y\rho$ and $g_z\rho$ and surface forces (2.63) are added according to general eqn (2.53) for the corresponding directions:

$$\rho \, Du/Dt = g_x \, \rho + (\partial\sigma_x/\partial x + \partial\tau_{xy}/\partial y + \partial\tau_{xz}/\partial z),$$

$$\rho \, Dv/Dt = g_y \, \rho + (\partial\tau_{xy}/\partial x + \partial\sigma_y/\partial y + \partial\tau_{yz}/\partial z), \qquad (2.64)$$

$$\rho \, Dw/Dt = g_z \, \rho + (\partial\tau_{xz}/\partial x + \partial\tau_{yz}/\partial y + \partial\sigma_z/\partial z).$$

Let us consider an ideal fluid, free of viscosity, for which all the shear stresses are equal to zero ($\tau_{xy} = \tau_{xz} = \tau_{yz} = 0$). Only normal stresses which are equal to one another remain. The negative value of any of these stresses is called the fluid pressure:

$$\sigma_x = \sigma_y = \sigma_z = -P. \qquad (2.65)$$

Thus, in a fluid without friction, the fluid pressure is equal to any of the normal stresses taken with a minus sign.

For a viscous fluid when analysing the surface forces, it would be useful to introduce the mean arithmetical value of the three normal stresses as:

$$-P = (1/3) \, (\sigma_x + \sigma_y + \sigma_z). \qquad (2.66)$$

This value which does not vary with the changing co-ordinates is called the fluid pressure.

The system of three equations comprises six stress components (σ_x, σ_y, σ_z, τ_{xy}, τ_{xz}, τ_{yz}). Therefore, the next problem is to establish the dependence of these components on deformation velocities, i.e., on components u, v and w.

From the normal stresses, let us isolate a value equal to pressure $-P$:

$$\sigma_x = -P + \sigma'_x; \; \sigma_y = -P + \sigma'_y; \; \sigma_z = -P + \sigma'_z. \qquad (2.67)$$

The remaining components σ'_x, σ'_y and σ'_z of normal stresses as well as shear stresses depend on viscosity and are expressed as displacement rates. Let us study the simplest case. The equation for tangential friction can be simply established by studying a plane laminar flow in which the rate u changes only in the direction of the y-axis (Fig. 2.12). In this case, friction arises only of the lateral boundaries of

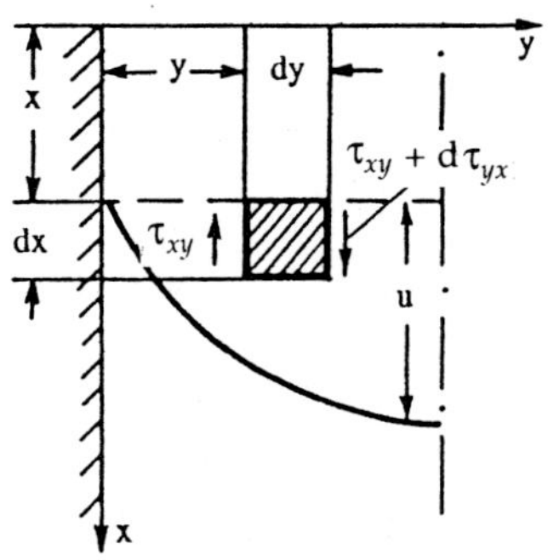

Fig. 2.12. Directions of shear stresses acting on the element of a fluid in one-dimensional flow.

the element. On the left face of the element, the velocity is less than in the element isolated and hence the force of friction is directed opposite to the flow and is equal to $\tau_{yx}\, dx\, dz$. At the right face of the element, the velocity is higher than in the element itself and hence in section $y + dy$, friction is oriented along the flow direction and equal to:

$$[\tau_{yx} + (\partial \tau_{yx}/\partial y)\, dy]\, dx\, dz.$$

The resultant of these forces is equal to the algebraic sum

$$[\tau_{yx} + (\partial \tau_{yx}/\partial y)\, dy]\, dx\, dz - \tau_{yx}\, dx\, dz = (\partial \tau_{yx}/\partial y)\, dV. \tag{2.68}$$

According to Newton's law, $\tau = \eta\,(\partial u/\partial y)$ and eqn (2.68) can be derived in the following form:

$$[\partial \tau_{yx}/\partial y]\, dV = \eta\,[\partial^2 u/\partial y^2]\, dV. \tag{2.69a}$$

In a general case, when velocity depends on the three co-ordinates and the fluid in incompressible ($\rho = \text{const}$), all the stress components are expressed as velocities in the following manner (Stokes' law):

$$\sigma'_x = 2\eta\, \partial u/\partial x;\ \tau_{xy} = \eta\,(\partial u/\partial y + \partial v/\partial x);$$

$$\sigma'_y = 2\eta\, \partial v/\partial y;\ \tau_{yz} = \eta\,(\partial v/\partial z + \partial w/\partial y); \tag{2.69b}$$

$$\sigma'_z = 2\eta\, \partial w/\partial z;\ \tau_{xz} = \eta\,(\partial u/\partial z + \partial w/\partial x).$$

2.9. Navier-Stokes' Equations

Let us study how the above equations are modified for an incompressible fluid. We isolate pressure P from normal stresses in the eqn of motion (2.64) according to eqns (2.67):

$$\rho\, Du/Dt = g_x\rho - \partial P/\partial x + (\partial \sigma'_x/\partial x + \partial \tau_{xy}/\partial y + \partial \tau_{xz}/\partial z),$$

$$\rho\, Dv/Dt = g_y\rho - \partial P/\partial y + (\partial \tau_{xy}/\partial x + \partial \sigma'_y/\partial y + \partial \tau_{yz}/\partial z) \text{ and}$$

$$\rho\, Dw/Dt = g_z\rho - \partial P/\partial z + (\partial \tau_{xz}/\partial x + \partial \tau_{yz}/\partial y + \partial \sigma'_z/\partial z). \tag{2.70}$$

We substitute in the equation of motion (2.70) the stress values expressed as velocity gradient (2.69b) and demonstrate this procedure by an example of direction along the x-axis at $\eta = \text{const}$ in accordance with eqn 2.53:

$$\partial \sigma'_x/\partial x + \partial \tau_{xy}/\partial y + \partial \tau_{xz}/\partial z = 2\eta\, \partial^2 u/\partial x^2 + (\partial/\partial y)\,[\eta(\partial u/\partial y + \partial v/\partial x)]$$

$$+ (\partial/\partial z)\,[\eta(\partial u/\partial z + \partial w/\partial x)] = \eta(\partial^2 u/\partial x^2 + \partial^2 u/\partial y^2 + \partial^2 u/\partial z^2). \tag{2.71}$$

Similarly, let us also find the equation for direction y- and z-direction. Let us express eqn (2.70) for the case of incompressible fluid ($\rho = \text{const}$). In this case, the continuity equation has the form of eqn (2.52). Since the temperature difference in an incompressible fluid is not significant, the coefficient of dynamic viscosity can be regarded as constant. Therefore, the equation of motion (2.70), considering eqns (2.69b) and (2.71) and continuity eqn (2.52), for a three-dimensional flow will have the following form:

64

$$\rho(\partial u/\partial t + u\,\partial u/\partial x + v\,\partial u/\partial y + w\,\partial u/\partial z)$$

$$= g_x\rho - \partial P/\partial x + \eta\,(\partial^2 u/\partial x^2 + \partial^2 u/\partial y^2 + \partial^2 u/\partial z^2);$$

$$\rho(\partial v/\partial t + u\,\partial v/\partial x + v\,\partial v/\partial y + w\,\partial v/\partial z)$$

$$= g_y\rho - \partial P/\partial y + \eta(\partial^2 v/\partial x^2 + \partial^2 v/\partial y^2 + \partial^2 v/\partial z^2);$$

$$\rho(\partial w/\partial t + u\,\partial w/\partial x + v\partial w/\partial y + w\partial w/\partial z)$$

$$= g_z\rho - \partial P/\partial z + \eta(\partial^2 w/\partial x^2 + \partial^2 w/\partial y^2 + \partial^2 w/\partial z^2); \tag{2.72}$$

$$\partial u/\partial x + \partial v/\partial y + \partial w/\partial z = 0. \tag{2.73}$$

These are the Navier-Stokes' eqns for an incompressible fluid derived in a rectangular co-ordinate system. If the mass forces are present, four unknown values P, u, v and w remain and four equations are available for determining them. To them should also be added the heat transfer eqn (2.38):

$$\partial T/\partial t + u\,\partial T/\partial x + v\,\partial T/\partial y + w\,\partial T/\partial z = a(\partial^2 T/\partial x^2 + \partial^2 T/\partial y^2 + \partial^2 T/\partial z^2). \tag{2.74}$$

Navier-Stokes' eqns (2.72) and (2.73) for an incompressible fluid and heat transfer eqn (2.74) can be expressed in vector form as follows:

$$\rho\,D\vec{u}/Dt = \vec{g}\,\rho - \operatorname{grad} P + \eta\nabla^2\,\vec{u}, \tag{2.75}$$

$$\operatorname{div}\vec{u} = 0, \tag{2.76}$$

$$DT/Dt = a\nabla^2 T + (\eta/c_p\rho)\,\Phi + Q_1/c_p\,\rho, \tag{2.77}$$

where $\nabla^2 = \partial^2/\partial x^2 + \partial^2/\partial y^2 + \partial^2/\partial z^2$ represents the Laplacian operator and $\Phi = 2[(\partial u/\partial x)^2 + (\partial v/\partial y)^2 + (\partial w/\partial z)^2] + (\partial v/\partial x + \partial u/\partial y)^2 + (\partial w/\partial y + \partial y/\partial z)^2 + (\partial u/\partial z + \partial w/\partial x)^2 - (2/3)\,(\partial u/\partial x + \partial v/\partial y + \partial w/\partial z)^2$ the dissipation function which represents viscous dissipation. The function describes the irreversible part of energy transport caused by shear stresses.

2.10. Boussinesq Approximation

Equation (2.75) has been derived for an impressible fluid. The density of fluid volume varies with temperature. The dependence of density on temperature is described by the equation of the state of the fluid. In turn, temperature variation at different points gives rise to density variation in them. In this case the resulting forces are the mass forces in the gravity field, often called buoyancy forces. Such a fluid should be regarded as compressible and all the effects associated with the density variation should be taken into consideration.

Buoyancy force in the gravity field is determined by the equation $\vec{F} = \rho\,\vec{g}$, where $\vec{g}$ is the gravity acceleration. Ignoring density variation implies ignoring the thermogravitational forces which, under natural convection, represent motive forces and thus convective flow cannot be described by the system of equations relevant for an incompressible fluid.

It can be seen that the temperature field is associated with the flow which, under natural convection, arises due to density variation associated with the change in temperature. Therefore, to obtain the distributions of velocity, temperature and pressure in space and time, equations of motion, continuity, heat transfer and state of matter will have to be solved simultaneously. This procedure essentially complicates the study of thermogravitational flows. For this reason, while studying natural convection, simplifying assumptions are made and more convenient approximation methods used.

The approximation introduced by Boussinesq is now considered. Local static pressure P in the equation of motion can be expressed by two terms: hydrostatic P_a and dynamic P_d pressures: $P = P_a + P_d$. Pressure P_a is equal to the hydrostatic pressure in the surrounding medium (fluid at rest): $P_a = \rho_a\, gx$, when the gravity acts in a positive direction along the x-axis (parallel to x-axis). Therefore, in the gravity field, grad $P_a = \partial P_a / \partial x = \rho_a\, g$ ($P_a = 0$ at $x = 0$ is adopted as the commencement of reckoning). Pressure P_d represents the dynamic component caused by flow and should be determined from the basic equations simultaneously with the temperature and velocity fields. In this case, the sum of mass and surface forces of pressure is represented as $\vec{F} -$ grad $P = \rho\vec{g} -$ grad $(P_a + P_d)$ $=$ $(\rho\vec{g} -$ grad $P_a) -$ grad P_d $=$ $(\rho - \rho_a)\,\vec{g} -$ grad P_d $= \Delta\rho\,\vec{g} -$ grad P_d. Equation of motion (2.75) is expressed as:

$$D\vec{u}/Dt = [(\rho - \rho_a)/\rho]\,\vec{g} - (1/\rho)\,\text{grad}\,P_d + (\eta/\rho)\,\nabla^2\,\vec{u} \tag{2.78}$$

Other eqns (2.77) and (2.51) have been given above. The major difficulty in solving the system of eqns (2.77), (2.78) and (2.51) is the need to take into account the change of density ρ with temperature in the continuity eqn (2.51). The problem is considerably simplified if the density change is considered only in the mass term $[(\rho - \rho_a)/\rho]\,\vec{g}$ and this variation in the continuity equation is ignored.

Let us study the correctness of such a simplification. It follows from the experiment that at relatively small temperature fluctuations, the relative variation of density is modest and only a small percentage of its value. For example, for water, the maximum temperature drop at atmospheric pressure is equal to $100°C$. For a coefficient of volume expansion of water

$$\beta = (-1/\rho)\,(\partial\rho/\partial T)_p = 2.2 \times 10^{-4}\,(1/°C),$$

the value of $\Delta\rho = \rho\,\beta\,\Delta T = 0.022\,\text{g/cm}^3$, i.e., only 2% of density value $\rho = 1\,\text{g/cm}^3$ and $\Delta\rho/\rho = 0.022$. Let us estimate how the relative variation of density influences (compared to the gradient of velocity) the continuity eqn (2.51). For this purpose, let us write:

$$\partial(\rho u)/\partial x = u\partial\rho/\partial x + \rho\,\partial u/\partial x.$$

For a case when the ambient fluid has constant temperature T_a and hence $\rho_a = \text{const}$, let us compare the relative change of density $(u[\partial\rho/\partial x])$ with relative change of velocity $(\rho\,[\partial u/\partial x])$. For this purpose, we evaluate the ratio of $u\,(\partial\rho/\partial x)$ to $\rho(\partial u/\partial x)$: $u(\partial\rho/\partial x)/\rho\,(\partial u/\partial x) \approx (\Delta\rho/\rho)\,(u/\Delta u)$.

Since the value of u in the medium varies from zero to u, u and Δu have an identical order of values. The density differences are determined from the equation

$\Delta\rho = \rho\beta\,(T - T_a)$; therefore, at $\beta(T - T_a) \ll 1$, the density variation in the continuity eqn (2.51) can be ignored compared to the other terms. Thus, the system of equations of motion, continuity and heat transfer will have the following form considering eqn (2.56):

$$\partial\vec{u}/\partial t + (\vec{u}\,\mathrm{grad})\vec{u} = \beta\,\vec{g}\,(T - T_a) - \mathrm{grad}\,P_d/\rho + (\eta/\rho)\,\nabla^2\vec{u}, \tag{2.79}$$

$$\mathrm{div}\,\vec{u} = 0, \tag{2.80}$$

$$\partial T/\partial t + (\vec{u}\,\mathrm{grad})\,T = (\lambda/c_p\rho)\,\nabla^2 T + Q_1/c_p\rho. \tag{2.81}$$

In the forgoing approximations, the temperature difference which is responsible for motion is taken into account only in the term describing the mass force. The density reduction caused by temperature drop is not considered in the other term of the equation of motion nor in the continuity equation. Let us also ignore the change of other parameters η, c_p and λ as well as heat generation due to friction in the equation of energy transfer. This approximation is called the *Boussinesq approximation*.

In eqn (2.79), let us analyse the terms which account for the thermal gravitational forces. Let us express $\beta g\,(T - T_a)$ as enthalpy i reckoned from temperature $T_a : i = c_p(T - T_a)$. In this case the mass term is recast and the following equation derived:

$$(T - T_a)\,c_p/[c_p/\beta g] = i/H, \tag{2.82}$$

where $H = c_p/\beta g$ has the dimension of length and every substance has its own H. It can be regarded as the characteristic thickness of the layer of substance. For example, for air, $\beta = 1/T$ and $H = c_p\,T/g = T/\gamma_{ad}$, where $\gamma_{ad} = \partial T/\partial x \approx g/c_p$ characterises the adiabatic temperature gradient. Let us determine the value of adiabatic gradient $\gamma_{ad} \approx g/c_p$ and the characteristic thickness of the layer of substance $H = 1/\beta\gamma_{ad} = c_p/\beta g$. For air at $T = 288°K$: $\gamma_{ad} = 9.8°K/km$ and $H = 32$ km; for water at $20°C$: $\beta = 2 \times 10^{-4}°K^{-1}$, $c_p = 4.2 \times 10^3$ J/kg°K, $\gamma_{ad} = 0.23°K/km$ and $H = 2000$ km; for the upper mantle: $\beta = 2 \times 10^{-5}°K^{-1}$, $c_p = 1.2 \times 10^3$ J/kg°K; $g/c_p = 8.2$ °K/km and $H = 6000$ km.

Boussinesq approximation is valid for layer thickness l much less than scale H, i.e., at $l/H \ll 1$, the density variation with the change of hydrostatic pressure can be ignored. Otherwise, the approximation of deep convection has to be used: in the continuity equation, the change of density with height is taken into account and potential, instead of general, temperature is used: $\theta = T + (\partial T/\partial x)_{ad}\,x$. Thus, the Boussinesq approximation is valid for a layer of thickness l if $(\partial T/\partial x)_{ad}\,l \ll \Delta T$. Considering that $H = 1/[\beta(\partial T/\partial x)_{ad}]$, this condition can be expressed as the ratio $l/H \ll \beta\Delta T$. The characteristic thickness of substance H is an interesting parameter not only for assessment of correctness of the Boussinesq approximation, but also for estimating the efficiency of the thermogravitational 'engine', i.e., assessing the amount of heat energy converted into mechanical energy. For the rate of heat transfer in a horizontal fluid layer heated from below and cooled from above, convection efficiency is determined as follows (Golitsin, 1980):

$$\text{Efficiency} = l/H. \tag{2.83}$$

By considering the upper the lower mantle of the Earth as horizontal layers heated from below and cooled from above, according to eqn (2.83), for the upper mantle

Efficiency $\approx 400/6000$ km $= 7\%$

and for the lower mantle

Efficiency $\approx 2100/6000$ km $= 33\%$.

Thus the lower mantle is the powerful generator of mechanical energy and global movements detected in the displacement of lithospheric plates, growth of mountains, young lithosphere at hot spots etc., are in all probability associated more with lower mantle convection than that of the upper mantle.

2.11. Boundary Conditions

Differential equations are derived on the basis of general principles of physics and all problems associated with this phenomenon are described by them. The system of Navier-Stokes' equations describes for example convection flows as well as heat transfer in the atmosphere, oceans, in the crystal growth system, Earth's mantle etc. In order to define unambiguously the problem under consideration, conditions of uniqueness or boundary conditions should be specified and temporal and spatial boundary conditions distinguished. Only spatial conditions are relevant in the case of steady-state processes. The basic boundary conditions are: shape and size of the heated surface, conditions of the fluid state, velocity, temperature and heat transfer conditions at the boundary. The boundary conditions can be specified in the form of numerical values of parameters, in the form of functions or as a differential equation.

Let us study the **boundary conditions for equations of motion (hydrodynamics)**. When analysing the problems associated with the flow of a viscous medium, the most important boundary condition is the rate of flow in the immediate vicinity of a solid bounding surface. If the immobile solid wall is impermeable to the fluid and does not absorb it, the velocity component normal to this surface (let it be v) is equal to zero ($v_w = 0$). Due to the seizure of the viscous fluid against the wall, the velocity along the wall is also equal to zero ($u_w = w_w = 0$).

If there is absorption or release of fluid at the boundary resulting from the physical or chemical processes, the component of velocity vector oriented normally to the wall is determined by the rate of this process. In this case, $v_w = v_n = m_n/\rho$, where m_n is the mass flow (kg/m^{2}s). The components of velocity gradient on the solid wall in the absence of mass of flow ($m_n = 0$) are equal to:

$$\partial u/\partial x = \partial w/\partial z = 0, \partial u/\partial y = \tau_w/\eta,$$

where τ_w is the shear stress at the wall. At the points of inflow and discharge of the fluid, the velocity and pressure are bound to be specified.

Boundary conditions for the energy conservation equation. Temporal conditions can be assigned (at some instant of time) such as the temperature distribution in the volume: $T = T_0(x, y, z)$. Spatial boundary conditions imply specifying heat

68

transfer conditions on the boundary surfaces. Three types of boundary conditions can be stated:

1) The first type of boundary conditions is the temperature distribution on the boundary surfaces over time, for example under stationary heat transfer and constant temperature on the walls, $T_w = $ const.

2) The second type involves the prescribed heat flux on the boundary surfaces as a function of co-ordinates and time, i.e., $q = -\lambda\,(\partial T/\partial y)_w$.

3) Type three represents the relation of wall temperature to temperature of the medium in terms of a given value of the coefficient of heat transfer through the fluid. In this case, the temperature at a given point on the wall surface is $T_w = T_l + q/\alpha$, where q is the specific heat flux (W/m^2) and α the coefficient of heat transfer. From the above equation, it can be seen that $q = \alpha\,(T_w - T_l)$. According to Fourier's law, this equation can be expressed as the temperature gradient at the solid wall:

$$- \lambda_w\,(\partial T/\partial y)_w = \alpha(T_w - T_l). \tag{2.84}$$

The latter represents the assigned boundary condition in the form of a differential equation.

Solving the problem for convective heat transfer by analytical methods has extremely restricted possibilities and solutions have been found for the simplest cases of laminar flows. A solution is possible by numerical methods using a computer but with limitations due to the difficulty of obtaining correct results, particularly for time-dependent three-dimensional flows.

To study turbulent flows, predominantly experimental methods are applied. Semi-empirical theories of turbulent flows and heat transfer are constructed based on these experimental results.

Because of limitations of using analytical methods and numerical simulation, laboratory experiments have become decisively important in the further study of thermogravitational convection. They are carried out to study not only the phenomenon under consideration, but also to obtain data for calculating other related phenomena. Experimental results can only be extrapolated to similar phenomena, however. Therefore, when setting up an experiment, one should know beforehand the values required to be measured in the experiment, how to process the results of measurements and which phenomena in the experiment are similar to those studied in nature.

2.12. General Concepts of Similarity

We have encountered the concept of similarity in geometry. For example, triangles are similar if corresponding angles are equal and the ratio of corresponding sides of triangles is constant, i.e., $l''_1/l'_1 = l''_2/l'_2 = l''_3/l'_3 = c$, where c is the coefficient of similarity. Using the properties of similar triangles, the height of a building or mountain, horizontal dimensions of objects, width of a river etc. can be determined without taking direct measurements.

The same is true for kinematic similarity, i.e., similarity of velocities and accelerations of fluid flows; dynamic similarity, i.e., similarity of forces causing similar motions; and thermal similarity, i.e., similarity of temperature fields and heat flows.

Physical phenomena are similar in which all the values characterising the phenomena under consideration are similar. For example, for thermal similarity of two fluid flows, there should be geometric similarity of canals with characteristic dimensions $l''/l' = c_1$, temperature similarity $T''/T' = c_T$, similarity of velocities $u''/u' = c_u$, similarity of viscosities $\eta''/\eta' = c_\eta$, similarity of densities $\rho''/\rho' = c_\rho$ and similarity of heat capacity $c''_p/c'_p = c_c$. Each physical value has its own similarity constant c whose subscript will show the physical property to which it is related.

In two similar processes, the values to similar properties at related space-time points differ from each other by a constant factor. Spatially similar points are those points which are geometrically similar. If vector values are considered, similar vectors should be similarly oriented in space.

Conversion factors (similarity constants) for different physical parameters may be distinct but they should not be selected arbitrarily: there should be some interrelations between them. Let us find this interrelation on the example of fluid flow. According to definition, velocity $u = l/t$. Let us apply this formula to two similar particles in similar flows which transverse similar routes:

for the first: $u' = l'/t'$;

for second: $u'' = l''/t''$.

By dividing them:

$$u''/u' = (l''/l')/(t''/t'). \tag{2.85}$$

If these flows are similar, for all similar points:

$$l''/l' = c_l, \; u''/u' = c_u \text{ and } t''/t' = c_t. \tag{2.86}$$

In eqn (2.85) let us substitute similarity constants instead of ratios, and obtain the following relation between them:

$$c_u \, c_t/c_l = 1. \tag{2.87}$$

By grouping the values pertaining to individual flows in eqn (2.85), we arrive at the following relation:

$$u't'/l = u''t''/l''.$$

This denotes that for similar flows, complex $H_0 = u\,t/l = $ idem (i.e., identical of the same thing). Thus, the main property of similarity between systems is the presence of dimensionless complexes which have the same numerical value for all similar systems. These complexes are called the criteria of similarity. The complex obtained above, $u\,t/l = H_0$, is called the criterion of homochronicity since time similarities can be established on its basis:

$$t''/t' = (u'/u'')\,(l''/l').$$

For such heat transfer processes, the boundary conditions, i.e., the uniqueness conditions, should also be similar.

70

Let us find the criterion of similarity for heat transfer conditions at the boundary specified by condition of type (2.84). Let us apply this equation to similar points of two similar systems:

for the first: $\lambda'_w (\partial T'/\partial y')_w = \alpha' \Delta T'$, (2.88)

for the second: $\lambda''_w (\partial T''/\partial y'')_w = \alpha'' \Delta T''$, (2.89)

where $\Delta T = (T_w - T_l)$.

According to the definition of the similarity, variables of the second system can be expressed as variables of the first system:

$$\lambda''_w = c_\lambda \lambda'_w; \quad \alpha'' = c_\alpha \alpha'; \quad T'' = c_T T'; \quad \Delta T'' = c_T \Delta T'; \quad y'' = c_l y'. \quad (2.90)$$

The values obtained in eqn (2.90) are substituted in eqn (2.89):

$$(- c_T c_\lambda/c_l) \lambda'_w (\partial T'/\partial y')_w = c_\alpha c_T \alpha' \Delta T'. \quad (2.91)$$

Then, instead of system (2.88) and (2.89), we obtain system (2.88) and (2.91). In eqns (2.88) and (2.91), the variables should be determined in the same manner. This is possible if the constants of similarity on the right and left sides of eqn (2.91) are equal, i.e., we obtain:

$$c_\alpha = c_\lambda/c_l \text{ or } c_\alpha c_l/c_\lambda = 1. \quad (2.92)$$

If in eqn (2.92), the ratio of similarity constants is substituted for the constants and all parameters of one system are rearranged to the left and of the other to the right side, we obtain:

$$\alpha' l'/\lambda' = \alpha'' l''/\lambda'' \text{ or } \mathrm{Nu} = \alpha l/\lambda = \text{idem}, \quad (2.93)$$

where Nu is the *Nusselt criterion* characterising the convective heat transfer and points out by how much the rate of convective heat transfer exceeds that of conductive heat transfer.

We found the criteria of similarity in the two examples discussed above because they exhibited analytical dependence between the variables. Therefore, if two phenomena are described by the same analytical dependence (they have the same physical and mathematical models), conditions can be established under which these two processes are similar. Even if we design a highly complex mathematical model but cannot integrate the corresponding equations, it can always be used to find the conditions of similarity of these processes. It is important to emphasise that the criteria of similarity derived from differential equations set up for an elementary volume are also applicable to the entire volume.

The first theorem of similarity can be formulated based on the above discussion: similar phenomena have identical criteria of similarity (Newton's theorem). The second theorem of similarity establishes the possibility of formulating an integral as a function of similarity criteria of a differential equation: physical dependence between the process variables can be represented in the form of a relationship between the similarity criteria $M_1, M_2, ..., M_n$:

$$f(M_1, M_2, ..., M_n) = 0.$$

By expressing the experimental results in the form of a criterial equation, we obtain a relationship that holds good for all similar processes.

A reciprocal theorem can also be constructed. Phenomena are similar if they have similar conditions of uniqueness and criteria derived from conditions of uniqueness are numerically identical. This theorem highlights the need for distinguishing criteria formulated from parameters appearing in the unique conditions and which are decisive. Criteria formulated from dependent variables are called non-decisive. Therefore, only the equality of decisive criteria provides the similarity condition of processes.

This theorem answers the questions posed before:

1) all the parameters involved in the similarity criteria should be determined experimentally;

2) the experimental results should be represented as a dependence between the criteria of similarity (in the form of a graph or equation) and

3) the experimental results should be extrapolated only to similar phenomena.

Let us turn to the application of the theory of similarity for analysing convective heat transfer. Let us examine individually the equations of motion and heat transfer which provide the conditions of mechanical and thermal similarity. The procedure of determination of the similarity conditions proposed by Mikheev (1947) seems to be clear from the mathematical and physical points of view. This procedure is used for analysing mechanical and thermal similarity.

2.13. Mechanical Similarity

Let us determine the conditions at which similar flows prevail in geometrically similar systems. We shall study the equations in the Boussinesq approximation for two similar systems but only for projection on the x-axis since calculation for other projections are analogous.

The equations for the first system may be written as:

$$\frac{\partial u'}{\partial x} + \frac{\partial v'}{\partial y'} + \frac{\partial w'}{\partial z'} = 0,$$

$$\frac{\partial u'}{\partial t'} + u'\frac{\partial u'}{\partial x'} + v'\frac{\partial u'}{\partial y'} + w\frac{\partial u'}{\partial z'} = \beta'g'_x\left(\frac{T}{T_a}\right) - \left(\frac{1}{\rho'}\right)\left(\frac{\partial P'}{\partial x'}\right)$$

$$+ v'\left(\frac{\partial^2 u'}{\partial x'^2} + \frac{\partial^2 u'}{\partial y'^2} + \frac{\partial^2 u'}{\partial z'^2}\right) \tag{2.94}$$

and for the second system:

$$\frac{\partial u''}{\partial x''} + \frac{\partial v''}{\partial y''} + \frac{\partial w''}{\partial z''} = 0, \quad \frac{\partial u''}{\partial t''} + u''\frac{\partial u''}{\partial x''} + v'\frac{\partial u''}{\partial y''} + w''\frac{\partial u''}{\partial z''}$$

$$= \beta''g''_x\left(\frac{T}{T_a}\right)'' - \left(\frac{1}{\rho''}\right)\left(\frac{\partial P''}{\partial x''}\right) + v''\left(\frac{\partial^2 u''}{\partial x''^2} + \frac{\partial^2 u''}{\partial y''^2} + \frac{\partial^2 u''}{\partial z''^2}\right), \tag{2.95}$$

where $v = \eta/\rho$ is the coefficient of kinematic viscosity. Similarity constants for the two similar systems will be:

$$\frac{x''}{x'} = \frac{y''}{y'} = \frac{z''}{z'} = c_l; \quad \frac{u''}{u'} = \frac{v''}{v'} = \frac{w''}{w'} = c_u;$$

$$\frac{t''}{t'} = c_t; \quad \frac{P''}{P'} = c_p; \quad \frac{g''_x}{g'_x} = c_g; \quad \frac{\beta''}{\beta'} = c_\beta;$$

$$\frac{(T - T_a)''}{(T - T_a)'} = c_T; \quad \frac{v''}{v'} = c_v.$$

The variables of the second system can be expressed in terms of the variables of the first system:

$$x'' = c_l x'; \quad u'' = c_u u'; \quad t'' = c_t t'; \quad P'' = c_p P'; \quad g''_x = c_g g'_x; \quad \rho''/\rho' = c_\rho;$$

$$\beta'' = c_\beta \beta'; \quad (T - T_a)'' = c_T(T - T_a)'; \quad v'' = v' c_v. \tag{2.96}$$

Substitution of the values (2.96) in eqn (2.95) gives:

$$\left(\frac{c_u}{c_l}\right)\left(\frac{du'}{\partial x'} + \frac{\partial v'}{\partial y'} + \frac{\partial w'}{\partial z'}\right) = 0, \tag{2.97}$$

$$\left(\frac{c_u}{c_t}\right)\left(\frac{\partial u'}{\partial t'}\right) + \left(\frac{c_u^2}{c_l}\right)\left(u'\frac{\partial u'}{\partial x'} + v'\frac{\partial u'}{\partial y'} + w'\frac{\partial u'}{\partial z'}\right) = c_\beta\, c_g\, c_T\, \beta'\, g'_x\, (T - T_a)'$$

$$-\left(\frac{c_p}{c_\rho c_l \rho'}\right)\left(\frac{\partial P'}{\partial x'}\right) + v'\left(\frac{c_v c_u}{c_l^2}\right)\left(\frac{\partial^2 u'}{\partial x'^2} + \frac{\partial^2 u'}{\partial y'^2} + \frac{\partial^2 u'}{\partial z'^2}\right) \tag{2.98}$$

Both systems can be expressed in terms of the variables of the first system. Equations (2.94), (2.97) and (2.98) possess equal solutions.

This is possible if the complexes composed from the constants of similarity in the equations of motion (2.98) are equal and they can be cancelled. Hence the relation between the constants of similarity is derived.

Continuity eqn (2.97) necessarily asserts and therefore does not impose additional limitations on selecting the constants of similarity.

To derive the relation between similarity constants we equate all the complexes appearing before the terms in eqn (2.98):

$$c_u/c_t = c_u^2/c_l = c_\beta c_g c_T = c_p/c_\rho c_l = c_v c_u/c_l^2. \tag{2.99}$$

If all the terms of eqn (2.99) are divided by any of these terms, the ratios will be equal to one.

Let us equate in pairs the terms of eqns (2.99) and find the criteria of similarity. Let us equate the first and second terms of eqns (2.99):

$$c_u/c_t = c_u^2/c_l; \quad c_u c_t/c_l = 1; \quad u't'/l' = u''t''/l'';$$

$H_0 = ut/l = \text{idem}$ represents the criterion of homochronicity. Let us equate the second and third terms of eqns (2.99):

$$c_u^2/c_l = c_\beta c_g c_T; \quad c_\beta c_g c_T (c_l/c_u^2) = 1; \quad \beta' g' T'l'/u'^2 = \beta'' g'' T''l''/u''^2;$$

$Fr = \beta g \Delta T\, l/u^2 =$ idem is the Froude criterion.

Let us equate the second and fourth terms of eqns (2.99):

$$c_u^2/c_l = c_p/c_\rho c_l; \quad c_p/c_\rho c_u^2 = 1; \quad P'/\rho' u'^2 = P''/\rho'' u''^2;$$

$Eu = P/\rho u^2 =$ idem is the Euler criterion.

From the second and fifth terms of eqns (2.99), we get:

$$c_u^2/c_l = c_v c_u/c_l^2; \quad c_u c_l/c_v = 1; \quad u' l'/v' = u'' l''/v'';$$

$Re = ul/v =$ idem is the Reynolds criterion.

By equating the terms of eqns (2.99) in different combinations, we can obtain criteria of a different type. The number of decisive criteria is equal to the difference between the number of independent process parameters and the total number of dimensions of these parameters.

One criterion composed from independent variables and from conditions of uniqueness, representing geometric properties of the system, physical properties of the bodies of the system, its initial state and boundary conditions will be decisive.

For example, under thermogravitational convection, velocity is accountable to temperature variations in the system. Therefore, the decisive process criteria will not be the Fr criterion, but complex $Gr = Fr\, Re^2 = \beta g\, \Delta T l^3/v^2$, the Grashof criterion. Each criterion of similarity has its own physical sense.

Thus the *homochronicity criterion* (Ho $= ut/l$) enables establishing identical instancy of time in which the systems will be similar to each other. It also characterises how rapidly the velocity field in the system changes.

The *Froude criterion* (Fr $= \beta g \Delta T\, l/u^2$) is the criterion of gravitational similarity and the ratio of gravity to inertia.

The *Euler criterion* (Eu $= P/\rho\, u^2$) represents the criterion of similarity of velocity fields and characterises the ratio pressure to inertia in the flow.

The *Reynolds criterion* (Re $= u\, l/v$) characterises the hydrodynamic regime of flow and is a measure of the ratio of inertia to molecular friction.

2.14. Thermal Similarity

Thermal similarity is possible in geometrically similar systems in which mechanical similarity has been fulfilled. Thermal similarity presupposes similarity of the temperature fields, heat flows and conditions of uniqueness. Let us examine two similar systems.

For the first system, the heat transfer equation will be as follows:

$$\partial T'/\partial t' + u'\partial T'/\partial x' + v'\partial T'/\partial y' + w'\partial T'/\partial z' = a'\,(\partial^2 T'/\partial x'^2$$

$$+ \partial^2 T'/\partial y'^2 + \partial^2 T'/\partial z'^2) + Q'/\rho' c_p', \tag{2.100}$$

and the heat transfer condition at the boundary will be:

$$-\lambda'\,(\partial T'/\partial y') = \alpha'\,\Delta T'. \tag{2.101}$$

74

Correspondingly, for the second system:

$$\partial T''/\partial t'' + u'' \, \partial T''/\partial x'' + v'' \, \partial T''/\partial y'' + w'' \, \partial T''/\partial z''$$

$$= a'' \, (\partial^2 T''/\partial x''^2 + \partial^2 T''/\partial y''^2 + \partial^2 T''/\partial z''^2) + Q''/\rho'' \, c_p{}'', \tag{2.102}$$

$$- \lambda'' \, (\partial T''/\partial y'') = \alpha'' \, \Delta T''. \tag{2.103}$$

We have already studies the similarity of boundary conditions (2.101) and (2.103). For similar processes, the following conditions are given:

$$x''/x' = y''/y' = z''/z' = c_l; \; t''/t' = c_t;$$

$$u''/u' = v''/v' = w''/w' = c_u;$$

$$T''/T' = c_T; \; a''/a' = c_a; \; c''_p/c'_p = c_c; \; \rho''/\rho' = c_\rho; \; Q''/Q' = c_Q. \tag{2.104}$$

The second system (2.102) and (2.103) can be expressed in terms of the variables of the first system:

$$(c_T/c_t) \, (\partial T'/\partial t') + (c_u c_T/c_l) \, (u'\partial T'/\partial x' + v\partial T'/\partial y' + w'\partial T'/\partial z')$$

$$= a' \, (c_a c_T/c_l^2) \, (\partial^2 T'/\partial x'^2 + \partial^2 T'/\partial y'^2 + \partial^2 T/\partial z'^2) + (c_Q/c_\rho c_c) \, (Q'/\rho' c_p'). \tag{2.105}$$

Since the solutions to eqns (2.100) and (2.105) should be the same, it follows that the coefficients derived from the constants of similarity in eqn (2.105) are equal:

$$c_T/c_t = c_u c_T/c_l = c_a c_T/c_l^2 = c_Q/c_\rho c_c. \tag{2.106}$$

For such thermal and convective systems, the following are the criteria of similarity:

$c_T/c_t = c_u \, c_T/c_l; \; c_u \, c_t/c_l = 1; \; u' \, t'/l' = u'' \, t''/l''$ or $u \, t/l = \mathrm{Ho} =$ idem is the criterion of homochronicity ;

$c_u \, c_T/c_l = c_a \, c_T/c_l^2; \; c_u \, c_l/c_a = 1; \; u'l'/a' = u'' \, l''/a''$ or $u \, l/a = \mathrm{Pe} =$ idem is the Peclet criterion;

$c_u c_T/c_l = c_Q/c_\rho c_c; \; c_Q c_l/c_u \, c_T \, c_\rho \, c_c = 1; \; Q' \, l'/u' \, T'\rho' \, c_p' = Q'' \, l''/u'' \, T'' \, \rho'' \, c_p'';$

$Q \, l/U \, T \rho \, c_p = IK =$ idem is the criterion characterising the relation between the internal heat generation and rate of convective transfer.

For such conductive systems in which the flow rates are equal to zero, the following criteria will be decisive:

$c_T/c_t = c_a \, c_T/c_l^2; \; c_a \, c_t/c_l^2 = 1; \; a' \, t'/l'^2 = a'' \, t''/l''^2$ or $at/l^2 = \mathrm{Fo} =$ idem is the Fourier criterion.

$$c_a c_T/c_l^2 = c_Q/c_\rho \, c_c; \; c_Q c_l^2/c_a \, c_T \, c_\rho \, c_c = 1; \; Q'l'^2/\lambda' \, T' = Q'' \, l''^2/\lambda'' \, T''$$

or $Ql^2/\lambda T = It =$ idem characterises the ratio of internal heat generation and conductive heat transfer.

For thermal similarity of convective systems, Nu, Pe and Ho are the decisive criteria.

Each criterion of similarity has its physical sense. Thus, the *Peclet criterion* ($\mathrm{Pe} = u \, l/a$) is a measure of the ratio of molecular and convective heat transfer in a flow as well as a criterion of thermal similarity.

The *Nusselt criterion* (Nu = $\alpha l/\lambda$) is a dimensionless coefficient of heat transfer characterising the relation between the rate of heat transfer and temperature field adjacent to the heat transfer surface (in the boundary layer of flow). In the conductive heat transfer regime, Nu = 1; for convective heat transfer Nu > 1. The Ho criterion has already been discussed.

Pe number can be expressed as the product of two criteria:

$$Pe = u\,l/a = (ul/v)(v/a) = Re\,Pr.$$

where Re is the criterion of hydrodynamic similarity (Reynolds number); $Pr = v/a$ is the *Prandtl criterion*, a measure of the similarity of temperature and velocity fields in the flow and also characterising the physical properties of the fluid. For example, for diatomic gases, Pr = 0.72 and triatomic gases, Pr = 0.8; for tetra-atomic and higher gases, Pr = 1; for molten metals, Pr << 1 and for the Earth's mantle Pr $\approx 10^{23}$.

When studying the rate of heat transfer, the main problem is determination of the coefficient of heat transfer α. Therefore, the criterial equation for convective heat transfer under forced convection has the following form:

$$Nu = f(Re; Pr). \tag{2.107}$$

In this gravity field, thermogravitational forces result from spatial changes in temperature. Therefore, Re and Gr criteria should be entered into criterial equations.

While studying mechanical similarity, it was pointed out that the Froude criterion represents a combination of Re and Gr criteria:

$$Fr = \frac{Gr}{Re^2}.$$

This complex characterises the relative role of natural convection compared to forced convection and hence the criterial equation in a general form will be:

$$Nu = f\left(\frac{Gr}{Re^2}; Pr\right). \tag{2.108}$$

At low values of Fr, the heat transfer is determined by the forced flow and hence

$$Nu = f(Re; Pr). \tag{2.109}$$

For thermogravitational flows (when Re $\rightarrow$ 0), the decisive criterion will be Gr.

$$Nu = f(Gr; Pr). \tag{2.110}$$

Finally, for gases of identical atomicity with equal Pr numbers

$$Nu = f(Gr). \tag{2.11}$$

Thus the theory of similarity enables derivation enables derivation of criteria of similarity from differential equations and determination of the general form of the criterial equation for similar processes. The theory of similarity, however, does not provide a general solution; it only helps to generalise the data of laboratory experiments or numerical simulation in a region in which the system under review are similar. Therefore, when applying the results of generalization, it must be checked whether or not conditions of similarity are satisfied.

2.15. Similarity Parameters in Natural Convection

We have studied the general rules for application of the theory of similarity. Let us now analyse some important actual cases using the method of recasting differential equations in a dimensionless form.

We shall examine a case of thermogravitational flows adjacent to a *vertical plate* having temperature T_w. The temperature of the fluid is T_a at infinity. The x-axis is oriented upwards and parallel to the gravity vector and the y-axis normal to the wall. Let us study a two-dimensional flow. Such flows do exist and have been studied experimentally. They are defined as flows of the boundary layer type. The boundary layer is studied below in detail.

The equations of free convection adjacent to the vertical plate will have the following form:

$$(\partial u/\partial t + u\, \partial u/\partial x + v\, \partial u/\partial y) = \beta\, g(T - T_a) - (1/\rho)\,(\partial P_d/\partial x)$$

$$+ (\eta/\rho)\,(\partial^2 u/\partial x^2 + \partial^2 u/\partial y^2),\, (\partial v/\partial t + u\, \partial v/\partial x + v\, \partial v/\partial y)$$

$$= -(1/\rho)\,(\partial P_d/\partial x) + (\eta/\rho)\,(\partial^2 v/\partial x^2 + \partial^2 v/\partial y^2),\, \partial u/\partial x + \partial v/\partial y = 0,$$

$$(\partial T/\partial t + u\, \partial T/\partial x + v\, \partial T/\partial y) = a\,(\partial^2 T/\partial x^2 + \partial^2 T/\partial y^2). \qquad (2.112)$$

where $a = \lambda/c_p\rho$ is the thermal diffusion coefficient. Henceforward, P will be used for P_d.

$\Delta T = T_w - T_a$ adopted as the temperature scale is expressed in a dimensionless form as $\overline{T} = (T - T_a)/\Delta T$, i.e., temperature is measured from T_a; the length of wall L can be adopted as the scale of length.

Velocity is the deciding parameter and hence the equation for it is found from the assumptions that convective terms and buoyancy forces are decisive, i.e.,

$$u\, \partial u/\partial x \approx \beta g\, \Delta T \text{ or } u^2 \approx \beta g\, \Delta T\, L$$

and $u_0 = (\beta g \Delta T\, L)^{1/2}$.

It would be correct to adopt the ratio a/L or v/L etc. as the velocity scale having the dimension of velocity. However, it is desirable to adopt values with a distinct physical sense for the phenomenon under consideration.

The value having the dimension of pressure and the sense of dynamic pressure $P_0 = \rho u_0^2$, i.e., $P_0 = \rho\,(\beta g\, \Delta T\, L)$ is accepted as the scale of pressure and $t_0 = L/u_0 = L/(\beta g\, \Delta T\, L)^{1/2}$ as the scale of time. The physical parameters of the fluid will be regarded as invariable and their value is taken for a fixed pressure and temperature.

Let us introduce the following dimensionless values and designate them by a bar:

$$\overline{u} = u/u_0 = u/(\beta g\, \Delta T\, L)^{1/2};\, \overline{v} = v/(\beta g\, \Delta T\, L)^{1/2};$$

$$\overline{P} = P/P_0 = P/\rho\beta\, g\Delta T\, L;\, \overline{t} = t\,(\beta g\, \Delta T\, L)^{1/2}/L;$$

$$\overline{x} = x/L;\, \overline{y} = y/L;\, \overline{T} = (T - T_a)/\Delta T. \qquad (2.113)$$

Let us rearrange eqn (2.112) into a dimensionless form. As an example, we recast one of the terms:

$$u\,\partial u/\partial x = (\beta g\,\Delta T\,L)^{1/2}\bar{u}\,(\partial\bar{u}/\partial\bar{x})\,[(\beta g\,\Delta t\,L)^{1/2}/L] = \beta g\,\Delta T\,\bar{u}\,(\partial\bar{u}/\partial\bar{x}).$$

In this case, the equation for x-axis will have the following form:

$$\beta g\,\Delta T\,(\partial\bar{u}/\partial\bar{t} + \bar{u}\,\partial\bar{u}/\partial\bar{x} + \bar{v}\,\partial\bar{u}/\partial\bar{y}) = \beta g\,\Delta T\,\bar{T} - \beta g\,\Delta T\,\partial\bar{P}/\partial\bar{x}$$

$$+ [v\,(\beta g\,L\,\Delta T)^{1/2}/L^2]\,(\partial^2\bar{u}/\partial\bar{x}^2 + \partial^2\bar{u}/\partial\bar{y}^2). \tag{2.114}$$

Equation for y-axis:

$$\beta g\,\Delta T\,(\partial\bar{v}/\partial\bar{t} + \bar{u}\partial\bar{v}/\partial\bar{x} + \bar{v}\partial\bar{v}/\partial\bar{y}) = -\,\beta g\Delta T\,\partial\bar{P}/\partial\bar{y}$$

$$+ [v\,(\beta g\Delta TL)^{1/2}/L^2]\,(\partial^2\bar{v}/\partial\bar{x}^2 + \partial^2\bar{v}/\partial\bar{y}^2). \tag{2.115}$$

Equation of continuity:

$$\partial\bar{u}/\partial\bar{x} + \partial\bar{v}/\partial\bar{y} = 0 \tag{2.116}$$

Equation of heat transfer:

$$[(\beta g\,\Delta T\,L)^{1/2}\,\Delta T/L]\,(\partial\bar{T}/\partial\bar{t} + \bar{u}\partial\bar{T}/\partial\bar{x} + \bar{v}T/\partial\bar{y})$$

$$= (a\Delta T/L^2)\,(\partial^2\bar{T}/\partial\bar{x}^2 + \partial^2\bar{T}/\partial\bar{y}^2). \tag{2.117}$$

Henceforward, the bars above the alphabetical notation are deleted without, however, forgetting that they represent dimensionless parameters.

After appropriate simplifications of eqns (2.114) to (2.117), we obtain

$$\partial u/\partial t + u\,\partial u/\partial x + v\,\partial u/\partial y = T^{\cdot} - \partial P/\partial x$$

$$+ (1/\mathrm{Gr}^{1/2})\,(\partial^2 u/\partial x^2 + \partial^2 u/\partial y^2), \tag{2.118}$$

$$\partial v/\partial t + u\partial v/\partial x + v\partial v/\partial y = -\,\partial P/\partial y + (1/\mathrm{Gr}^{1/2})$$

$$\times\,(\partial^2 v/\partial x^2 + \partial^2 v/\partial y^2), \tag{2.119}$$

$$\mathrm{div}\ \vec{u} = 0, \tag{2.120}$$

$$\partial T/\partial t + u\partial T/\partial x + v\partial T/\partial y = (1/\mathrm{Gr}^{1/2}\,\mathrm{Pr})\times(\partial^2 T/\partial x^2 + \partial^2 T/\partial y^2), \tag{2.121}$$

where $\mathrm{Gr} = \beta g\,\Delta T\,L^3/v^2$ is the Grashof number and $\mathrm{Pr} = v/a$ is the Prandtl number. Heat transfer at the boundary between the fluid and the solid wall proceeds only by heat conduction and the boundary condition has the following form:

$$q = -\,\lambda\,(\partial T/\partial y)_w. \tag{2.122}$$

In a dimensionless form. eqn (2.122) will be as follows:

$$q = - \lambda\, (\partial \overline{T}/\partial \overline{y})_w\, \Delta T/L \text{ or } \alpha L/\lambda = - (\partial \overline{T}/\partial \overline{y})_w, \tag{2.123}$$

where $\alpha = q/\Delta T$.

We established before that $\alpha l/\lambda = \text{Nu}$ is the Nusselt criterion representing the dimensionless coefficient of heat transfer.

The solution to the system of eqns (2.118) to (2.121) under condition (2.123) depends on the values of the criteria. For geometrically similar systems having identical values of Gr and Pr criteria, solutions to eqns (2.118) to (2.121) under condition (2.123) are identical. Thus, we have established the condition of similarity. For free convective heat transfer at a vertical surface, the determining criteria are Gr, Pr and Nu derived for the boundary condition. The following will be the general form of the criterial equation:

$$\text{Nu} = f(\text{Gr}; \text{Pr}). \tag{2.124}$$

In practical problems, it is usually necessary to first know the quantity of heat transferred from the body to a fluid. This quantity of heat can be expressed as the coefficient of heat transfer $\alpha = q/\Delta T$ which is determined for each point or for the surface as a whole as an average value. Usually, for the similar systems considered above, the experimental or numerical results are expressed in the form of a criterial equation:

$$\text{Nu} = f(\text{Pr})\, \text{Gr}^n.$$

For laminar convection, $n = 1/4$.

2.16. Boundary Layer Approximation

Let us assess the terms in the equations of motion and convective heat transfer at a vertical surface. We apply the approximation of the boundary layer to the equations of thermogravitational convection. The boundary layer term was coined by L. Prandtl in 1904 for a flow forced over the surface. The basic concept of this approximation is as follows: under forced flow over the surface the effect of viscosity is confined to a thin layer adjacent to the surface while the flow outside of this region is almost inviscid (Fig. 2.13).

Seizure of fluid against the surface is assumed, i.e., $u_w = 0$. Therefore, under forced flow, the flow velocity varies from zero on the surface to the velocity of free flow outside the boundary layer. The convective and diffusive components of energy and impulse transport have the same order in the boundary layer and hence the equations should be considered in their entirety. Nonetheless, because of the small thickness of the boundary layer compared to the distance in the flow direction or the body size, significant simplifications can be made. The main one is the assumption that values of gradients oriented normally to the surface are much greater than those obtained under forced flow. Pressure in the boundary layer is approximately equal to the pressure outside the boundary layer and elliptical equations of a two-dimensional flow rearrange to parabolic. Thus the problem

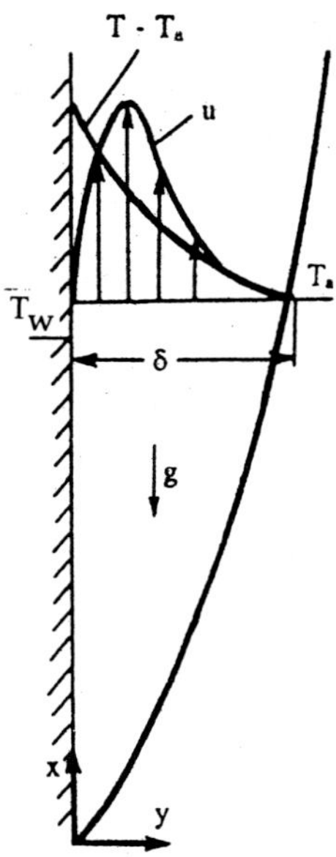

Fig. 2.13. Boundary layer at a vertical flat plate in thermogravitational convection.

of determining the flow is broken into two parts: modelling the regions in which the flow is inviscid and modelling the flow in a thin layer in which convective transport of momentum and energy occurs. The pressure field obtained by solving the first problem is used later for solving the problem for the boundary layer and determining the fields of velocity and temperature and hence the wall-friction and heat transfer at the wall. These problems have been well studied in Shlikhting's monograph *Theory of Boundary Layer* (1969).

The basic ideas used in the method of boundary layer approximation for thermogravitational convection are similar to those used in the case of forced flow. The major difference is that pressure in the region outside the boundary layer is independent of the conditions of external flow but represents the hydrostatic pressure, and the longitudinal velocity (u) outside the boundary layer is equal to zero.

It has been assumed that longitudinal flow is restricted to a thin zone adjacent to the heat transfer surface. Temperature variation occurs only in this boundary layer, as confirmed experimentally. This suggests that the gradients under forced flow along the surface are much less than those normally oriented to it.

Let us study a two-dimensional flow under conditions of thermogravitational convection of a heated (or cooled) isothermal vertical surface (T_w = const). The fluid is also isothermal (T_a = const) (see Fig. 2.13).

Let us write the equation of free convection in Boussinesq approximation for a laminar flow regime at a vertical plate in a dimensionless form in terms of eqn (2.113):

$$\partial u/\partial x + \partial v/\partial y = 0,$$
$$\text{(1)} \qquad \text{(1)} \tag{2.125}$$

$$\partial u/\partial t + u\partial u/\partial x + v\partial u/\partial y = T - \partial P/\partial x$$
$$\text{(1)} \qquad \text{(1)} \quad \text{(1)} \qquad (\delta)\,(1/\delta) \quad \text{(1)}$$

$$+ (1/Gr^{1/2})\,(\partial^2 u/\partial x^2 + \partial^2 u/\partial y^2)$$
$$\text{(1)} \qquad (1/\delta^2) \tag{2.126}$$

$$\partial v/\partial t + u\partial v/\partial x + v\partial v/\partial y = - \partial P/\partial y + (1/Gr^{1/2})$$
$$(\delta) \qquad \text{(1)}\ (\delta) \quad (\delta)\ \text{(1)}$$

$$\times (\partial^2 v/\partial x^2 + \partial^2 v/\partial y^2)$$
$$(\delta) \qquad (1/\delta) \tag{2.127}$$

$$\partial T/\partial t + u\partial T/\partial x + v\partial T/\partial y = (1/Gr^{1/2}\,Pr)$$
$$\text{(1)} \qquad \text{(1)} \quad \text{(1)} \ (\delta)\,(1/\delta_T)$$

$$\times (\partial^2 T/\partial x^2 + \partial^2 T/\partial y^2).$$
$$\text{(1)} \qquad (1/\delta_T^2) \tag{2.128}$$

Suppose the variation of u and T across the boundary layers δ and δ_T in thickness.

By this thickness of dynamic boundary layer (δ) is meant the region in which velocity changes form zero at the wall to values equal to 1% of maximum velocity. at the outer border of the boundary layer. Similarly, by the thickness of thermal (temperature) boundary layer δ_T is meant the region of variation of dimensionless temperature form 1 at the wall to values 0.01 at the boundary. We assume that the thickness of the thermal δ_T and dynamic δ boundary layers is much smaller than the linear size L. In this case, the dimensionless thickness of the dynamic layer $\delta/L \ll 1$, and of the thermal layer, $\delta_T/L \ll 1$.

Let us estimate the order of the values of terms involved the equations of free convection. Its maximum value, the longitudinal component of velocity u_m is accepted as the velocity scale. In this case the dimensionless value of longitudinal velocity u/u_m is of the order 1 since the velocity changes from zero to the maximum value. Co-ordinate x also varies from zero to 1 and the order of x is about 1. Let us evaluate the order of value v. Velocity u changes along x as along y from zero to 1. Therefore, the order of the value of derivative $\partial u/\partial x$ is also equal to one. It follows from continuity eqn (2.125) that the order of derivative $\partial v/\partial y$ is equal to 1. Since the order of y is δ, v also has order $\delta \ll 1$. The order of value $\partial^2 u/\partial x^2 \cong 1$ and of the second derivative $\partial^2 v/\partial y^2 \cong 1/\delta \gg 1$. In eqns (2.125) to (2.128), the order for each term is given below it: (1), (δ) and (1/δ). It follows from eqn (2.126) that $\partial^2 u/\partial y^2 \gg \partial^2 u/\partial x^2$ and thus term $\partial^2 u/\partial x^2$ can be ignored.

In the boundary layer, the terms responsible for convective transfer are of the same order as the terms in the case of viscous friction. An interesting conclusion emerges form this fact. Since convective terms ($u\,\partial u/\partial x + v\,\partial u/\partial y$) are of the order 1, it follows that $(1/Gr^{1/2})\,(\partial^2 u/\partial y^2) = 1$, $1/Gr^{1/2}\,\delta^2 \cong 1$ and $\delta \cong 1/Gr^{1/4}$. The dimensional value $\delta \cong L/\mathrm{Gr}^{1/4}$. Since $\delta << 1$, the boundary layer approximation is valid for high values of the Gr criterion.

It follows from the heat transfer eqn (2.128) that $\partial^2 T/\partial x^2 << \partial^2 T/\partial y^2$ and that the first term can be ignored. Since convective transfer ($u\,\partial T/\partial x + v\,\partial T/\partial y$) is comparable with conductive transfer, the order of the conductive term is equal to that of the convective term and $1/(\mathrm{Pr}\,\mathrm{Gr}^{1/2}\,\delta_T^2) \cong 1$; hence, $\delta_T \cong L/\mathrm{Pr}^{1/2}\,\mathrm{Gr}^{1/4}$. We have found a general relation for the thickness of the thermal boundary layer, i.e., the relation between δ and δ_T: $\delta_T/\delta \cong 1/\mathrm{Pr}^{1/2}$. Hence it follows that $\delta_T < \delta$ for very viscous fluids ($\mathrm{Pr} >> 1$) and $\delta_T > \delta$ for liquid metals.

In eqn (2.127) for y-axis, the terms are of order δ. Since $\partial P \cdot \partial y \cong \delta$, $\Delta P \cong \delta^2$. Equation (2.127) itself is of order δ and can be ignored compared to eqn (2.126). In eqn (2.126) $\partial P/\partial x \cong \delta^2/L \approx \delta^2$, and this term can likewise be ignored. The non-stationary term in eqn (2.126) $\partial u/\partial t$ can be taken as being of the same order as the convective term and term $\partial T/\partial t$ of eqn (2.128) can be considered as of the same order as terms of the convective heat transfer.

Following the evaluation of terms of the equations of free convection, we obtain the approximate system of equations:

$$\partial u/\partial x + \partial v/\partial y = 0, \tag{2.129}$$

$$\partial u/\partial t + u\partial u/\partial x + v\partial u/\partial y = T + (1/\mathrm{Gr}^{1/2})\,\partial^2 u/\partial y^2, \tag{2.130}$$

$$\partial T/\partial t + u\partial T/\partial x + v\partial T/\partial y = (1/\mathrm{Gr}^{1/2}\mathrm{Pr})\,\partial^2 T/\partial y^2. \tag{2.131}$$

We have derived equations of the boundary layer for heat transfer at a vertical wall under conditions of free convection. As can be seen, this system is considerably simplified compared to the initial eqns (2.125) to (2.128).

2.17. Approximate Solutions to Thermogravitational Convection Equations

The heat transfer equations given above were analysed in an effort to describe the mechanism of heat transfer and to elucidate the basic criteria characterising the process. However, solving these equations is an extremely sophisticated task even in simple cases such as flow at an isolated vertical plate.

Some methods are available for approximately solving the equations, e.g., the method of successive approximations and the integral method.

The method of successive approximations suggested by M.E. Shvets (1954) can be applied. Let us determine the heat transfer at a vertical isothermal plate at T_w placed in a large liquid parcel at temperature T_a.

The following are the boundary conditions:

at $y = 0, u = v = 0, T = T_w,$

$$\text{at } y \rightarrow \infty, u \rightarrow 0, T \rightarrow T_a. \tag{2.132}$$

As before, the scales of velocity, temperature and length adopted are: $u = (\beta g\,\Delta T\,L)^{1/2}$, $\Delta T = T_w - T_a$ and L. The dimensionless temperature is the tempera-

82

ture measured from T_a; $\overline{T} = (T - T_a)/(T_w - T_a)$; henceforward, the bar is deleted since only dimensionless values are used in all the calculations.

Velocity v is derived from the continuity equation:

$$v = -\int_0^y (\partial u/\partial x)\, \partial y.$$

The substitution of velocity v into eqns (2.130) and (2.131) gives for a case of stationary heat transfer:

$$u\partial u/\partial x - (\partial u/\partial y)\int_0^y (\partial u/\partial x)\, \partial y = T + \mathrm{Gr}^{-1/2}\, \partial^2 u/\partial y^2, \tag{2.133}$$

$$u\partial T/\partial x - (\partial T/\partial y)\int_0^y (\partial u/\partial x)\, \partial y - \mathrm{Gr}^{-1/2}\, \mathrm{Pr}^{-1}\, \partial^2 T/\partial y^2. \tag{2.134}$$

Hydrodynamic (δ) and thermal (δ_T) boundary layers are assumed to persist. From the estimations made, for $\mathrm{Pr} = 1$, $\delta \cong \delta_T$. The problem will be studied for these conditions. Let us similarly assume finite thickness δ, it being a function of x. To determine the thickness of the boundary layer, let us accept the conditions that at the outer border of the boundary layer, $(\partial T/\partial y)_\delta = 0$ and $(\delta u/\partial y)_\delta = 0$. Therefore, the boundary conditions for the approximations made before will be:

at $y = 0$, $T = 1$ and $u = v = 0$;

at $y = \delta$, $T = 0$, $u = 0$, $\partial u/\partial y = 0$, $\partial T/\partial y = 0$. \qquad (2.135)

The first approximation to heat transfer eqn (2.134) is found by ignoring convective heat transfer:

$$\partial^2 T/\partial y^2 = 0. \tag{2.136}$$

The solution to eqn (2.136) at boundary conditions (2.135) will be:

$$T = 1 - y/\delta. \tag{1.137}$$

The first approximation for determining u will be found from the equation of motion (2.133) by ignoring the convective terms and taking into consideration eqn (2.137):

$$\mathrm{Gr}^{-1/2}\, \partial^2 u/\partial y^2 = -T \text{ or } \mathrm{Gr}^{-1/2}\, \partial^2 u/\partial y^2 = y/\delta - 1. \tag{2.138}$$

By integrating this equation under the boundary conditions of eqn (2.135):

$$\mathrm{Gr}^{-1/2}\, u = y^3/6\delta - y^2/2 + c_1 y + c_2.$$

At $y = 0$, $u = 0$ and $c_2 = 0$; at $y = \delta$, $u = 0$ and $c_1 = \delta/3$. The solution for velocity will then have the following form:

$$u = \mathrm{Gr}^{1/2}\, \delta^2\, [(1/6)\,(y/\delta)^3 - (1/2)\,(y/\delta)^2 + (1/3)\,(y/\delta)]. \tag{2.139}$$

Let us find the second approximation to temperature. For this purpose, we use the values of u and T found in eqns (2.137) and (2.139) and their derivatives:

$$\partial T/\partial y = -1/\delta;\ \partial T/\partial x = \delta' y/\delta^2;\ \partial u/\partial x = \mathrm{Gr}^{1/2}\,(y\delta'/3 - y^3\,\delta'/\partial\delta^2),$$

$$\int_0^y (\partial u/\partial x)\,\partial y = \delta'\mathrm{Gr}^{1/2}\,(y^2/6 - y^4/24\delta^2), \tag{2.140}$$

where $\delta' = d\delta/dx$. Let us substitute the values obtained into the left side of heat transfer eqn (2.134):

$$(\mathrm{Ra}\ \delta')^{-1}\partial^2 T/\partial y^2 = 3y^4/24\delta^3 - y^3/2\delta^2 = y^2/2\delta. \tag{2.141}$$

Here, Ra $=$ Pr Gr is the Rayleigh criterion. Integration of eqn (2.141) gives:

$$(\mathrm{Ra}\ \delta')^{-1}\partial T/\partial y = y^5/40\delta^3 - y^4/8\delta^2 + y^3/6\delta + c_1. \tag{2.142a}$$

After the second integration, we obtain:

$$(\mathrm{Ra}\ \delta')^{-1}T = y^6/240\delta^3 - y^5/40\delta^2 + y^4/24\delta + c_1 y + c_2. \tag{2.142b}$$

Considering that at $y = 0$, $T = 1$, we obtain $c_2 = (\mathrm{Ra}\ \delta')^{-1}$ and at $y = \delta$, $T = 0$, we have $c_1 = -1\ \mathrm{Ra}\ \delta'\delta - \delta^2/48$.

In this case, the equation for temperature (2.142b) will have the following form:

$$T = 1 - y/\delta + \mathrm{Ra}\,[\delta'\,(y^6/240\,\delta^3 - y^5 . 40\delta^2 + y^4/24\delta) - (\delta'\delta^2/48)y]. \tag{2.143}$$

Determination of the second approximation to u is similar and can be omitted. Let us find the thickness of the boundary layer from condition $(\partial T/\partial y)_\delta = 0$; we find form eqn (2.143) that

$$-1/\delta + 11\ \mathrm{Ra}\ \delta^2\delta'/240 = 0 \ \text{and}\ \delta^3\delta' = 240/11\ \mathrm{Ra}. \tag{2.144}$$

We integrate eqn (2.144) considering that $\delta' = d\delta/dx$ and $\delta = 0$ at $x = 0$;

$$\delta = (960x/11\ \mathrm{Ra})^{1/4}. \tag{2.145}$$

From eqn (2.144), we determine δ' and substituting it into eqn (2.143), we obtain:

$$T = 1 - (1/11)\,[16(y/\delta) - (y/\delta)^6 + 6(y/\delta)^5 - 10(y/\delta)^4]. \tag{2.146}$$

From this the local heat flux can be found. Considering that $\mathrm{Nu} = -(\partial T/\partial y)_w$ (see eqn (2.123)) and δ is determined using eqn (2.145), we obtain from eqn (2.146) that

$$\mathrm{Nu} = \alpha L/\lambda = -(\partial T/\partial y)_w = 16/11\delta = 0.48\,(\mathrm{Ra}/x)^{1/4}. \tag{2.147}$$

Equation (2.147) for dimensioned value x will have the following form:

$$\mathrm{Nu}_x = \alpha x/\lambda = 0.48\,\mathrm{Ra}_x^{1/4}. \tag{2.148}$$

According to numerical solutions, for Pr > 1, $\mathrm{Nu}_x = 0.503\,\mathrm{Ra}_x^{1/4}$; this indicates good agreement with our solution.

In dimensional terms, eqn (2.145) will have the following form:

$$\delta = 3.06\, x/\mathrm{Ra}_x^{1/4}, \tag{2.149}$$

and the thickness of the boundary layer depends only on the Rayleigh criterion, as confirmed by the numerical solutions for $\mathrm{Pr} > 3$ (Dzhaluriya, 1983). According to eqn (2.146), the temperature profile depends only on y/δ. This suggests that, at $\mathrm{Pr} > 3$, temperature is a function of argument $y/\delta \cong (y/x)\,\mathrm{Ra}_x^{1/4}$.

Thus, for the problem under study, velocity and temperature depend only on similarity variable $\xi \cong (y/x)\,\mathrm{Ra}_x^{1/4}$ and equations in partial derivatives (2.129) to (2.31) can be recast to the form of ordinary differential equations in variable ξ.

The integral method and the method of similarity variable have been outlined in Dzhaluriya's book (1983).

2.18. Processing of Experimental Results

We shall now analyse in greater detail the experimental results on heat transfer in free convection.

It follows from the above analysis that the criterial equation of free convective heat transfer will have the following form:

$$\mathrm{Nu} = f(\mathrm{Gr};\mathrm{Pr}). \tag{2.150}$$

Quite often eqn (2.150) is represented in the form of a power function, as numerical simulation and laboratory experiments suggest:

$$\mathrm{Nu} = c\,\mathrm{Pr}^n\,\mathrm{Gr}^m, \tag{2.151}$$

where c, n and m are constants for a given heat transfer regime; they differ for laminar and turbulent regimes of convection. These coefficients may also differ for geometrically different heat transfer surfaces (flat and cylindrical layers, flat and cylindrical surfaces placed in a large fluid parcel).

Thermogravitational flows are initiated by density variation in the fluid due to temperature, which also alters the physical parameters. Therefore, when processing experimental results, it is important to average the physical parameters or to adopt a controlling temperature whereby the physical parameters can be found.

We do not have prior knowledge of the process conditions and hence a controlling temperature cannot be uniquely determined. The simplest and widely used method is to select the so-called average temperature in the boundary layer as a governing factor:

$$\overline{T} = 0.5\,(T_w - T_a), \tag{2.152}$$

where T_w and T_a are the temperatures of the wall and fluid outside the boundary layer. For a flat layer, $\overline{T} = 0.5\,(T_1 - T_2)$, where T_1 and T_2 are the temperatures of the heat transfer surfaces.

If the temperature profile across the boundary (or flat) layer is known, it is best to adopt the average temperature across the boundary layer:

$$\overline{T}_\delta = \frac{1}{\delta} \int_0^\delta T \, dy. \tag{2.153}$$

Selection of the determining temperature depends on a given case. It depends on the nature of variation of the physical properties of the fluid with temperature, range of temperature variation over the averaging area and on the required accuracy. Specialised experiments and analyses of data available suggest that when processing experimental results, it is best to adopt the average fluid temperature $\overline{T}_\delta$ (eqn 2.153). Selecting the average temperature according to eqn (2.153) takes well into consideration the effect of temperature on the fluid properties.

Since selection of the decisive temperature is not unique, the empirical criterial equations should be used with caution. The average temperature should be adopted in precisely the same manner as when deriving the equations. Therefore, when processing experimental results, the selected average temperature should be indicated and appropriate notations made in the form of indexes for the criterial, e.g., Ra_T or $Ra_{\delta, av}$.

The longitudinal co-ordinate can be used as a determining linear site. In the case of heat transfer in flat or cylindrical layers, the layer thickness and the ratio between it and the longitudinal co-ordinate should be adopted as controlling sites.

The diameter or vertical site of the rod should be used as crucial site in the case of a horizontal cylinder, sphere or horizontal rod of different cross-sections placed in a large volume.

Ra, Gr and Nu numbers are calculated from the experimental results conforming to the selected parameters of average temperature, determining geometric size and governing temperature drop. Temperature drop across layers of different geometry as well as the temperature difference across the boundary layer is measured in the course of the experiment and, when determining the criteria, is adopted for the steady heat transfer regime.

Coefficient c and degrees n and m are determined by graphic plotting of the experimental results in the form of dependence (2.151). When plotting eqn (2.151) in a logarithmic form, a straight line is obtained. Taking a logarithm of eqn (2.151) for a fixed Pr number gives:

$$\lg \text{Nu} = \lg c + m \lg \text{Gr}. \tag{2.154}$$

This represents an equation for a straight line. The value of m represents a tangent of the slope of straight line to the axis denoting the Gr number. A family of curves is obtained on the graph in the case of two arguments (Pr and Gr).

The criterial equation found experimentally is applicable in the range of Pr and Gr numbers corresponding to the data. Therefore, for each criterial equation, the limits of its applicability, decisive temperature and linear sizes are indicated. The equation derived can only be used for similar processes in the range of its applicability.

3

Thermogravitational Flows in a Horizontal Layer and Some Geological Applications

3.1. Conditions of Initiation of Convection Flow

Heat transfer under thermal convection is crucial in the thermal regime of the atmosphere and oceans of the Earth, atmosphere of other planets as well as the convective stellar zones. There are grounds to believe that convection flows are of prime importance in the heat transfer regime of the upper and lower mantle of the Earth.

Cellular convection in a horizontal fluid layer represents one of the types of flows arising under conditions of free convection.

Let us consider a horizontal fluid layer heated from below and cooled from above. The temperature gradient set up in this layer is called an inverse gradient since the liquid at the bottom of the layer will be lighter than at the top due to thermal expansion. The lighter liquid will tend to ascend while the cold, heavy one descends. However, this natural tendency in the fluid will be restrained by its intrinsic viscosity. Therefore, the inverse temperature gradient should exceed a certain level before attainment of an unstable state.

Thus, transition from a stable to an unstable state may occur only on exceeding the critical temperature gradient. Movement on loss of stability has a cellular character in the form of rolls and polygonal cells.

Benard (1901) carried out the earliest experiments on studying natural convection in a horizontal layer. The conditions of initiation of the flow in a horizontal layer were first studied theoretically by Rayleigh (1916) who showed that the stability loss is characterised by the Rayleigh criterion:

$$\mathrm{Ra} = \frac{\beta g \, \Delta T l^3}{a \nu} > \mathrm{Ra_{cr}}$$

(i.e., stability loss happens as the critical value $\mathrm{Ra_{cr}}$ is exceeded), where l is the layer thickness and $\Delta T = T_1 - T_2$ the temperature drop across the layer (Fig. 3.1). At $\mathrm{Ra} > \mathrm{Ra_{cr}}$, stability is lost over the whole volume of the fluid.

Let us turn to determining the critical value $\mathrm{Ra_{cr}}$. At $\mathrm{Ra} < \mathrm{Ra_{cr}}$ in a horizontal layer, heat is transferred by thermal conduction (see Fig. 3.1a). In the absence of convection, heat transfer eqn (2.74) rearranges to:

$$\frac{\partial^2 T_0}{\partial y^2} = 0. \tag{3.1}$$

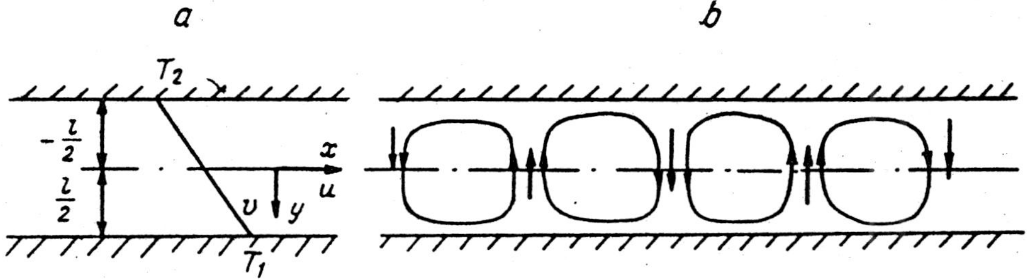

Fig. 3.1. Fluid flow in a horizontal layer.

The boundary conditions are:

$$\text{at } y = \frac{-l}{2}, \; T = T_2; \text{ at } y = \frac{l}{2}, \; T = T_1. \tag{3.2}$$

Integrating doubly eqn (3.1), we obtain

$$T_0 = c_1 y + c_2. \tag{3.3}$$

According to eqns (3.2) and (3.3), the integration constants will be:

$$c_2 = \frac{(T_2 + T_1)}{2}; \; c_1 = \frac{-(T_2 - T_1)}{l},$$

and in the absence of convection, there is a linear temperature profile:

$$T_0 = \frac{(T_1 + T_2)}{2} - \frac{y\,(T_2 - T_1)}{l} \tag{3.4}$$

Experimentally the onset of instability for $\Delta T_{cr} = T_1 - T_2$ is evidenced as convection rolls, i.e., the flow is two-dimensional (see Fig. 3.1b).

At the steady state the temperature profile across the layer will correspond to eqn (3.4). Condition $\Delta T > \Delta T_{cr}$ corresponds to onset of convection and the temperature profile will differ from eqn (3.4). Close to the stability threshold, temperature distribution will differ from (3.4) by an infinitely small value, T':

$$T' = T - T_0 = T - \frac{(T_2 + T_1)}{2} + \frac{y\,(T_2 - T_1)}{l} \tag{3.5}$$

The flow velocity close to the stability threshold will also be infinitely small (u', v'). Let us substitute the actual temperature profile (eqn 3.5) in the free con-

88

vection equations in two-dimensional approximation. In this case these equations will have the following form:

$$\rho_0 \, (\partial u'/\partial t + u' \, \partial u'/\partial x + v' \, \partial u'/\partial y)$$

$$= - \, \partial P'/\partial x + \eta \, (\partial^2 u'/\partial x^2 + \partial^2 u'/\partial y^2), \tag{3.6}$$

$$\rho_0 \, (\partial v'/\partial t + u' \, \partial v'/\partial x + v' \, \partial v'/\partial y)$$

$$= - \partial P'/\partial y - \rho_0 \, \beta g T' + \eta \, (\partial^2 v'/\partial x^2 + \partial^2 \, v'/\partial y^2) \tag{3.7}$$

$$\partial u'/\partial x + \partial v'/\partial y = 0, \tag{3.8}$$

$$\partial T'/\partial t + u' \, \partial T'/\partial x + v' \, \partial T'/\partial y + v' \, (T_1 - T_2)/l$$

$$= a \, (\partial^2 T'/\partial x^2 + \partial^2 T'/\partial y^2). \tag{3.9}$$

As the values of T', u' and v' are small, the quadratic terms relative to the disturbances $u' \, \partial u'/\partial x$, $u' \, \partial T'/\partial x$ etc., can be ignored in eqns (3.6) to (3.9). In this case we obtain a system of equations which determine the variation of small perturbations of temperature T', velocities u' and v' and pressure P':

$$\rho_0 \, \partial u'/\partial t = - \, \partial P'/\partial x + \eta \, (\partial^2 u'/\partial x^2 + \partial^2 u'/\partial y^2), \tag{3.10}$$

$$\rho_0 \, \partial v'/\partial t = - \, \partial P'/\partial y - \rho_0 \, \beta g T' + \eta \, (\partial^2 v'/\partial x^2 + \partial^2 v'/\partial y^2) \tag{3.11}$$

$$\partial u'/\partial x + \partial v'/\partial y = 0, \tag{3.12}$$

$$\partial T'/\partial t + v' \, (T_1 - T_2)/l = a \, (\partial^2 T'/\partial x^2 + \partial^2 T'/\partial y^2). \tag{3.13}$$

Equations (3.10) to (3.13) characterising the development of small perturbations are linear and the theory used to determine the stability threshold called the 'linear theory of stability'.

For $\Delta T < \Delta T_{cr}$, the system remains in a stable state. This is because the small thermal perturbations present (or fluctuations of buoyancy causing the fluctuation of liquid volumes) are balanced by dissipation processes due to friction and heat dissipation. The problem of studying the stability of a system with given parameters is essentially an attempt to determine the response of the system to minor disturbances: whether the system quiets down after disturbances or departures from the initial state.

The system is stable in the first case but unstable in the second. By subjecting the system to various types of perturbations, it is possible to determine under what types it is unstable and under what types it is stable. The neutral or indifferent stability state represents the transition from the stable to unstable states. Finding the stability of a system is one of the main problems in the study of hydrodynamic stability.

When studying the stability of a horizontal layer heated from below, it can be assumed that all the parameters of this system are constants except one, e.g., temperature drop across the layer. The system can then be studied for its transition from stable to unstable states for different values of ΔT by imposing small perturbations on the system each time.

Two types of transient states can be recognised from the response of the system to small perturbations. In the first case, transition from a stable to a new steady state occurs and a change in form of stability is observed from one steady state to another. In the second case, transition occurs through a neutral state of the system to a new one in the form of oscillatory instability with certain frequency characteristic.

The following is the sequence of solving the stability problem. The steady state of the system is studied first. In the present case this is represented by eqn (3.5). The general system of equations, i.e., (3.6) to (3.9), is then written. In these equations, non-linear quadratic terms relative to disturbances are ignored and a linear system of equations is obtained, i.e., eqns (3.10) to (3.13). The concept of stability is extended to all possible forms and wavelengths of small perturbations T', u' and v'. In practice, these perturbations are represented in the form of simple harmonic waves and consequently stability is studied relative to each wavelength individually. The stationary state in the horizontal layer is determined only by the temperature variation with the transverse co-ordinate (eqn 3.5). Arbitrary disturbances can be represented in the form of two-dimensional periodic waves. In fact, experiments show that, in a horizontal layer heated from below, longitudinal rolls are formed close to the stability threshold.

A system will remain stable if it is stable against all perturbations of all wavelengths λ. It will be unstable if it is unstable even to the disturbances of a single wave. It thus follows that the neutral state will be characterised by instability of one wavelength, i.e., the critical wavelength λ_{cr}, and will be stable to all other waves, $\lambda < \lambda_{cr}$. The parameters of the system corresponding to the aforesaid limiting stability state will be critical values.

Let us turn to solving the stability problem of a horizontal layer heated from below.

The boundary conditions to equations for minor disturbances (3.10) to (3.13) will be as follows: as indicated before, walls are isothermal and there is no fluid flow through the surface:

$$T' = 0, \ v' = 0 \ \text{at} \ y = \pm \ l/2. \tag{3.14}$$

If the fluid layer is bounded by rigid walls, adherence conditions will be satisfied i.e., there is no slippage of fluid along the surface (no-slip boundaries):

$$u' = 0 \ \text{at} \ y = \pm \ l/2 \tag{3.15}$$

If the surfaces are free, with slippage of fluid along them, shear stresses τ'_{xy} are absent and then at $y = \pm \ l/2$, according to eqn (2.69b),

$$\tau'_{xy} = \eta \left(\partial u'/\partial y + \partial v'/\partial x \right) = 0. \tag{3.16}$$

Since $v' = 0$ on the free interface for all values of x, $\partial v'/\partial x = 0$; it follows from eqn (3.16) that at $y = \pm \ l/2$:

$$\partial u'/\partial y = 0. \tag{3.17}$$

An analytical solution to the linear problem can be obtained for free interfaces for which boundary conditions (3.14) and (3.17) are applicable. Let us introduce the current function ψ which satisfies continuity eqn (3.12):

$$u' = - \partial\psi'/\partial y; \quad v' = \partial\psi'/\partial x. \tag{3.18}$$

By substituting eqn (3.18) in (3.10) and (3.11), we obtain

$$- \rho_0 \, \partial^2\psi'/\partial t \, \partial y = - \partial P'/\partial x - \eta \, (\partial^3\psi'/\partial y \, \partial x^2 + \partial^3 \, \psi'/\partial y^3), \tag{3.19}$$

$$\rho_0 \, \partial^2\psi'/\partial t \, \partial x = - \partial P'/\partial y - \rho_0 \, \beta g T' + \eta(\partial^3 \, \psi'/\partial x^3 + \partial^3 \, \psi'/\partial x \, dy^2). \tag{3.20}$$

Let us eliminate pressure from (3.19) and (3.20) and, for this purpose, differentiate eqn (3.19) for y and eqn (3.20) for x and subtract one from the other. We then obtain:

$$\rho_0 \, (\partial^3 \, \psi'/\partial t \, \partial y^2 + \partial^3\psi'/\partial t \, \partial x^2)$$

$$= - \rho_0 \, \beta g \, \partial T'/\partial x + \eta \, (\partial^4 \, \psi'/\partial y^4 + 2\partial^4 \, \psi'/\partial x^2 \, \partial y^2 + \partial^4 \, \psi'/\partial x^4). \tag{3.21}$$

Equation (3.13) is rewritten taking into account the current function (3.18):

$$\partial T'/\partial t + [(T_1 - T_2)/l] \, \partial\psi'/\partial x = a \, (\partial^2 \, T'/\partial x^2 + \partial^2 T'/\partial y^2). \tag{3.22}$$

Thus the problem consists in solving the linear system of eqns (3.21) and (3.22) relative to ψ' and T' with constant coefficients.

Disturbances of the current and temperature functions are represented in the from of a two-dimensional wave of horizontal length λ and maximum amplitudes T'_0 and ψ'_0 satisfying boundary conditions for free surface (3.14) and (3.17). The time variation is represented by value ω:

$$\psi' = \psi'_0 \, \cos\left(\frac{\pi y}{l}\right) \sin\left(\frac{2\pi x}{\lambda}\right) e^{\omega t},$$

$$T' = T'_0 \, \cos\left(\frac{\pi y}{l}\right) \cos\left(\frac{2\pi x}{\lambda}\right) e^{\omega t}. \tag{3.23}$$

For positive values of ω instabilities are amplified with time and the layer heated below is convectively unstable: at negative temperatures, instabilities are damped out and the layer becomes stable against small disturbances.

Let us substitute the values ψ' and T' (eqn 3.23) in the equations of motion (3.21) and heat transfer (3.22):

$$(2\pi/\lambda) \, \rho_0 g \, \beta T'_0 = \psi'_0 \, [(4\pi^2/\lambda^2 + \pi^2/l^2)^2 \, \eta + \omega \rho_0 \, (\pi^2/l^2 + 4\pi^2/\lambda^2), \tag{3.24}$$

$$(\omega + a\pi^2/l^2 + a \, 4\pi^2/\lambda^2)T'_0 = - \, [2\pi \, (T_1 - T_2)/\lambda l] \, \psi'_0. \tag{3.25}$$

By dividing eqn (3.25) by (3.24), the amplitudes of perturbations ψ'_0 and T'_0 are eliminated:

$$(\omega + a\,\pi^2/l^2 + a\,4\pi^2/\lambda^2)/(2\pi/\lambda)\ \rho_0 g\ \beta$$

$$= -\,[2\pi\,(T_1 - T_2)/\lambda l]/[\eta(4\pi^2/\lambda^2 + \pi^2/l^2)^2$$

$$+ \omega\,\rho_0\,(\pi^2/l^2 + 4\pi^2/\lambda^2)]. \tag{3.26}$$

We are interested in neutral perturbations. As pointed out before, the experiment fixes the formation of stable convection rolls implying that a change in form of stability arises, i.e., the development of steady flows after the loss of stability, and hence neutral perturbations are characterised by condition $\omega = 0$.

It follows from eqn (3.26) at $\omega = 0$ that the critical number $\mathrm{Ra_{cr}} = [\beta g\,(T_1 - T_2)l^3]/av$ depends only on wave number $K = 2\pi l/\lambda$:

$$\mathrm{Ra_{cr}} = (K^2 + \pi^2)^3/K^2. \tag{3.27}$$

Relation (3.27) is an equation of the neutral boundary of stability. From it can be found the minimum value of $\mathrm{Ra_{cr}}$ by differentiating it with respect for wave number K and equating the derivative to zero:

$$d\,\mathrm{Ra_{cr}}/dK = 6\,(K^2 + \pi^2)^2/K - (K^2 + \pi^2)^3\,2/K^3 = 0. \tag{3.28}$$

From eqn (3.28) it can be seen that $K_{cr} = \pi/\sqrt{2}$. By substituting the value found for K_{cr} into eqn (3.27), we find the minimum critical value:

$$\mathrm{Ra_{cr}} = 27\,\pi^4/4 = 657.5. \tag{3.29}$$

At $\mathrm{Ra} > \mathrm{Ra_{cr}}$, the onset of convection is observed in the horizontal layer heated from below. The horizontal size of two rolls will then be equal to $\lambda = (2\sqrt{2})\,l$, where $\lambda = 2\,l_1$ and l_1 the horizontal size of the roll.

The most real conditions are the boundary conditions of seizure against the rigid surfaces (3.14) and (3.15) which can be solved only numerically. The results of such investigations have been presented in Chandrasekhar's monograph (1961). In this case, $\mathrm{Ra_{cr}} = 1707.8$ and $K_{cr} = 3.117$, i.e., $\lambda = 2.016\,l$. Figure 3.2 shows the dependence of the value of $\mathrm{Ra_{cr}}$ on K for a fluid layer bounded by rigid surfaces. The region $\mathrm{Ra} < \mathrm{Ra_{cr}} = 1707.8$ is stable against perturbations. At high values of $\mathrm{Ra} > \mathrm{Ra_{cr}}$, there is a wavelength range (wave numbers) at which convection flows arise. This region is enclosed between the branches of curves 1 (see Fig. 3.2).

3.2. Horizontal Cell Size and Stability

Theoretical investigations of secondary flows and numerical simulation of the structure of two-dimensional flows were carried out assuming a strict periodicity of convection rolls. Experiments showed that even at a minor super-

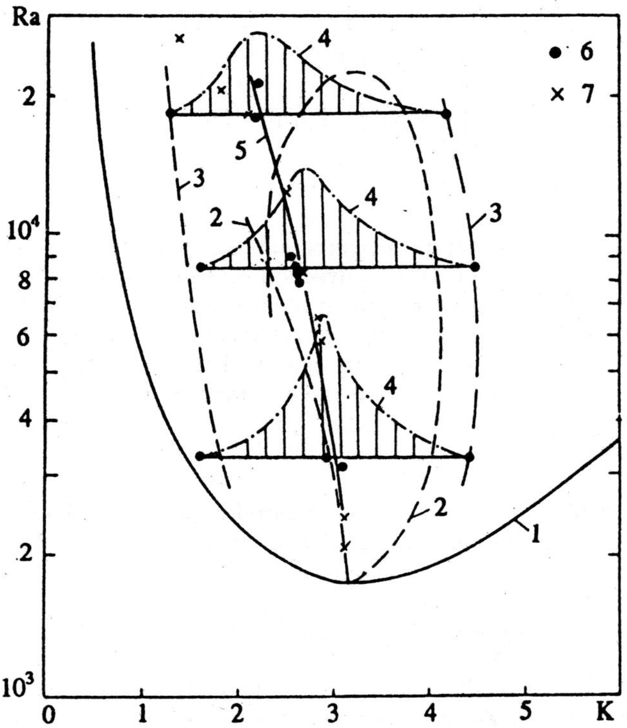

Fig. 3.2. Diagrams showing the stability of convection rolls: 1—neutral curve of the evolution of convection flows (Chandrasekhar, 1961); 2— curve limiting the stability region of rolls for high values of Prandtl number (Pr → ∞) (Busse, 1967); experiments carried out by Berdnikov and Kirdyashkin (1978), Pr = 16; 3—boundary of roll stability; 4—curve of probability distribution of wave numbers K; 5 and 6—average value of K; 7—average value of K according to Krishnamurti (1970).

criticality, the cellular structure is only quasi-periodic with a wide spectrum of wavelengths (Berdnikov and Kirdyashkin, 1978). This aspect did not initially attract attention while analysing the experimental results. Figure 3.3a, b depict the boundaries of cellular flows taken from photographs of cells in a horizontal plane (Berdnikov and Kirdyashkin, 1978). Visual observations showed that as the Rayleigh number increases, a qualitative change in steady spatial forms of cells occurs. In the region $5 \times 10^3 \leq Ra \leq 2 \times 10^4$ (Pr = 14), the quasi-two-dimensional steady roll structure is slowly transformed, initially into a three-dimensional (see Fig. 3.3b) and later even into an oscillating structure. The three-dimensional state is seen even in the curvatures of rolls and their ends are terminated. Moreover, the motion of fluid particles in the horizontal section is not rectilinear but helical, i.e., secondary flows arise (see Fig. 3.3b). Figure 3.2 shows the results of measurements of horizontal size of longitudinal rolls. They are depicted in the form of probability distribution curves for different wave numbers K; here, $K = \pi \, l/l_1$. The zone of roll convection, according to experiments, is bounded by curves 3. There is no convection outside this region. Within the zone of convection rolls, however, the most probable wave number and its average value can be estimated (points 6 and 7).

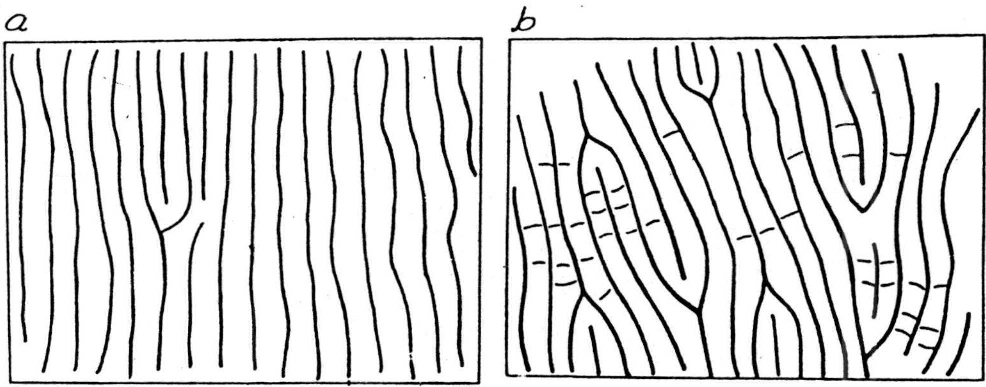

Fig. 3.3. Steady convection rolls.
a—Ra = 4.4 × 10³, l = 3.15 mm; b—Ra = 1.14 × 10⁴, l = 3.15 mm (Berdnikov and Kirdyashkin, 1978).

The probability distribution curves (see Fig. 3.2, curves 4) show that the steady-state flow has a broad spectrum of wavelengths. As Ra rises, this spectrum broadens and the most probable value of l_1 increases. The distribution is not symmetrical relative to the most probable value of l_1. For example, at Ra = 18,000 the range of horizontal cell dimensions will be: $l_1 = (0.7 \text{ to } 2.4) \, l$ and the average horizontal roll size $l_1 = 1.25 \, l$.

Figure 3.2 (curve 2) also shows the results of Busse's (1967) theoretical investigations of roll stability assuming that the cellular structure is strictly periodic in the horizontal plane. The experiment shows that the onset of the spatial form of cellular flow and the distribution of horizontal roll sites in the horizontal plane has a random character. Therefore, differences are noticed between the experimentally determined region of convection rolls and Busse's theoretical curve.

3.3. Flow Regimes and Heat Transfer Rate

As Ra number increases, the flow patterns become complex. Roll flows transform into three-dimensional flows with polygonal structures (Fig. 3.4), which in turn lose stability. Finally, the growth of instabilities leads to a regime called the turbulent flow.

94

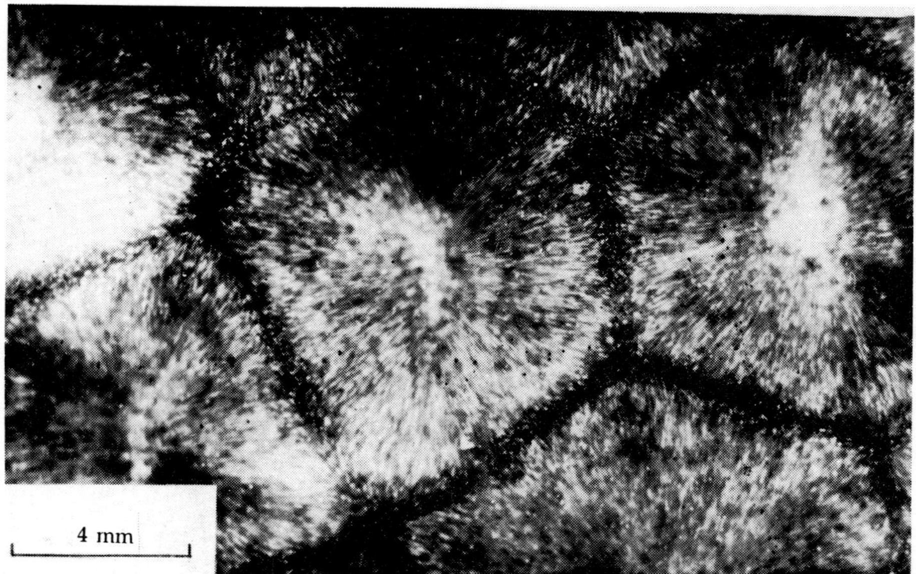

Fig. 3.4. Polygonal form of flow. Pr = 16, l = 4 mm and Ra = 22,600.

It is very difficult to trace in detail the sequence of structural changes of convection flows. It can be done by classifying the flow regimes into the following types arbitrarily:

a) steady two-dimensional;
b) steady three-dimensional
c) unsteady three-dimensional, and
d) turbulent.

These regimes depend on Rayleigh (Ra) and Prandtl (Pr) numbers.

Figure 3.5 presents the chart of flow regimes according to observational data for convection in a horizontal fluid layer heated from below. Experiments pursued by A.G. Kirdyashkin and A.A. Kirdyashkin (in litt.) include the examination of short-period local Rayleigh number and heat flux fluctuations behaviour and flow velocity measurements.

Curve (demarcation) VI is defined by growth of high-frequency fluctuations of local Rayleigh number and local heat flux q_l through the layer (see Sec. 3.6, Fig. 3.14a). Furthermore, short-period fluctuations superimpose on long-period ones associated with large-scale changes of cell structure (Fig. 3.14b, 3.15a). Boundary VI is determined by Rayleigh number $Ra_{VI} \approx 1.7 \times 10^5$.

Transition to a developed turbulent regime, labelled VII, is determined by the relation between average amplitude of local heat flux high-frequency fluctuations and average value of local heat flux A/q_l. The onset of developed turbulence is associated with sufficiently great value of relative average amplitude $A/q_l \approx 0.35$ (Fig. 3.15b).

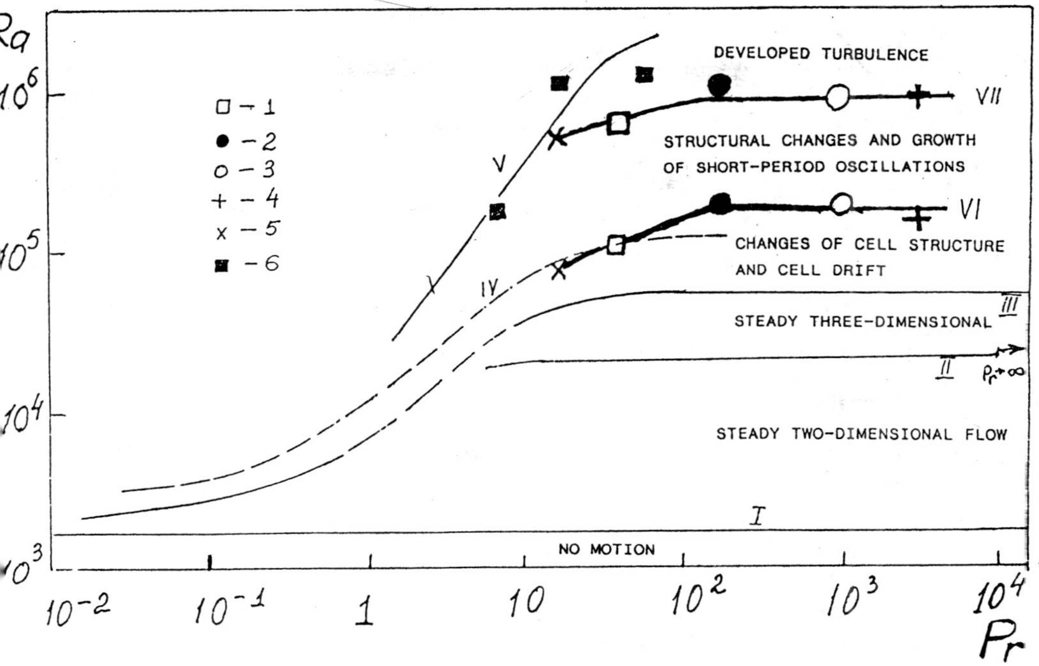

Fig. 3.5. Demarcations between flow regimes (regime diagram). A.G. Kirdyashkin and A.A. Kirdyashkin's observations (to be published 1998): 1—Pr = 35–44; 2—Pr = 165–205; 3—Pr = 858–1194; 4—Pr = (2.5–4.1) × 10^3; data obtained by Berdnikov et al. (1987): 5— Pr = 14–16; transition to turbulence observed by Willis and Deardorff (1967): 6 and curve V; transitions observed by Krishnamurti (1970, 1973): curves I–IV.

Turbulence observations can be explained in terms of intermittency. Intermittency represents a relation between the present time of short-period fluctuations in a given realisation and the total record length. Line VI corresponds to an intermittency coefficient $\varphi = 0$, that is, curve VI marks the transition from long-period fluctuations to the onset of short-period fluctuations of local Rayleigh number and heat flux.

However, amplitude of fluctuations is small, whereas the intermittency $\varphi = 1$. Therefore, the turbulent regime is not developed for $\varphi = 1$. For example $\varphi = 1$. for Ra = 1.8 × 10^5 (Pr = 43); Ra = (4.4 – 6.0) × 10^5 (Pr = 166); Ra = (3.0 – 3.6) × 10^5 (Pr = 940); Ra = 3.2 × 10^5 (Pr = 3.6 × 10^3). The line labelled VII is determined by $Ra_{VII} \approx 10^6$, exceeding Rayleigh numbers which correspond to $\varphi = 1$. Rayleigh numbers corresponding to $\varphi = 1$ in Willis and Deardorff's observations (1967) exceed our data. One reason is that temperature drop across the layer increased or decreased progressively in the course of their experiments, i.e., the boundary conditions were unsteady. In our experiments temperature difference in the layer was constant. Another probable cause lies in the numerical method of intermittency calculation in the Willis and Deardorff observations. Thus the boundary of turbulent regime V is significantly refined (Fig. 3.5).

As can be seen from Figure 3.5, there is only a slight variation in turbulent flow boundary for Pr > 15.

Different heat transfer laws are found depending on the flow regime. Close to the stability threshold in the range $Ra_{cr} < Ra < 3.5 \times 10^3$, dependence of the Nusselt criterion $Nu = \alpha l / \lambda$ on Ra is as follows:

$$Nu = 1 + 1.43 \left(\frac{1 - Ra_{cr}}{Ra} \right) \tag{3.30}$$

In the range $4 \times 10^3 < Ra < 10^5$:

$$Nu = 0.2 \, Ra^{1/4} \tag{3.31}$$

In a turbulent flow regime at $Ra > 10^5$:

$$Nu = 0.1 \, Ra^{1/3}. \tag{3.32}$$

It can be seen from eqn (3.32) that the coefficient of heat transfer α is independent of the layer thickness, i.e.,

$$\frac{\alpha}{\lambda} = 0.1 \left(\frac{\beta g \Delta T}{a\nu} \right)^{1/3} \tag{3.33}$$

This highlights the characteristic internal size or length scale inherent in heat transfer itself as a determining factor. Kirdyashkin (1966) experimentally determined that a vertical layer arises adjacent to the heat transfer surface in the regime of developed turbulent convection and its thickness is determined from the condition $Ra_{h.1} = 3100$, i.e., the thickness of this layer depends on the physical properties of the fluid and temperature drop across the layer, where

$$Ra_{h.1} \frac{\beta g \, \Delta T_{h.1} l_{h.1}^3}{a\nu}.$$

In this case, $l_{h.1} = (3100 \, a\nu / \beta g \, \Delta T_{h.1})^{1/3}$, where $\Delta T_{h.1} = 0.5 \, \Delta T$ for a layer enclosed between two plates and $\Delta T_{h.1} = (T_1 - T_\infty)$ for a large fluid parcel with the horizontal plate heated from below, where T_∞ is the temperature in the fluid layer at a distance from the horizontal surface far exceeding $l_{h.1}$.

3.4. Temperature and Velocity Fields in Convection Cells

Temperature changes in a horizontal layer heated from below, at $Ra \leq Ra_{cr}$, indicate a linear temperature profile corresponding to a conduction regime of heat transfer. At $Ra > Ra_{cr}$, temperature profiles in the ascending ($x/l_1 = 0$) and descending ($x/l_1 = 1$) flows deviate from a straight line and a region of constant temperature is noticed in them even at $Ra = 8.68 \times 10^3$ (Fig. 3.6).

At other values of x/l_1 temperature profiles have a reverse gradient. Figure 3.7 shows the temperature profiles in a polygonal cell. Reversal of heat flux is also noticed outside the descending and ascending flows. Thermal boundary layers occur at the walls.

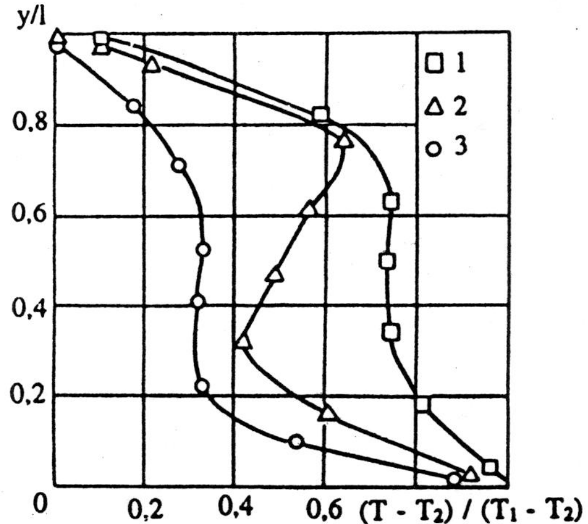

Fig. 3.6. Temperature profiles in a roll (Berdnikov and Kirdyashkin, 1978).
$1—x/l_1 = 0.065$ (ascending flow); $2—x/l_1 = 0.735$; $3—x/l_1 = 1$ (descending flow).
$Ra = 8.7 \times 10^3$; $l = 3.15$; $\Delta T = 3.75°C$; $T_2 = 23.52°C$.

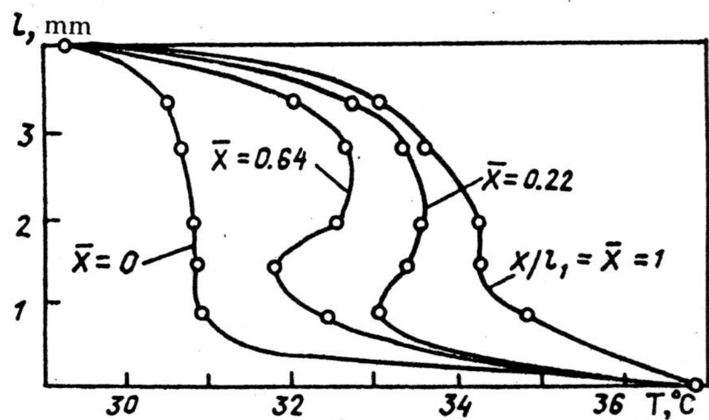

Fig. 3.7. Temperature profiles in a polygonal cell (Kirdyashkin, 1966).
$Ra = 41,200$ and $Pr = 15$.

The temperature profile measurements in the descending and ascending flows
for different Ra values enabled establishing a simple law of temperature variation
in the ascending and descending flows and the temperature difference between
them (Fig. 3.8). Temperature profiles in the ascending and descending flows are
similar in co-ordinates $(T - T_0)/\Delta T$ and y/l, where T_0 is the average temperature
in the layer. For the greater part of the layer thickness, the relative difference in
temperature between descending and ascending flows is constant and equal to
$T_{x=1} - T_{x=0} = 0.5 \ \Delta T$.

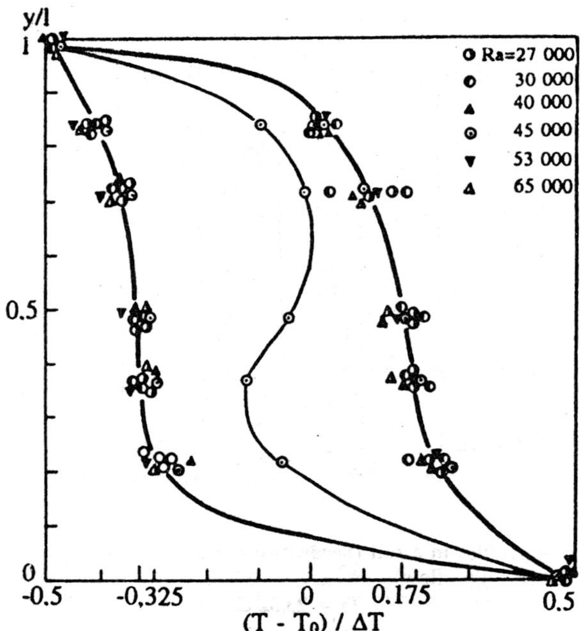

Fig. 3.8. Relative temperature profiles in the descending and ascending flows of a polygonal cell through the thickness of a fluid layer (Kirdyashkin, 1966).

Velocity measurements in the cell were done by the stroboscopic visualisation method: the fluid layer was illuminated by a pulsed source of light. Aluminium particles suspended in the layer were photographed on a stationary photographic film in the form of tracks (Fig. 3.9). By knowing the time interval between the flashes and the scale of photographing, the velocity field of two-dimensional flow in the cell could be determined.

Figure 3.10 depicts the velocity field in a convection roll while Figure 3.11 shows the same property in a polygonal cell. Measurements were carried out in a wide range of Ra numbers. Figure 3.12 shows the dependence of the maximum value of horizontal velocity (u_{max}) on Ra number where $Re_m = u_m l/v$ is the Reynolds number calculated from u_{max} and $Pr = v/a$ is the Prandtl number determined from the average temperature in the layer $\overline{T} = (T_1 + T_2)/2$. In this graph, the continuous line represents the theoretical values of Re_m Pr obtained by Chandrasekhar (1961) by the finite amplitudes method in the vicinity of Ra_{cr}.

$$Re_m \ Pr = 0.24 \, (Ra - Ra_{cr})^{1/2}. \tag{3.34}$$

Good agreement is evidenced between the experimental and theoretical findings also at Ra $\gg$ Ra_{cr}. A similar pattern emerges also at Ra $\gg$ Ra_{cr} from the solution obtained by applying the boundary layer theory: $Re_m \approx Ra^{1/2}$.

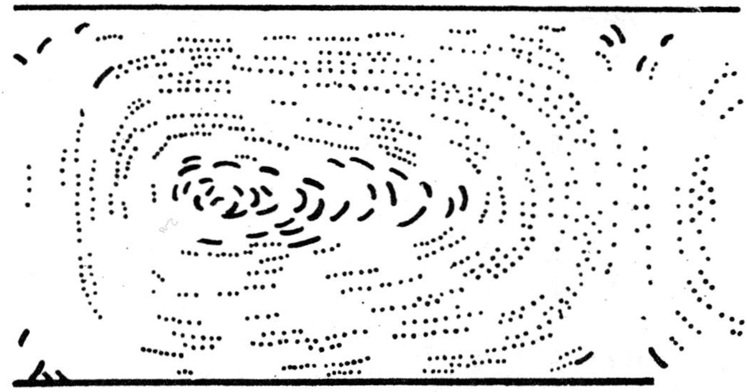

Fig. 3.9. Stroboscopic flow lines in the form of tracks (Berdnikov and Kirdyashkin, 1978).

3.5. Effect of Prandtl Number on Flow Structure and Heat Transfer in Thermogravitational Convection

Let us set up the equation of free convection in Boussinesq approximation (eqns (2.79) to (2.81)) in a dimensionless form for the following scales: velocity a/l; temperature ΔT; length l; time l^2/a and pressure $a^2\rho/l^2$

$$\left(\frac{1}{\mathrm{Pr}}\right)\left[\frac{\partial \vec{u}}{\partial t} + (\vec{u}\ \mathrm{grad})\vec{u}\right] = \mathrm{Ra}\ \vec{j}\ T - \left\{\frac{1}{\mathrm{Pr}}\right\}\mathrm{grad}\ P + \nabla^2\ \vec{u}, \tag{3.35}$$

$$\frac{\partial T}{\partial t} + (\vec{u}\,\mathrm{grad})\ T = \nabla^2 T \tag{3.36}$$

and

$$\mathrm{div}\ \vec{u} = 0, \tag{3.37}$$

where $\vec{j}$ is the unit vector of gravity.

Let us analyse the general properties of eqns (3.35) to (3.37) for two cases wherein $\mathrm{Pr} \ll 1$ and $\mathrm{Pr} \gg 1$. In the case of steady-state heat transfer, $\partial\vec{u}/\partial t = 0$ and $\partial T/\partial t = 0$.

Fluids with $\mathrm{Pr} \ll 1$ should be classified as highly heat-conducting materials, for example liquid metals. At $\mathrm{Pr} \ll 1$, convective term $(\vec{u}\,\mathrm{grad})\ T$ in the heat transfer eqn (3.36) can be ignored, i.e., heat transfer essentially proceeds by thermal conduction. The equation of motion (3.35) should be used in toto.

Let us examine a liquid with $\mathrm{Pr} \gg 1$. This is a highly viscous fluid. Such substances include mantle material for which $\mathrm{Pr} \sim 10^{22}$ to 10^{23}. The flow velocities in highly viscous fluids are low. Such flows are called creep movements. In the case of creep, convection terms can be ignored in the equation of motion (3.35) $(\vec{u}\,\mathrm{grad})\ \vec{u}$.

100

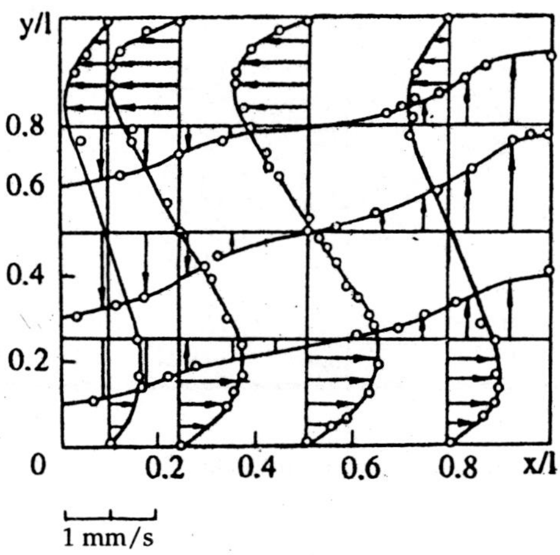

Fig. 3.10. Velocity field in a convection roll:
$Ra = 8.8 \times 10^3$; $l = 3.15$ mm; $l_1 = 3.6$ mm; $u_{max} = 0.77$ mm/s; $\overline{T} = 33.7°C$; $Pr = 15$ ethanol (Berdnikov and Kirdyashkin, 1978).

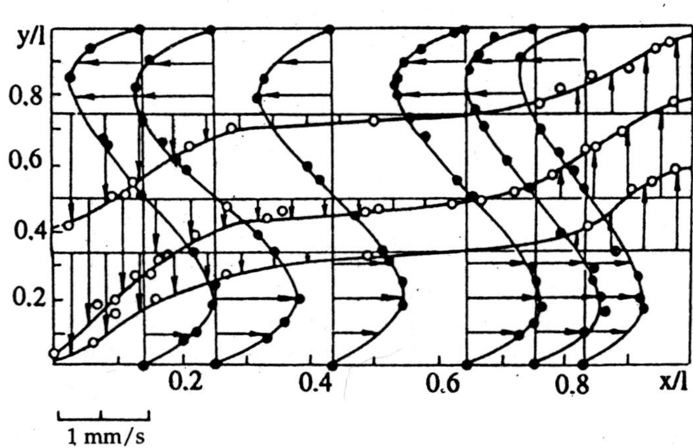

Fig. 3.11. Velocity field in a polygonal cell.
$Ra = 1.9 \times 10^4$; $l = 4.07$ mm; $u_{max} = 0.89$ mm/s; $\overline{T} = 31.7°C$; $v_{max} = 1.66$ mm/s ethanol (Berdnikov and Kirdyashkin, 1978).

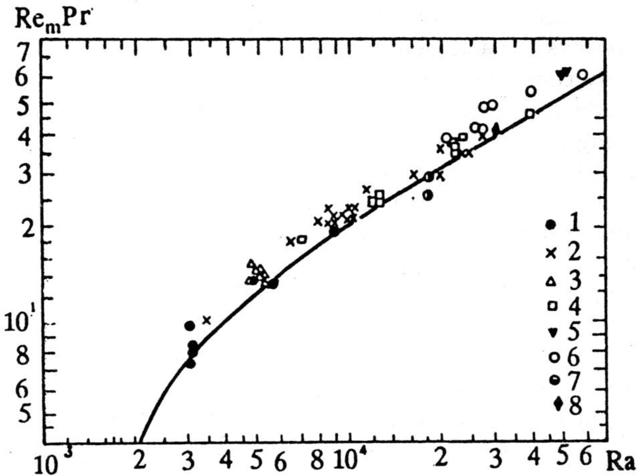

Fig. 3.12. Reynolds number Re_m plotted against Rayleigh number Ra for Pr $\approx$ 16 according to the experiments of Berdnikov and Kirdyashkin (1978).
Rolls: 1 — l = 2.5 mm; 2 — l = 3.15 mm; 3 — l = 3.6 mm; 4 — l = 4 mm; 5 — l = 5.1 mm; *hexagons*: 6 — l = 4 mm; 7 — l = 3.6 mm; *rectangles*: 8 — l = 4 mm; *continuous line*: Chandrasekhar's solution (1961).

Under steady boundary conditions, unsteady flow occurs due to stability loss such that an onset of turbulent flow is observed as the Ra number increases. In this case, the unsteady term can likewise be ignored in the equation of motion: $\partial \vec{u}/\partial t = 0$. Terms on the left side of eqn (3.35) for Pr $\gg$ 1 are much smaller than the viscous terms $\nabla^2 \vec{u}$ and the gravity term $\vec{j}$ Ra T can be ignored for this reason. The term representing the pressure gradient $(1/Pr)$ grad P should not be ignored.

In a three-dimensional case, the following system is obtained by eliminating pressure from the three scalar equations in which the equation of motion (3.35) is resolved and the unsteady and convective terms at Pr $\gg$ 1 ignored:

$$\text{Ra}\ \frac{\partial T}{\partial y} + \frac{\partial}{\partial y}(\nabla^2 u) - \frac{\partial}{\partial x}(\nabla^2 v) = 0, \tag{3.38}$$

$$\frac{\partial}{\partial z}(\nabla^2 v) - \frac{\partial}{\partial y}(\nabla^2 w) = 0, \tag{3.39}$$

$$\frac{\partial T}{\partial t} + u\ \frac{\partial T}{\partial x} + v\ \frac{\partial T}{\partial y} + w\ \frac{\partial T}{\partial z} = \nabla^2 T \tag{3.40}$$

and

$$\frac{\partial u}{\partial x} + \frac{\partial v}{\partial y} + \frac{\partial w}{\partial z} = 0. \tag{3.41}$$

It can be seen from the above system of equations that at Pr >> 1, Rayleigh number Ra represents the sole determining criterion of unsteady thermal convection. The unsteady convection at Pr >> 1 appears as thermally induced since the unsteady term $\partial T/\partial t$ is present only in the heat transfer eqn (3.40) and the unsteady term $\partial u/\partial t$ is negligibly small in the equation of motion. The heat transfer equation remains in a complete form since the convective term $(\vec{u}\ \text{grad})\ T$ plays the main role in heat transfer and the term $\nabla^2 T$ accounting for conductive heat transfer constitutes the determining factor adjacent to the heat transfer surface independent of the Pr number.

The criteria equation (heat transfer law) for Pr >> 1 is of the form:

$$Nu = f(Ra). \tag{3.42}$$

For free convection at a vertical heat transfer surface at constant temperature difference across the boundary layer, eqn (3.42) holds for Pr > 3 (Dzhaluriya, 1983):

$$Nu = 0.503\ Ra^{1/4}. \tag{3.43}$$

Thus, when studying convection at a vertical plate, convective terms can be ignored even for Pr > 3. For a horizontal fluid layer heated from below, eqn (3.42) holds even for Pr > 1, as can be seen from eqns (3.31) and (3.32). Apart from experimental results, Figure 3.5 presents the results of three-dimensional numerical modelling of thermogravitational convection in a spherical shell for Pr $\rightarrow \infty$ (Zhang and Yuen, 1995). The solutions were obtained only for three values of Ra numbers and correspond to the flow regimes shown in Figure 3.5. For the Earth's mantle, Pr $\approx 10^{23}$ and inertia can be ignored. It can be seen from the regime diagram (Fig. 3.5) that the stability boundaries of steady flows are independent of the Pr number even for Pr > 5. Boundaries of unsteady flow regimes (Fig. 3.5, IV, VI, VII) are independent of Pr for Pr > 15. This shows that the Rayleigh number for Pr > 15 is decisive in unsteady flows. This is supported by the heat transfer law under a turbulent flow regime in a horizontal layer (3.32) and for a turbulent regime at a vertical plate (Kirdyashkin, 1979; Paolucci, 1990);

$$Nu = 0.098\ Ra^{1/3}$$

Thus for experimental modelling of convection in the lower mantle, it is appropriate to use a fluid with Pr > 15 as the working medium.

3.6. Time-scales of Unsteady Flows

As can be seen from the system of eqns (3.38) to (3.41), unsteady flow is thermally induced since the non-stationary term remains only in the heat transfer equation. The unsteady state is mainly due to thermal and not hydrodynamic inertia since the convective terms in the equation of motion are negligibly small. The change in velocity is determined by temperature variation with time and a correlation should be seen between them. Increases in temperature drop $(T_1 - T_2)$ give rise to enhancement of velocity.

One cannot predict the frequency characteristics of temperature fluctuations and hence velocity in an unsteady convection regime from the time-average steady boundary conditions.

A decisive criterion of unsteady convection, i.e., homochronicity criterion, was derived earlier (in Sec. 2.14):

$$\text{Ho} = u\ t_0/l. \tag{3.44}$$

In the case of fluctuations of temperature drop relative to the average value in a horizontal layer, velocity can be determined using eqn (3.34):

$$u = \frac{0.24\,a\ (\text{Ra} - \text{Ra}_{cr})^{1/2}}{l}\ .$$

The velocity scale adopted $u \approx a(\text{Ra} - \text{Ra}_{cr})^{1/2}/l$ is substituted in eqn (3.44):

$$\text{Ho} = \frac{t_0\,a}{l^2}\ (\text{Ra} - \text{Ra}_{cr})^{1/2} = \text{Fo}\ (\text{Ra} - \text{Ra}_{cr})^{1/2}\ , \tag{3.45}$$

where $\text{Fo} = t_0\,a/l^2$ is the Fourier criterion inherent to unsteady conductive heat transfer. At $\text{Ra} \gg \text{Ra}_{cr}$, eqn (3.45) will have the following form:

$$\text{Ho} = \frac{t_0\,a}{l^2}\ \text{Ra}^{1/2} = \text{Fo}\ \text{Ra}^{1/2}. \tag{3.46}$$

For the known value of Ho, the characteristic time value t_0 can be determined from eqn (3.46):

$$t_0 = \frac{l^2 \text{Ho}}{a\text{Ra}^{1/2}}\ . \tag{3.47}$$

The characteristic time-scale of unsteady heat transfer under free convection can be determined by laboratory experiments.

Now we shall consider the transition from unsteady initial regime to steady or time-average steady flow wherein oscillating instabilities have already appeared. A steady convection regime existed priorily, as determined from the constant heat flux, average constant temperature drop and steady flow pattern. On attaining these conditions, the cooler or upper heat exchanger was lifted and the fluid mixed manually. The heat exchanger was subsequently quickly installed and readings from the digital voltmeters simultaneously entered in the computer. Cellular patterns were likewise obtained.

At $\text{Ra} < 4 \times 10^4$ in the initial regime, fluctuations in temperature drop across the layer, hence of Ra number and heat flux (Fig. 3.13), and change-over to steady cellular convection were observed. These fluctuations are caused by the descent of cold and ascent of hot flows in a time comparable to that of one 'cycle' or 'revolution' of the convection cell. These cause a decrease in the Ra number and hence in the rate of fluid flow, which in turn causes an increase in temperature difference.

Thus in the period of transition to the steady regime oscillatory 'unwinding' of the cell occurs and the unsteady regime bears a thermally induced character

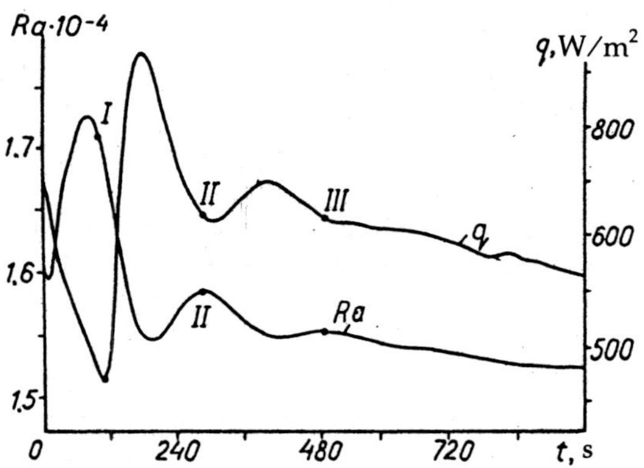

Fig. 3.13. Development of convection in a horizontal glycerine layer:
$Pr = 3 \times 10^3$; $\overline{Ra} = 1.54 \times 10^4$; $l = 17.8$ mm.

since it is controlled by the variation of temperature drop across the layer. This is indicated by the correlation of variation of temperature difference in the layer and the change in local specific heat flux: curves $Ra(t)$ and $q(t)$ are in opposite phases (see Fig. 3.13). The extreme values of these curves correlate well and the maximum value of Ra corresponds to the minimum value of q. This also confirms the analysis of free convection equations, which suggests that the unsteady state is thermally induced for $Pr \gg 1$. On the curves of Figure 3.13 can be distinguished characteristic points associated with transition to the steady convection regime. The time corresponding to point 1 accounts for the onset of the flow pattern. Convection cells are circular in form with indistinct outlines while the system of ascending flows is only forming. In this time interval convection is weak, heat transfer proceeds essentially by conduction and hence heat flux diminishes with time.

An increase in velocity is observed for the time interval $t_I < t < t_{II}$. This causes an increase in heat flux and some reduction in temperature drop. At point II cellular structure is already formed and later precisely the same cycle of fluctuations with smaller amplitudes is repeated. This is completed at point III at $t = t_0$. Slow structural changes in convection cells occur later.

It was experimentally found that at $Pr = (3 \text{ to } 3.3) \times 10^3$ and $1.5 \times 10^4 < Ra < 4 \times 10^4$, the value of Ho for glycerine calculated from the time of transition to the steady regime t_0 is equal to 16.5 ± 2. In an unsteady convection regime for $Ra > 4.1 \times 10^5$, convection is oscillatory and it is difficult to determine time t_0 of the time-average steady regime.

Long-period fluctuations of local heat flux q_l associated with cell drift are observed for $Ra > Ra_{III}$ (see Sec. 3.3, Fig. 3.5) in the horizontal layer (Fig. 3.14a). Short-period, high-frequency fluctuations in the convection cell are superimposed on long-period fluctuations for $Ra > Ra_{VI}$ (Fig. 3.5), hence short period fluctuations are intermittent (Fig. 3.14b). Furthermore, high-frequency fluctuations extend over the whole time of the experiment (the whole record length, Fig. 3.15a). Short-period

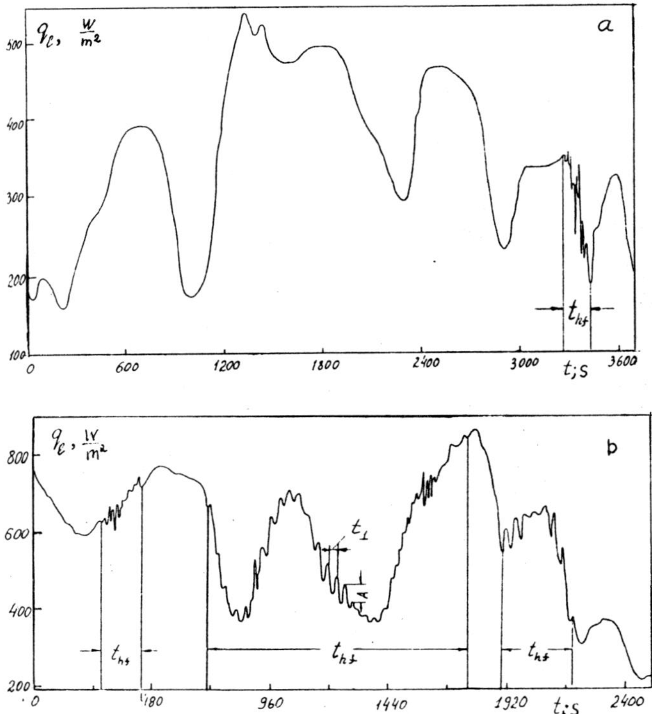

Fig. 3.14. Variation of local heat flux with time (A.G. Kirdyashkin and A.A. Kirdyashkin, 1991): a—Ra $= 2.1 \times 10^5$, Pr $= 190$, $l = 12.4$ mm; b— Ra $= 2.6 \times 10^5$, Pr $= 181.5$, $l = 12.4$ mm.

fluctuations grow in amplitude with increasing Rayleigh number. By this is meant that the fluctuations extend over all convection cells. It is difficult to distinguish long-period fluctuations in a developed turbulent regime (Fig. 3.15b).

Average periods of short-period temperature fluctuations (or local heat flux adjacent to the wall) t_1 were determined in a turbulent convection regime. The results were used to calculate the homochronicity criterion:

$$\overline{\text{Ho}_1} = t_1 a\,(\text{Ra} - \text{Ra}_{\text{cr}})^{1/2}/l^2. \tag{3.48}$$

The functions of normalised spectral density of local heat flux $R_q\,(f)$ and normalised spectral density of local Ra number, $R_{\text{ra}}\,(f)$, were calculated in experiments with vacuum oil VM-4 at Pr $= 1.1 \times 10^3$ to 1.3×10^3. The fundamental frequency of temperature fluctuations (average period of fluctuations of local values of Ra(t) and $q(t) - (t_1)$) could be determined quite accurately in all the experiments at Pr $= 1.1 \times 10^3$ to 1.3×10^3 as can be seen from Fig. 3.16 (at $l/B = 1/3.2$, where B is the horizontal size of the layer). The average periods of high-frequency fluctuations t_1 estimated from the initial experimental curves Ra (t) and $q(t)$ differ insignificantly from the characteristic periods calculated from the spectra of local Rayleigh number and heat flux. For VM-4 (Pr $= 1.1 \times 10^3$ to 1.3×10^3) at Ra $= 4.4 \times 10^5$, 8×10^5 and

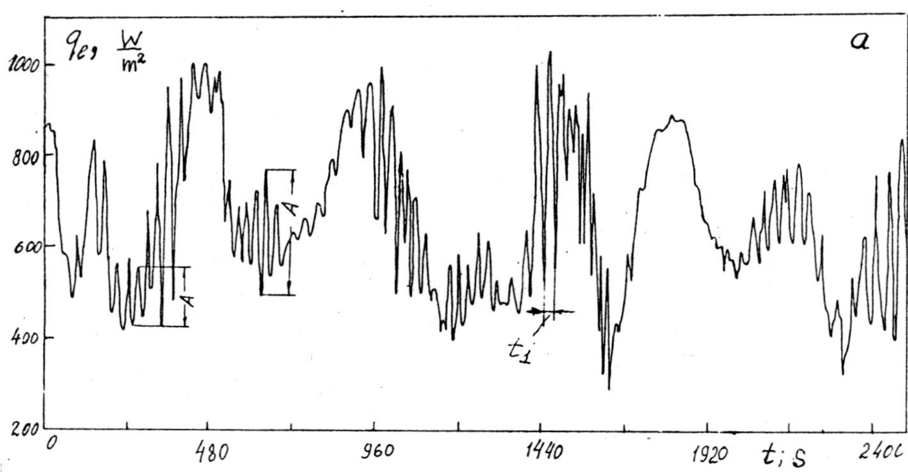

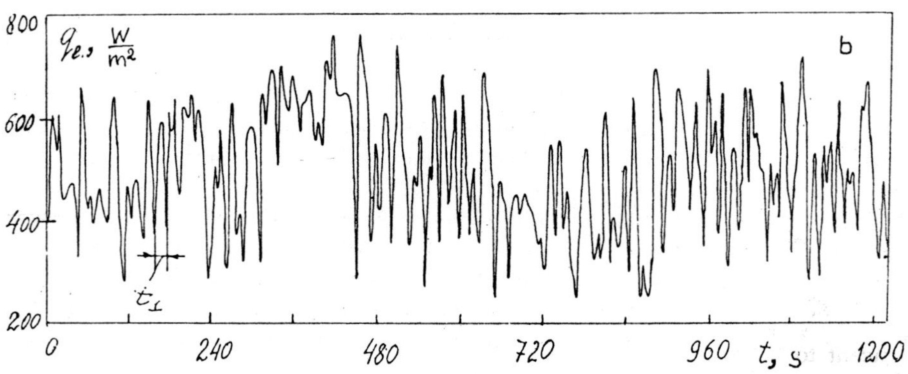

Fig. 3.15. Variation of local heat flux with time (A.G. Kirdyashkin and A.A. Kirdyashkin, 1998): a—Ra 6×10^5, Pr = 165.5, $l = 14.8$ mm; b—Ra = 3×10^6, Pr = 182, $l = 27.3$ mm .

1.6×10^6, Ho_1 values are respectively equal to 6.3, 7.0 and 7.4. For glycerine (Pr = 2.2×10^3 to 3×10^3) at Ra = 3.4×10^5, 4.8×10^5, 6.5×10^5, 7.7×10^5 and 1.06×10^6, the respective values of Ho_1 are equal to 6.1, 6.4, 6.2, 6.2 and 6.2. For hexadecane (Pr = 28) at Ra = 1.6×10^5 and 1.95×10^5, Ho_1 values are respectively equal to 4.2 and 6.7. According to Gollub and Benson (1980), for Pr = 2.5 to 5, short-period fluctuations arise at Ra > 5.9×10^4 and Ho_1 values 7.0, 6.9, 7.0 and 5.4, these values corresponding to Ra = 5.95×10^4, 7.7×10^4, 7.97×10^4 and 1.1×10^5.

Thus, Ho_1 variation is insignificant as Pr number varies from 2.5 to 3×10^3, i.e., for different liquids of low as well as high viscosity. This denotes that, even at $Pr > 2.5$, the dimensionless time scale Ho_1 is independent of Pr number. Convection flows in the lower mantle are characterised by values $Ra = 5 \times 10^5$ to 5×10^6 and correspond to a turbulent flow. Since the Pr number exerts almost no influence on Ho_1 in the region $Ra = 3 \times 10^5$ to 5×10^6 as can be seen from the experimental results and analysis of Navier-Stokes' equations for high-viscosity flows, the value of Ho_1 obtained can be used for calculating the periods of temperature fluctuations of lower mantle flows from eqn (3.47): $t_1 = l^2 Ho_1/aRa^{1/2}$. At $Ho_1 = 6.5$, $l = 2.1 \times 10^6$ m, $a = 2 \times 10^{-6}$ m^2/s and $Ra = 10^6$, the period of temperature oscillations in the lower mantle convection cells is $t_1 = 450$ million years.

The above period is comparable with the longest of known geological cycles, also called Wilson cycles or megacycles (Khain, 1994) and determined by the cyclic formation and disintegration of supercontinents, 'from Pangaea to Pangaea'. In experiments at $Ra = 106$, apart from short-period fluctuations, long-period fluctuations associated with structural changes and drift of convection cells were also observed. The value of the homochronicity criterion calculated from these fluctuations is $Ho_2 \sim 50$ which is an order of magnitude greater than Ho_1 and hence $t_2 \sim 10 \, t_1$. In other words, the period of complete reshaping of convection cells in the lower mantle at present-day viscosity and heat flow is comparable with the age of the Earth.

These results of experimental modelling for the lower mantle agree with the results of numerical simulation (Zhang and Yuen, 1995) carried out within the framework of the model of a spherical shell based on solving a system of equations of thermal convection at $Pr \rightarrow \infty$ (see Fig. 3.5).

Experiments were carried out in a regime of turbulent free convection under stationary boundary conditions without temperature variation at the heat transfer surfaces with time. Under actual geodynamic conditions, the structure of lower mantle flows depends on conditions in the upper mantle and at the core-mantle boundary (CMB). These conditions vary with the changing structure of upper mantle flows (with the 'jumping' of subduction zones, continental drift, displacement of oceanic ridges etc.) and are rigorously reflected in the structure of lower mantle flows. It may be anticipated that the peculiarities of CMB phenomena, which are poorly known, influence the structure of lower mantle flows.

The above discussion suggests that the structure of lower mantle convection while being appreciably controlled by the boundary conditions in the upper mantle and at the CMB, influences the boundary conditions in turn.

The effect of thermal jet separating from the heated surface on transfer has not been adequately studied in a horizontal layer heated from below in a turbulent free convection flow regime. The effect of local heat plumes above local heat sources on convective cellular flow has been studied even less. One can only predict the influence of these types of flow on heat transfer in a horizontal layer. When the local sources are few and of low intensity, it may be deduced that heat plumes are independent of the mantle convection or that the role of cellular flows is decisive. In the presence of a powerful local heat source, the cellular flows ascend at the site of its disposition. In the presence of a plethora of local heat sources, heat plumes will play a decisive role.

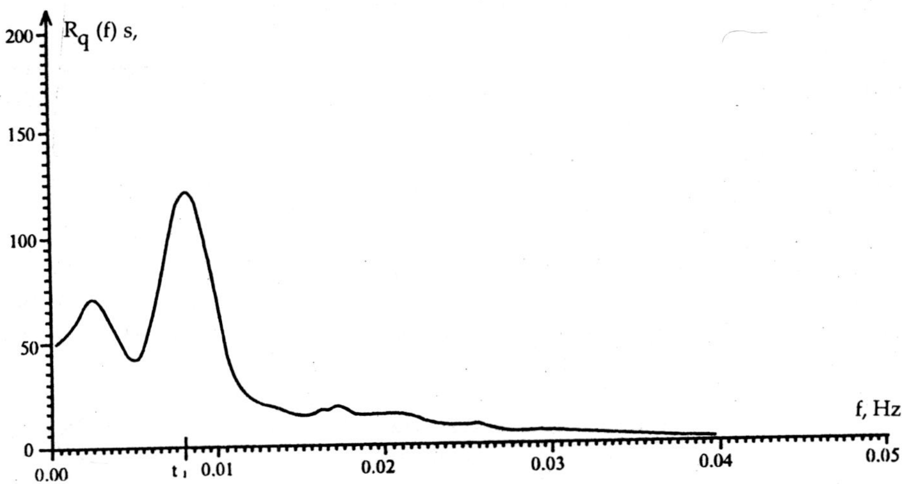

Fig. 3.16. Fluctuation spectra of local heat flux for Pr = 1.1 × 10³ and Ra = 1.6 × 10⁶; characteristic fluctuation period t_1 = 133 s.

The interaction between unsteady heat sources and cellular flow remains to be seen. The characteristic period of plume action is ≈ 10² less than the time-scale of mantle convection. In the total absence of the action of intense plume fields, initiation of convection flows and their formation extend over a hundred million years. Thus, at this period formation of cellular flows in the lower mantle is influenced by upper mantle processes which represent the boundary conditions for convection in the lower mantle.

The interplay of mantle plumes and thermogravitational flows in the lower and upper mantle are modelled in Chapter 4.

3.7. Convective Heat and Mass Transfer in Diamond Recrystallisation

Let us study diamond recrystallisation as an example of the practical application of the above principles. Other examples of larger scale geodynamic processes will be dealt with in Chapters 4 and 5.

Recrystallisation of diamond in high-pressure metal melts occurs under conditions of high solubility in the hot zone and reduced solubility in the cold zone. This causes the transport of carbon in the direction opposite to the temperature gradient. Considering the importance of studying this mechanism, A.G. Kirdyashkin, I.I. Fedorov, A.I. Chepurov and Yu.M. Borzdov experimentally studied the mass transfer rate of carbon and proposed a model for quantitatively describing this process (Kirdyashkin et al., 1986).

Experiments were carried out in a high-pressure multiple-plunger apparatus of the type detachable sphere by the method of Sobolev et al. (1983). The initial specimens were gathered by the standard scheme: carbon source, fine-grained graphite of high quality in the hottest central zone; a metal layer below; and the diamond seed further down. As graphite completely transformed into diamond within the first few minutes of the experiment, carbon transport proceeded as follows: diamond → carbon solution in the metal melt → diamond. Eutectic alloy $Mn_{0.6}Ni_{0.4}$ placed in an ampoule in the form of a pressed tablet 5 mm in diameter and 2 mm in height served as the carbon solvent. Experiments were carried out at pressure from 5.5 to 6 GPa and temperature ~ 1350°C. Their duration, reckoned from the moment of saturation of the melt with carbon, ranged from 0.5 to 9 h. The temperature gradient in the working zone of the cell filled tightly with talc was less than 15°K/mm.

The volume in which diamond crystallised on the seed can be considered an inclined flat layer filled with liquid metal (Fig. 3.17). The top wall (carbon source) was heated more while the bottom one cooled. The diamond seed crystal of size x_0 was centred on the bottom.

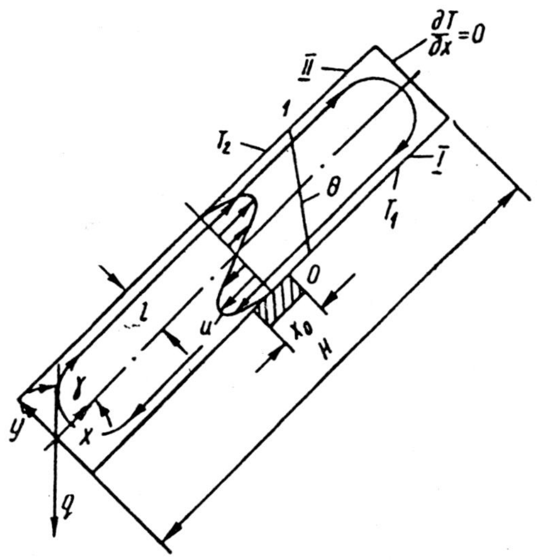

Fig. 3.17. Fluid flow in the high-pressure cell.

In the flat fluid layer closed at the butt-ends, with a different orientation about gravity adjacent to isothermal plates of varying temperature, an ascending flow arose at the heated surface and a descending one at the cooled heat transfer surface. At a different orientation of the flat layer, when the heated plate was placed above the cooled one, a free convection flow was observed in the central region with respect to the length of the plates for Ra < 1.63 × 10³ and H/l >> 1. Under these conditions the temperature profile across the layer was a linear function of y/l while the velocity profile was a cubic parabola (Kirdyashkin and Mukhina, 1971):

110

$$u = \left[\frac{\beta g \Delta T l^2}{6\nu}\right]\left[\left(\frac{y}{l}\right)^3 - \frac{y}{l}\right] \cdot \cos \gamma$$

and

$$\Theta = \frac{(T - T_1)}{\Delta T} = 0.5\left(1 + \frac{y}{l}\right). \tag{3.49}$$

Here, $Ra = \beta g\,\Delta T l^3/a\nu$ is the Rayleigh number: $\Delta T = T_2 - T_1$ the temperature drop between the heat transfer surfaces; l the half-thickness of the layer; y the co-ordinate normal to the heat transfer surfaces; and γ the slope of the layer relative to the vertical ($\gamma > 0$ when the heated surface was aligned above the cooled one).

Numerical simulation carried out for the Prandtl number $Pr \to 0$ revealed that eqns (3.49) will be satisfied for $Gr = \beta g\,\Delta T l^3\nu^{-2} < 8 \times 10^3$, where $Gr = Ra/Pr$ is the Grashof criterion. In our case, the liquid metal heat carrier had $Pr = 0.07$.

The results of experimental estimation of the Rayleigh number in the diamond crystallisation regime showed that $Ra < 1.6 \times 10^3$ and $Gr < 8 \times 10^3$, i.e., the thermogravitational flow in the cell where diamond crystallises is described by eqn (3.49). The effect of concentration gravity on the thermogravitational flow was ignored for the test conditions because of low mass transfer on the seed. Let us examine the steady-state regime of flow which was established in the liquid metal system comparatively rapidly relative to the entire crystallisation period. It may be noted that crystallisation rate is restricted by the rate of mass transfer in the concentration boundary layer and not by the kinetics of surface processes.

On surface II comprising the diamond charge (see Fig. 3.17), the liquid metal is saturated with carbon conforming to the temperature and pressure of the liquid metal system in the neighbourhood of the wall. The convecting liquid metal flows around surface I because of the closed flow line in laminar circulatory flow. On surface I the carbon crystallisation rate on the diamond seed is determined by mass transfer in the concentration boundary layer. Therewith, carbon concentration in the metal at the diamond surface decreases to the saturation concentration corresponding to the temperature of diamond surface T_1.

The scheme of hydrodynamics and mass transfer in the concentration boundary layer δ_c adjacent to the diamond surface is shown in Figure 3.18.

In the conditions under consideration, $Pr \ll Sch$, where $Sch = \nu/D$ is the Schmidt criterion ($Sch = 250$) and D the diffusion coefficient. In this case, the thickness of the concentration boundary layer δ_c is much less than l and its development proceeds in the region in which a linear approximation of velocity profile $u = \tau_w y/\eta$ can be adopted. Here, τ_w is the wall friction and η the coefficient of dynamic viscosity.

Shear stress on the wall is determined by eqn (3.49):

$$\tau_w = \eta\left(\frac{\partial u}{\partial y}\right)_w = \frac{(\eta\beta g \Delta T l \cos \gamma)}{6\nu} \tag{3.50}$$

The two-dimensional equation of the concentration boundary layer is expressed in the following manner: $u\,\partial\bar{c}/\partial x + v\,\partial\bar{c}/\partial y = D\,\partial^2\bar{c}/\partial y^2$. Since flow of type (3.49)

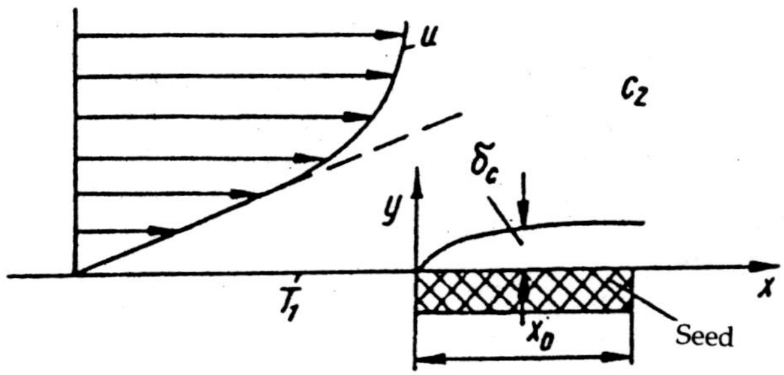

Fig. 3.18. Mass transfer in the concentration boundary layer.

is plane-parallel, the normal component of velocity v = 0. Thus, taking into consideration the preceding equations, we obtain

$$By \frac{\partial \bar{c}}{\partial x} = \frac{\partial^2 \bar{c}}{\partial y^2} , \qquad (3.51)$$

where $B = \tau_w/D\eta$; $\bar{c} = (c - c_2)/(c_1 - c_2)$; c_1 is the carbon concentration of saturation in the liquid metal at the diamond crystal surface at temperature T_1; c_2 the carbon concentration of saturation in the outer part of the boundary layer of liquid metal at temperature T_2.

The boundary conditions for eqn (3.51) are: if $y = 0$, $\bar{c} = 1$; if $y \geq \delta_c$, $\bar{c} = 0$. By integrating eqn (3.51) for the given boundary conditions using the substitution suggested by Shvets (1954), we derive

$$A = y \left(\frac{B}{x}\right)^{1/3} = y \left(\frac{\tau_w}{D\eta\,x}\right)^{1/3} \qquad (3.52)$$

After rearrangement of eqn (3.51) we obtain the follow equation in new terms:

$$\frac{\partial^2 \bar{c}}{\partial A^2} + \left(\frac{A^2}{3}\right)\left(\frac{\partial \bar{c}}{\partial A}\right) = 0. \qquad (3.53)$$

The solution to eqn (3.53) satisfying the boundary conditions $\bar{c} = 1$ for $A = 0$ and $A = \infty$ takes the form:

$$\bar{c} = 1 - \int_0^A \exp\left(-A^3/9\right) dA \Big/ \int_0^\infty \exp\left(-A^3/9\right) dA, \qquad (3.54)$$

where

$$\int_0^\infty \exp\left(-A^3/9\right) dA = (1/3)^{1/3} \int_0^\infty e^{-\xi} \xi^{(1/3 - 1)} d\xi = \Gamma(1/3)/3^{1/3} = 1.855 \ (\Gamma(1/3)$$

is the gamma function).

The local values of mass transfer coefficient of the crystallising diamond surface (α_c) are calculated as:

$$\alpha_c = -D(\partial \bar{c}/\partial y)_{y=0} = -D\,(\partial\bar{c}/\partial A)_{A=0}\,(\partial A/\partial y)_{y=0}. \tag{3.55}$$

We find from eqn (3.54) that $(\partial\bar{c}/\partial A)_{A=0} = -1/1.855$ and from eqn (3.52) we obtain $(\partial A/\partial y)_{y=0} = (\tau_w/D\,\eta x)^{1/3}$. By substituting these values into eqn (3.55) and allowing for eqn (3.50), the local values of mass transfer coeffficients can be determined:

$$\alpha_c = (D/1.855)\,(\beta\,g\Delta\,Tl\,\cos\,\gamma/6v\,Dx)^{1/3}. \tag{3.56}$$

The average value of mass transfer coefficient is calculated using the following equation:

$$\bar{\alpha}_c = (1/x_0)\int_0^{x_0}\alpha_c\,dx = 0.455\,(D/x_0)\,(x_0^2\,\beta g\,\Delta Tl\,\cos\,\gamma/vD)^{1/3}. \tag{3.57}$$

Equation (3.57) can be expressed in a dimensionless form as follows:

$$\mathrm{Nu}_c = \bar{\alpha}_c x_0/D = 0.455 k_0^{2/3}(\cos\gamma)^{1/3}\,\mathrm{Gr}^{1/3}\,\mathrm{Sch}^{1/3}, \tag{3.58}$$

where Nu_c is the dimensionless mass transfer coefficient; $k_0 = x_0/l$; $\mathrm{Gr} = \beta g\,\Delta Tl^3 v^{-2}$; and $\mathrm{Sch} = v/D$.

The mass transport rate of carbon in the course of diamond crystallisation (j_c) is determined as:

$$j_c = \bar{\alpha}_c\,\Delta c = \bar{\alpha}_c\,(\partial c/\partial T)\,\Delta T. \tag{3.59}$$

To calculate the mass transfer rate using eqns (3.57) to (3.59), numerical values of parameters involved in the equations under the adopted experimental conditions had to be found. The diffusion coefficient of carbon in nickel melt at high pressure has been given (Strong et al., 1967) as: 1.4×10^{-3} mm^2/s at 1350°C. Values of $\partial c/\partial T$ and β calculated for $\mathrm{Mn}_{0.6}\mathrm{Ni}_{0.4}$ according to the values reported by Ershova et al. (1981), are 10^{-3} mg $\cdot$ mm$^{-3}\cdot$ deg^{-1} and 6×10^{-4} deg^{-1}. The kinematic viscosity coefficient of melt $\mathrm{Mn}_{0.6}\mathrm{Ni}_{0.4}$ at atmospheric pressure according to Baum (1979) is 0.55 mm^2/s. Bridgeman (1935) pointed out that pressure rise to 12 kb increases the kinematic viscosity coefficient of mercury by 30%. On this basis, we accepted that $v \approx 1$ mm^2/s in eqn (3.50). The slope of cell to gravity vector in the apparatus was 45°. Substituting these values in eqn (3.59) gives the relation:

$$j_c = 9 \times 10^{-6}\,(\Delta T)^{4/3}\,x_0^{-1/3}, \tag{3.60}$$

where j_c is in mg $\cdot$ mm$^{-2}\cdot$ s^{-1}; ΔT in °C; and x_0 in mm.

The transfer rate (3.60) depends on the typical crystal size which increases in the course of growth. By expressing the size as surface area and considering that the latter rises linearly as the test period is extended, we derive an equation for calculating the average velocity of carbon transport:

$$\bar{j}_c = 10^{-5}\,[(S_k^{11/6} - S_0^{11/6})/(S_k^2 - S_0^2)]\,(\Delta T)^{4/3}, \tag{3.61}$$

where S_0 and S_k are the initial and final values of crystal surface area in mm^2.

Within the accuracy of calculations, $(S_k^{11/6} - S_0^{11/6}) / (S_k^2 - S_0^2) \approx 1$ can always be adopted and relation (3.61) represented as $j_c = 10^{-5} (\Delta T)^{4/3}$. At temperature drop $\Delta T = 5$ to 10°C, the average mass transfer rate is $(0.9$ to $2.2) \times 10^{-4}$ mg $\cdot$ mm$^2 \cdot$ s^{-1}. This accords well with the experimental values, especially for the crystal face growing normally towards the cooler part, for which eqn (3.57) was derived. Therewith the average velocity of transfer was equal to $(0.6$ to $2.5) \times 10^{-4}$ mg $\cdot$ mm$^{-2} \cdot$ s^{-1}.

For comparison, we calculated the rate of purely diffusive transport of carbon without convection when the transported carbon was deposited uniformly on the substratum in the cold part of the ampoule. This process in the diffusive model is described by the following equation:

$$j_c = D \; \partial c / \partial y = D(\partial c / \partial T) \, (\Delta T / L), \tag{3.62}$$

where $L = 2l$ is the layer thickness.

The value of j_c calculated using eqn (3.62) under the test conditions $(\Delta T = 10°C)$ was equal to 7×10^{-6} mg $\cdot$ mm$^{-2} \cdot$ s^{-1} which is much less than the mass transfer rate of carbon $(5 \times 10^{-5}$ to 5×10^{-4} mg $\cdot$ mm$^{-2} \cdot$ s$^{-1})$ determined in experiments in which a fine-grained diamond layer was formed in the cold zone.

Thus, mass transfer of carbon in diamond recrystallisation with an inclined high-pressure cell occurs by thermogravitational convection. The transfer rate is described with adequate accuracy by eqns (3.57) to (3.59).

The high-pressure cells are set up horizontally in most Russian (anvil with recess) and foreign (belt) apparatus. According to eqns (3.49), convection does not occur when the top wall is more heated and wall surfaces are isothermal. However, for an inclined cell a horizontal temperature gradient appears and hence convection is observed. It may be noticed that, in this case, there is no stability threshold and convection is evident at any horizontal temperature gradient, even at infinitesimal gradients.

Experiments show that unlike diamond synthesis, controlling by carbon diffusion through a film of the metal melt, mass transfer of carbon in the recrystallisation processes depends on the design features of the high-pressure cell and can be controlled by thermogravitational convection.

4

Geodynamic Processes in the Mantle and Their Modelling

In this and the next chapter, the main problems of deep-level geodynamics are dealt with systematically on the basis of the introductory material and the principles of physical modelling discussed in the preceding chapters. Deep-level convection flows represent the main factor responsible for the developmental dynamics of the Earth and for most of the surface processes studied in geology.

Interactions among the following four groups of deep-level factors are reflected in the structure of the Earth's surface envelopes: 1) structure of convection flows in the lower mantle and their interaction with the outer liquid core of the Earth, 2) mantle plumes ascending from the boundary of the lower mantle or penetrating the entire mantle, 3) convective flows in the asthenosphere interacting with (a) lower mantle flows, (b) mantle plumes, (c) descending flows in the subduction zones and (d) non-homogeneous base of lithospheric plates—all these interactions cause, in particular, local mantle cells (diapirs) interconnected with rifts, marginal seas and local magmatic structures and 4) lateral heterogeneity and stratification (tectonic stratification) of the lithosphere that can be divided primarily into the Earth's crust and lithospheric mantle formed from the asthenosphere. The most important inhomogeneity of the lithosphere is represented by continents with thick (up to 60 km) crust having an acidic composition and a high level of radioactive heat generation serving as thermal insulators for the mantle heat and thick (up to 250 to 300 km under Africa and Asia) lithospheric mantle.

The upper envelopes (Earth's crust, lithospheric upper mantle and asthenosphere) are combined together under the name 'tectonosphere' (*Tectonosphere of the Earth*, 1978; Dobretsov, 1981) as interrelated tectonic, magmatic and metamorphic processes occur here and a great number of such processes can be studied directly. The deepest rocks of the lithospheric upper mantle and partly of the asthenosphere are encountered in the form of deep-seated xenoliths in kimberlites and alkali basalts (Sobolev, 1974). The composition and state of asthenospheric upper mantle are determined from the intensity and composition of basaltic and ultrabasic magmatism as well as from geophysical data.

As pointed out before, the composition and structure of the deeper envelopes of the Earth including the transition layer and lower mantle can be inferred only indirectly using geophysical (seismic, gravitational, thermal and electromagnetic) data, experimental and numerical modelling and by analogy with meteorites.

We shall now attempt to estimate the interaction of the major factors enumerated above on the basis of the proposed global model of convection in the Earth's mantle (Kirdyashkin and Dobretsov, 1991; Dobretsov and Kirdyashkin, 1993a) and available information. In many cases the known estimates have been supplemented with our experimental and theoretical data, specially on the spatial seales and velocity of lower mantle convection flows, stability mechanism of ascending mantle streams (plumes) and effect of continental masses and subduction zones on the upper mantle asthenospheric convection.

The presentation is made on the principle of 'from general to the specific', commencing with a discussion of the problems of convection flows in the upper and lower mantles, mantle plumes and their formation at the core-mantle and upper-lower mantle interfaces.

In the next (fifth) chapter the geodynamic processes occurring at the lithosphere-asthenosphere boundary and giving rise to plate tectonics are discussed, i.e., spreading and formation of oceanic crust; subduction and origin of island arc crust; and zones of collision and formation of folded belts. In conclusion, a model of the evolutionary development of folded belts and the Earth as a whole is presented. Interrelation and the cyclic nature of geodynamic processes are estimated.

4.1. Mantle Convection Models

Although models of thermogravitational convection are commonly used for estimating deep-level processes in the Earth, some Russian authors regard the gravitation-concentration mechanism as justified (Keondzhyan, 1980; Monin and Sorokhtin, 1982; Sorokhtin and Ushakov, 1991). Trubitsyn and Kharybin (1988) demonstrated that this mechanism can be manifest under very specific conditions of the distribution of heterogeneities in the mantle as well as under specific boundary conditions. Gravitation-concentration global flows may arise if the particle sizes are less than critical, which in turn depend on the differences in densities and physical properties of the viscous liquids. With reduction in particle size, initiation of global concentration flows also becomes problematic and flows arising as a result of Rayleigh-Taylor thermogravitational instability may become more real. The conditions of prolonged formation of the Earth's core impose great restrictions (Vityazev, 1991).

Two types of models of mantle thermogravitational convection can be distinguished. According to the first model, convection is presumed all along the thickness of the mantle from the lithosphere (30 to 100 km) to the core-mantle boundary (2890 km) (Pekeris, 1936; Molnar et al., 1979; Peltier and Jarvis, 1982; Davies, 1984; Bercovici et al., 1989; Davies and Richards, 1992). The second model assumes that convection proceeds in two layers (in the upper and lower mantles) with no significant mass transport at their boundary (Richter and Parsons, 1975; Richter and McKenzie, 1981; Dobretsov, 1980, 1981; Christensen and Yuen, 1984, 1985; Cserepes and Rabinowicz, 1985; Cserepes et al., 1988; Irvine, 1988, 1991; Kirdyashkin and Dobretsov, 1991; Zhao et al., 1992; Dobretsov and Kirdyashkin, 1993; Honda, 1995).

116

Much of the geological and geophysical data available have been explained from alternative standpoints within the framework of these models. For example, the immense horizontal size of many lithospheric plates may suggest the great depth of convection cells encompassing the entire mantle (Molnar et al., 1979; Davies, 1984, 1992). The modelling of asthenospheric flows (Kirdyashkin, 1989; see also Chapter 5) demonstrates the existence of elongated and thin asthenospheric cells, however. These estimates agree with the results of experimental studies of the two-layer convection model given below (Kirdyashkin and Dobretsov, 1991; Dobretsov and Kirdyashkin, 1993). Our experiments showed that thin and elongated cells may occur in the asthenosphere as a result of the horizontal temperature gradient under the influence of major lower-mantle cells or lower-mantle plumes flowing along the boundary of the upper and lower mantle (see below).

Another set of arguments relates the special features of subduction zones with the nature of the upper-lower mantle boundary. Proportional variation of the length of the subduction zone at the surface and subduction rate can be explained by thermal assimilation of subducting plates and by the absence of a barrier at the bottom of the upper mantle (Molnar et al., 1979). The same dependence can also be explained by the accumulation of heavy restites at the bottom of subduction zones (Fukao et al., 1994) as well as by intense mantle flows along the subduction barrier (Russo and Silver, 1996).

Stress distribution in the subducting plates and grouping of deep-focus earthquakes in them suggest the presence of a barrier at a depth of 650 to 700 km (Irvine, 1988, 1991). The distribution of earthquake foci in the extremely extended Izu-Bonin and Mariana subduction zones (Van der Hilst and Seno, 1993) is depicted in Figure 4.1. It can be seen in the sections that the seismicity zones can be divided into shallow (up to 200 to 300 km) and deep-focus (up to 660 km) zones. The latter at a depth of 550 to 660 km reveal change of slope as the probable result of the above-mentioned barrier at a depth of 650 to 700 km. This turnaround can be seen distinctly in sections c, d and h. Other characteristics of these sections, especially the absence of deep-focus earthquakes in sections e, f and g are discussed while studying the subduction mechanism in Chapter 5.

The possible low-viscosity zone located below 700 km points to the necessary existence of two convective layers while the absence of such a zone confirms the whole mantle convection model (Peltier and Jarvis, 1982). The presence of such a low-viscosity zone is confirmed by the data of seismic tomography (Fukao et al., 1994; Su et al., 1994) and is a probable explanation for maximum coherence of hot spots (mantle plumes) with the low-velocity regions of the upper mantle at depths of 700 to 1200 km (Cadec et al., 1995). Similarly, the presence of a cold, high-viscosity layer at a depth of 400 to 650 km detected in most regions from the data of detailed seismic tomography (Spakman, 1990; Spakman et al., 1993; Fukao et al., 1994) as well as seismic reflections from the boundary at a depth of 660 km and its relief are likewise indicative of a two-layer convection model (Shearer and Masters, 1992). Finally, based on an analysis of geoid anomalies and surface topography (Cazenave et al., 1988; Ricard and Wuming, 1991; Cazenave and Thoraval, 1994), the conclusion was reached in Chapter 1 that there is viscous stratification in the mantle.

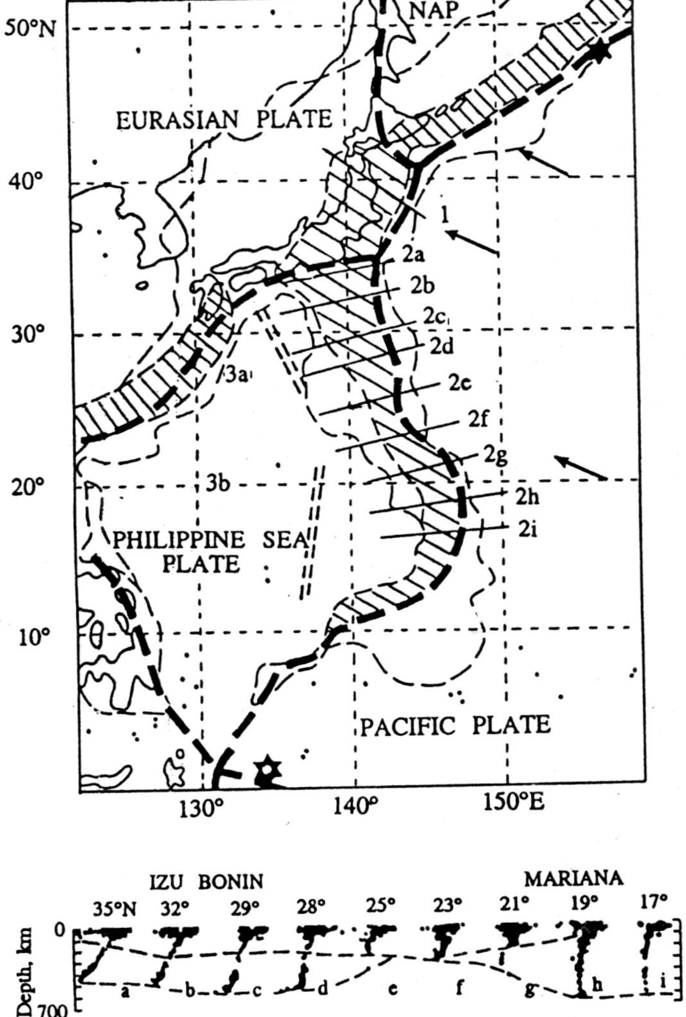

Fig. 4.1. Seismicity of Kuril-Japan, Izu-Bonin and Mariana arcs (Van der Hilst and Seno, 1993): thick lines—plate boundaries; NAP—Sea of Okhotsk plate; arrows show the directions of movement of the Pacific Ocean plate; hatched area—projection of seismic focal zone with maximum number of earthquakes; outer broken line indicates the border of the zone of infrequent earthquakes. Also indicated are isolated earthquakes outside this line. Section through the arc of Japan along line 1 is shown in Fig. 5.43. Sections of seismic focal zone through Izu-Bonin and Mariana arcs (2a to 2i) are depicted at the bottom of the Figure. The double broken line in the Philippine Sea shows the spreading centre of Bonin basin.

118

Thus attempts at reproducing and refining mantle convection models have been undertaken primarily on the basis of seismotomographic models. These models have continued to be refined (Dziewonski and Anderson, 1981; Dziewonski, 1984; Dziewonski and Woodhaus, 1989; Inoue et al., 1990; Spakman, 1990; Shearer, Masters et al., 1992; Spakman et al., 1993; Fukao et al., 1994; Su et al., 1994). The main difficulty here is that all the lateral variations of seismic wave velocities in the mantle fall in the range of ± 2% of the standard model, as for example PREM (Montagner and Anderson, 1989) or model S12 WM 13 (Su et al., 1994; Cadec et al., 1995). A great body of seismic data, special programmes and tests and supercomputers are used for reliable ascertainment of seismic velocities. Some doubts about the reliability of identified anomalies nevertheless persist; further, the resolution of seismotomography is still quite low: 200 km along the depth and 2500 km along the lateral direction. Therefore, correlation of seismotomographic anomalies with independently derived factors such as surface profile, geoid anomalies, heat flux and geological structures is of great importance. This aspect has been partly dealt with above and we shall return to it later. Let us recall here only one of the latest successful attempts in which seismotomographic anomalies have been inter-correlated as well as correlated with three important types of geological structures: mid-oceanic ridges, subduction zones and hot spots (Cadec et al., 1995).

Figures 4.2 and 4.3 depict the distribution of cold and dense masses in the mantle corresponding to the positive anomalies of V_p and probable descending flows in the lower mantle based on the data of seismic tomography (Fukao et al., 1994). At depths of 700 to 1300 km, cold zones correlate with cold dense anomalies at depths of about 200 km and these in turn with present-day subduction zones or subduction zones that existed 100 to 200 million years ago (see sections marked 1 and 2 and black sections). However, in the case of sections 3 and 4 in Figure 4.2a and for the lower part of the mantle (Fig. 4.2b), cold masses do not correspond to the subduction zones and form isolated bodies under eastern Asia, North America, Antarctica and northern part of the Pacific Ocean. A similar patchy distribution of heated masses in Figure 4.2b suggests that this structure is independent of the seismotomographic structures of the intermediate layer and upper mantle.

The actual distribution of dense (+ ΔV) and less dense (− ΔV) masses at different levels in the mantle (see Figs. 1.3, 4.2 and 4.3) suggests two-layer type mantle convection, interpreting the dense masses as cold descending flows and the less dense ones as hot ascending flows. Such a multilayer and complex convection pattern is shown in Figure 4.4 using the data of seismic tomography (Fukao et al., 1994; Maruyama, 1994). We have added the upper convective cells of a different size but they are evident from the distribution of masses and motions shown in Figures 4.3 and 4.4. The thin-layered structure of the upper half of the mantle (to depth 1000 to 1300 km) and a more complex spotted pattern of heterogeneities in the lower mantle have engaged attention in other contemporary seismotomographic models (Su et al., 1994; Cadec et al., 1995).

Most geochemical data and recent isotopic studies show that the upper and lower mantle represent two different geochemical reservoirs with highly restricted mass transfer between them (Dobretsov, 1980, 1981; Allegre, 1982; Allegre et al., 1986, 1995; Zonenshain and Kuz'min, 1983, 1993a, b; O'Nions and Tolstikhin, 1996).

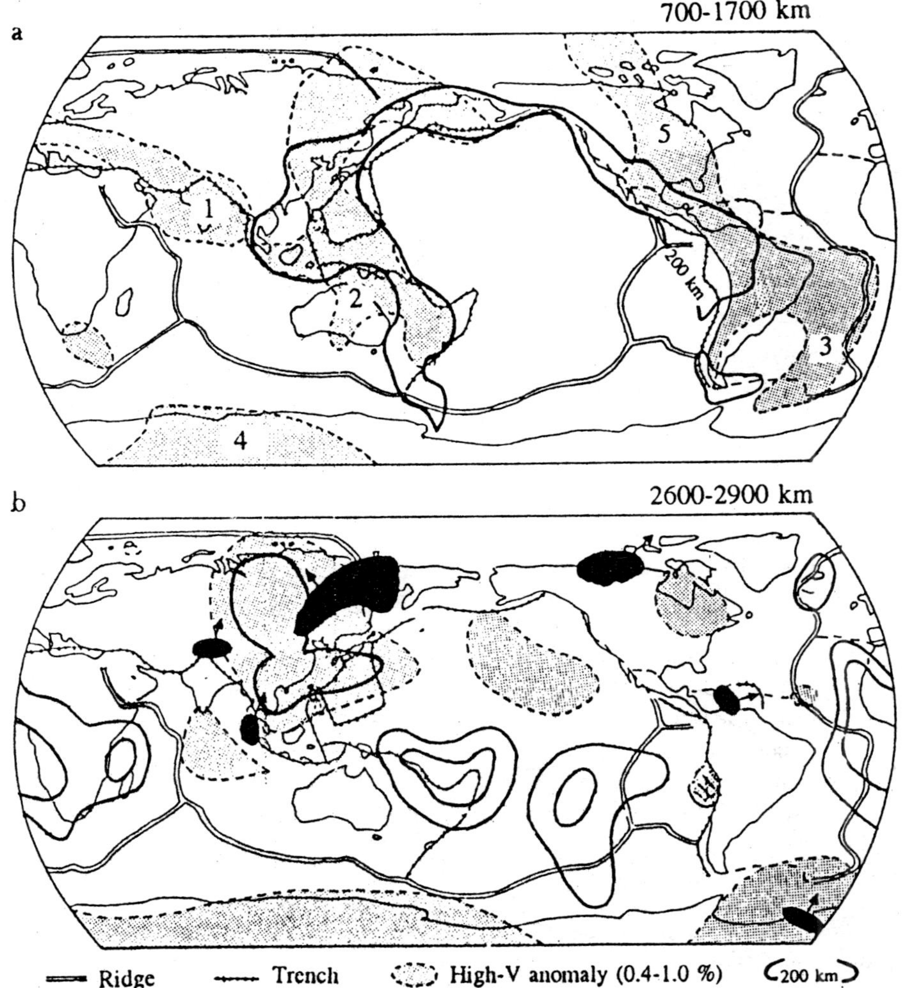

Fig. 4.2. Seismotomographic representation of anomalies in the lower mantle (Fukao et al., 1994, modified). Positive anomalies of P waves at depth (a) 700 to 1300 km and (b) 2600 to 2900 km are hatched. Thick isolines show the probable ascending flows in the upper mantle at depth (a) 200 to 300 km and (b) in the lower mantle. Black regions depict sections that underwent prolonged subduction for 180 million years and correspond to maximum volumes of subducted material.

In an overwhelming number of cases, geochemical data support the two-layer convection model while most geophysical data can be interpreted in favour of whole mantle convection (see above). At the International Geological Congress held in Washington in 1989, this aspect was recognised as the major contradiction between

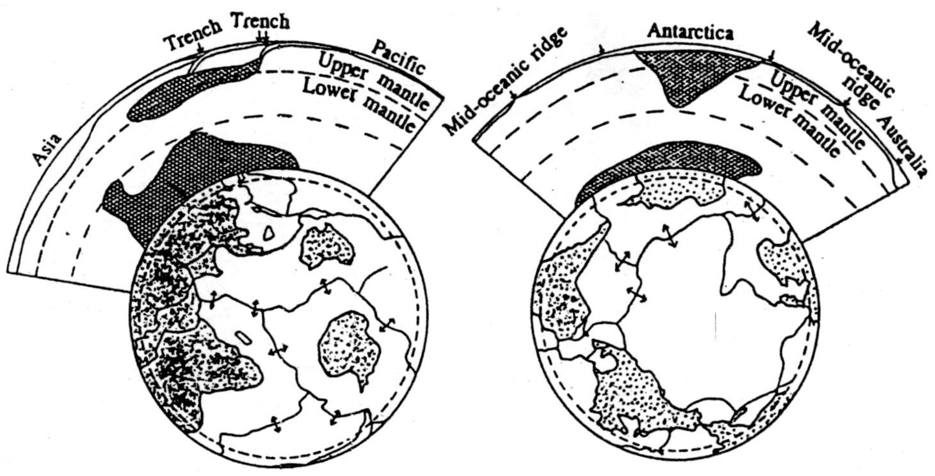

Fig. 4.3. Model of accumulation of heavy subducted plate material under eastern Asia (left) and Antarctica (right) (Fukao et al., 1994). Legend same as in Fig. 4.2.

deep-level geophysical and geochemical data. The latest review on the isotopic pattern of rare gases He, Ne and Ar (O'Nions and Tolstikhin, 1996) serves to illustrate clearly the independence between the upper and lower mantle geochemical reservoirs. These authors reached the conclusion that the mass transport between the upper and lower mantles throughout the Earth's history, after the early stages of its formation, did not exceed 10% and most probably ranged from 1 to 3%. This picture emerges from the estimation that the balance of rare gases (He, Ne and Ar) and their isotopes is ensured if the average flux from the lower mantle is $\leq 3 \times 10^{13}$ kg per million years over 4.35 billion years or 3% of present-day slab mass flux of 10^{15} kg per million years. The present-day mass flux carried by the plumes is estimated at 2.56×10^{14} kg per million years (Sleep, 1990), which is ten times more than the above estimated mass transport between the lower and upper mantles, i.e., only 10% of present-day plumes can be of lower mantle origin (see below). Finally, these relations may undergo change in the history of the Earth but differences of 1 or 2 orders represent an extremely important subject for discussion.

It was on the basis of geochemical data Zonenshain and Kuz'min (1983, 1993a, b) delineated the hot fields and demonstrated their independence from the lithospheric plate boundaries and their stable position on the Earth's sphere over the preceding 120 to 150 million years, i.e., after the formation of the Atlantic and active movement of lithospheric plates. These considerations as well as a comparison with the geophysical data led L.P. Zonenshain and M.I. Kuz'min to the two-layer convection model and, correspondingly, to the two-level tectonics of the Earth (Fig. 4.5). According to Zonenshain and Kuz'min (1993a, p. 126), a distinction

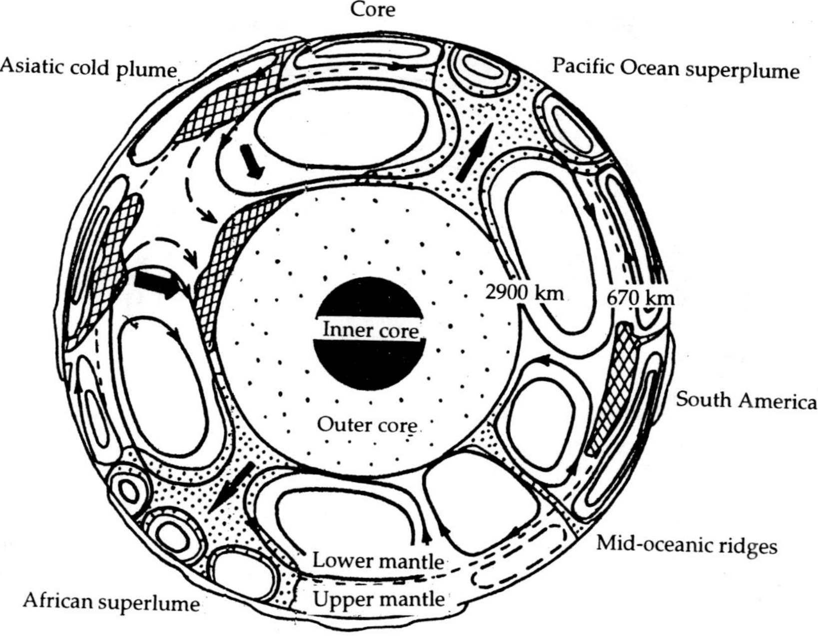

Fig. 4.4. Model of present-day Earth and scheme of thermogravitational flows in the mantle (Maruyama, 1994, modified).

should be made between the deep-level lower mantle processes, or the tectonics of hot fields, and the processes at the upper level or plate tectonics. The system of convection flows in the upper mantle according to the developed concept should be distinguished from the lower mantle convection.

Alternative models were proposed even as the numerical model of thermal convection in the mantle was being developed. The two-layer model of mantle convection was first formulated (Richter and McKenzie, 1981; Christensen and Yuen, 1984) assuming that the boundary between the layers at depth 650 km is chemical or a combination of chemical and phase boundaries, viscosities of the layers being roughly equal. In this case thermal coupling plays a dominant role at the interface flows in opposite directions arise in the cells and horizontal cell sizes remain constant.

However, if the viscosity and thickness of the layers change, different types of interactions may occur at the interface and the horizontal sizes of the cells may differ in the two layers (Honda, 1982; Cserepes and Rabinowicz, 1985; Cserepes et al., 1988). Numerical simulation in two-dimensional approximation (Cserepes and Rabinowicz, 1985) showed that the structure of two-layer convection at Rayleigh criterion values Ra = 3×10^6 to 3×10^7 depends on the ratio l_1/l_2 and v_1/v_2, where l_1 and l_2 are the thickness of the upper and lower layers and

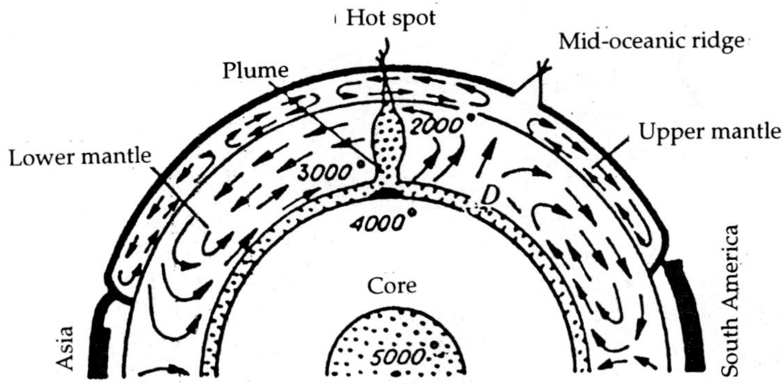

Fig. 4.5. Section of the Earth's sphere along the equator through the Pacific Ocean. Layer D and the plume rapidly ascending from it are depicted. This plume intersects the convection mantle flows (Zonenshain and Kuz'min, 1993b).

v_1 and v_2 the kinematic viscosities of the layers (see Chapter 3, the Rayleigh criterion which depends on β, the coefficient of volume expansion, ΔT the temperature difference between the layer boundaries, l the layer thickness and a and v the coefficients of thermal diffusion and of viscosity). In this case the effective Rayleigh number was determined for v_1 = min and $l = l_1 + l_2$. At $l_1/l_2 = 2$ to 4, Ra = 3×10^6 and $v_1 = v_2$, the cells in both cases had the same horizontal sizes; ascending and descending flows in the upper and lower layers occurred in opposite directions. This suggests identical direction of motion at the interface, which is determined by the viscous coupling of the layers. On raising the viscosity ratio $\gamma = v_1/v_2 > 1$, one large cell in the lower layer corresponded to two or three cells in the upper layer but viscous coupling remained up to $\gamma = 2.5$. At much higher values of Ra = 3×10^7, the upper and lower cells were equal in horizontal size up to $\gamma = 20$ to 25. Flows in opposite directions along the interface, suggesting a predominantly thermal interface, appeared at viscosity ratio $\gamma > 100$. At $l_1/l_2 = 1/3$, $\gamma = 300$ and Ra = 3×10^6, the upper and lower cells were identical in horizontal size, showed the same flow direction in the cells, but flows of opposite directions at the boundary. Our experiments showed identical findings (see below, Fig. 4.7; Kirdyashkin and Dobretsov, 1991).

Three-dimensional numerical simulation of mantle convection has been under development in recent years (Cserepes et al., 1988; Bercovici et al., 1989; Zhao et al., 1992; Yuen et al., 1994; Zhang and Yuen, 1995). In earlier studies, viscosity was treated as constant in the different layers. At Prandtl numbers Pr = $v/a \to \infty$ and viscosity ratios $\gamma = 1$ or 10, viscous coupling was observed (Cserepes et al., 1988) and square cells formed in the lower layer; cells of the same size formed in the upper layer but flowed in the opposite direction. At $\gamma \geq 100$, a thermal boundary was observed, i.e., circulation of the fluid in the cells in the same direction, flows along the boundary in opposite directions and the convective structure of the lower layer more or less duplicated in the upper layer. A similar flow structure was observed in our experiments at low $\gamma = 75$ (see below). At intermediate values of $\gamma = 25$, the flow directions in the rolls of the upper layer become perpendicular to the direction in the lower layer (Cserepes et al., 1988).

In more complex three-dimensional models, stratification of convection is manifest under conditions of varying stratified viscosity or in the presence of phase transitions in the mantle. Turbulence is also manifest due to complex unstable cells, and unstable ascending and descending flows or plumes, cylindrical as well as drop-shaped (Zhao et al., 1992; Tackley et al., 1993; Yuen et al., 1994; Zhang and Yuen, 1995). Three-dimensional modelling with varying viscosity has also shown (Zhang and Yuen, 1995) that in all methods of computation, first the descending jets arise, followed by ascending flows only when the former have reached the core boundary. Probably, the inflow of cold material changes the instability of transition D_2 in the mantle, which initiates the ascent of plumes to the surface. Cold anomalies in the model of Zhang and Yuen (1995) reach maximum at a depth of 140 to 150 km from the core surface (i.e., within layer D_2) and may constitute in temperature one-half the total temperature difference in the mantle, i.e., up to 1000°C. The number and localisation of plumes in a spherical shell are stabilised by the stratification of variable density on lowering the effective Rayleigh number. The internal heat sources in the mantle increase the number of plumes but exert little influence on their stability at viscosity stratification.

Modelling of mantle convective flows inevitably poses the question of the applicability of experiments in a rectangular box since the depth of the mantle is almost half that of the Earth's radius. Although models of convection in cylindrical co-ordinates allowing for the Earth's curvature appeared more than twenty years ago (Hsui et al., 1972), work on the effect of curvature of the layer on the convection regime has appeared only relatively recently (Jarvis, 1994). The main conclusion drawn from this research is that the average temperature, Nusselt number and dependence of the Rayleigh critical number on aspect ratio for rectangular layers and layers bent along the cylindrical surface are extremely proximate under conditions of equality of aspect ratio of the layers. Some differences in geophysical characteristics (heat flow at the surface, topography, gravity etc.) of the layers arise due to the varying length of horizontal boundaries in a cylindrical case.

Summarising the above discussion, density, viscosity and thus mantle convection, may be regarded as stratified; this stratification can be described by more complicated models than the two-layer convection ones. Complication is due to two factors:

a) viscous stratification and phase transitions cause an instability and stratification of whole mantle convection; for example, migrating ascending and descending plumes arise with 'penetrating convection' between them (Yuen et al., 1994; Zhang and Yuen, 1995) and

b) in the Earth's history, two-layer convection changed into whole mantle convection, or vice versa, perhaps many times (Haggerty, 1994; Honda, 1995).

An attempt at comparing the different models using the parametrical approach (Honda, 1995) is depicted in Table 4.1. The variable criteria adopted were parameter β in equation $Nu = \alpha\, Ra^\beta$ (see Chapter 3), Urey ratio, $U = q_x/q_{total}$ (where q_x is the radiogenic heat), and f the concentration of radiogenic heat sources in the lower mantle. It can be seen from this Table that the whole mantle and

124

two-layer models are possible at different parameters of β and U while the model of transition from the two-layer model to whole mantle convection hardly differs from that of whole mantle convection, i.e., the mantle 'has no memory' of the preceding convection pattern after a billion years. Present-day knowledge of possible parameters of β, U and f as well as the above-discussed seismotomographic models corresponds more to the two-layer model of mantle convection. Therefore, transition from whole mantle to two-layer convection or a more complex variant is quite likely: transition from two-layer convection to whole-mantle convection about 2.5 billion years ago as suggested by Maruyama (1994) and, later, transition again to two-layer convection about a billion years ago according to Honda's (1995) conclusions.

Table 4.1. Criteria of whole-mantle and two-layer convection (Honda, 1995)

Model	$β^*$	$U = q_x^{**}/q_{total}$	Remarks
Whole mantle	< 0.1	~ 0.4	Unrealistic β
	~ 0.3	~ 0.8	Unrealistic U (?)
Layered mantle	~ 0.3	≥ 0.4	Depleted lower mantle at $f^{***} ≈ 0.1175$
Transition from layered to whole	~ 0.3	~ 0.8	Same as for whole mantle model
Transition from whole to layered	~ 0.3	~ 0.4	Same as for layered mantle model

*β—constant for Nusselt equation Nu = $α/Ra^β$, usually β = 0.25 to 0.33.
$^{**}q_x$—radiogenic source of heat flow; q_{total} is total heat flux.
$^{***}f$—concentration of radiogenic heat source in the lower mantle (if all the radiogenic heat sources are in the lower mantle, $f = 1$; otherwise, $f = 0$).

It remains to be pointed out in conclusion that studies of free convection in the mantle using numerical simulation contributed greatly to evaluation of the basic criteria when selecting alternative models. The results of numerical modelling, however, depend on a great number of general assumptions and correctness of approximation of equations and boundary conditions. Thus, physical modelling of two-layer convection is all the more necessary because experimental studies are extremely fragmentary and not sufficiently detailed for quantitative analysis (Nataf et al., 1981).

The correlation of numerical calculations with our laboratory experiments is discussed below.

4.2. Experimental Studies of Two-layer Mantle Convection

The two-layer system of fluids in our experiments (Kirdyashkin and Dobretsov, 1991; Dobretsov and Kirdyashkin, 1993b) was formulated as follows (Fig. 4.6): a bath with heat exchangers (1, 2) and transparent lateral walls (7) was first filled with the heavy liquid, glycerine (8) to the required height and hexadecane (9) added later. The upper heat exchanger 1 was then lowered along graduated scales (6) to ensure continuous contact with the surface of the upper hexadecane layer.

For visualisation and measurement of flow velocity, aluminium particles 5 to 15 µm in size were placed in the liquids. These particles remained in the flow in the course of the experiment. The liquids with the particles in them were transparent throughout the horizontal direction. Photographing and telemonitoring were carried out through the front wall by illumination with optical blades 4 mm in thickness placed in the same vertical plane laterally through wall 7) and on top (through heat exchanger 1).

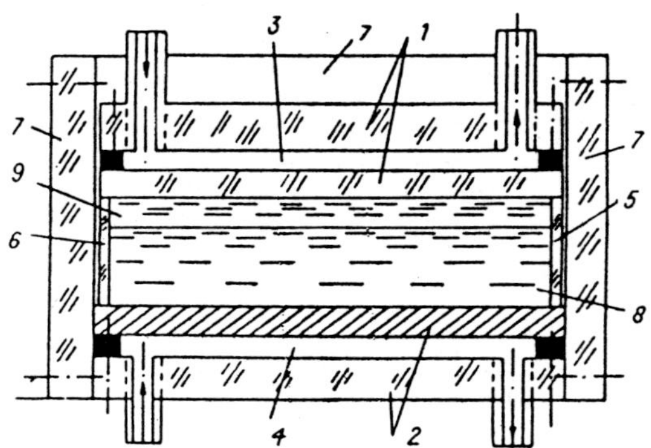

Fig. 4.6. Experimental unit:
1, 2—heat exchangers; temperature-controlled water: 3—cooling and 4—heating; 5, 6—plates specified the height of the liquid layer; 7—transparent walls of the container; 8—glycerine; 9—hexadecane.

Experimental results pertain to layer thicknesses $l_1/l_2 < 1$ and viscosity ratio $v_2/v_1 = 73.5$. In a non-steady regime, cellular flows arose in the upper liquid layer, their horizontal dimensions being comparable to the thickness of the upper thin layer l_1. In the lower viscous layer of the liquid, the horizontal dimensions of unsteady cellular flows were also comparable to its thickness l_2. In a steady-state regime of thermogravitational convection in the lower layer of the more viscous liquid, the horizontal sizes of cellular convective flows (L) corresponded to the structure of cellular flows in an independent horizontal liquid layer heated from below and cooled on top, in which the horizontal cell sizes were: $L/l_2 < 1.8$. In the upper thin layer, the horizontal dimensions of convective cells corresponded to the horizontal size of cellular flows in the lower layer, $L/l_1 \gg 1.8$ and the flow was plane-parallel in character (Fig. 4.7).

An accurate accord was observed between the ascending flows in the upper and lower layers. Similarly, descending flows in the lower layer were located under the descending flows in the upper layer. As the Ra number increased consequent on the change of l_2 or T_2, three-dimensional cellular flows arose in the lower layer. In the upper layer, the horizontal sizes of cells as also their spatial disposition fully corresponded to the structure of cellular flow in the lower layer.

Rolls could be oriented arbitrarily relative to the lateral wall through which the photographs were taken. To generate two-dimensional rolls with their axes

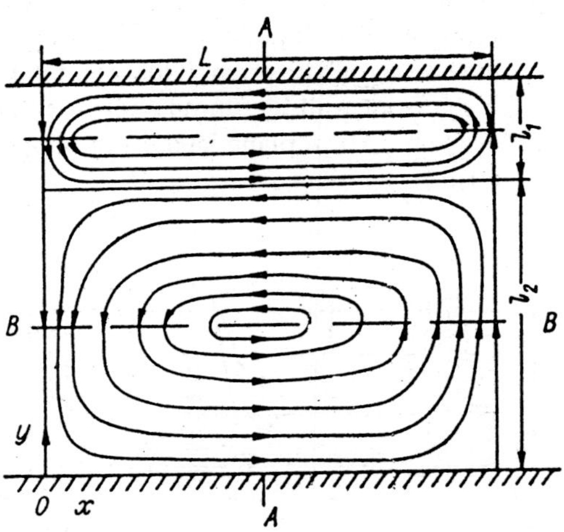

Fig. 4.7. Flow lines in a two-layer system of liquids hexadecane and glycerine determined from a videofilm. Velocity profiles shown in Fig. 4.8 have been determined in sections A–A and B–B.

oriented normal to the lateral surface, two parallel thin-walled copper tubes 5 mm in diameter were placed close to the cooling surface. These tubes were situated perpendicular to the lateral surface and water cooled to the temperature of the cooling water of the upper heat exchanger (1) pumped through them. These cooling pipes were used to fix the zones of descending flow in the rolls, and the roll axes were oriented perpendicular to the plane of the lateral wall through which photographing and videographing were carried out.

It must be pointed out that the tubular coolers could not determine the flow structure in a two-layer system. There were always correlations between the descending flows in the upper and lower layers and also between the ascending flows even in the absence of tubular coolers. This is a consequence of the nature of free-convective flows in the two-layer system when the thickness of the upper liquid layer of lesser viscosity is much less than the thickness of the lower, more viscous layer. Tubular coolers simulated the subduction zones in the upper mantle and were used for rapid transition to a steady state and for spatial orientation of rolls. This feature helps conclude that subduction zones might also play an important role in mantle convection and descending flows are formed under them in the lower mantle. By varying the distance between the cooling tubes, the distance was adjusted between the descending flows of rolls and thus the horizontal roll sizes. Experiments demonstrated that on changing the dimensions of rolls, they can remain stable at $L/l_2 < 1.75$. At $L/l_2 > 1.8$, the ascending flow becomes unstable and the 'breakdown' of rolls and formation of polygonal cells are possible.

The flow lines in the two-layer system of liquids obtained from the videofilm are shown in Figure 4.7. Using the videofilm, flow velocity profiles were found in sections A–A and B–B (Fig. 4.8) for glycerine ($l_2 = 19$ mm) and hexadecane ($l_1 = 7$ mm). In the hexadecane layer, the flow is plane-parallel with the exception of the region of descending and ascending flows; in the lower layer, the horizontal dimensions of cells are comparable with the thickness of this layer and the flows are slow. Flows above and below the liquid interface are oriented in opposite directions, that is counterflow occurs at the boundary.

Experimentally measured temperature profiles in different vertical sections, parallel to the roll axes where the temperature of the upper heat exchanger T_1 was adopted as the reference temperature, are shown in Figure 4.9. A horizontal temperature gradient arose due to counterflow adjacent to the boundary. As a result, it became difficult to determine the exact temperature drop for the lower and upper layers. The overall temperature drop (ΔT) in the two-layer system could be determined reliably: it is equal to 12°C (see Fig. 4.9). The temperature difference between the descending and ascending flows in section $y = l_2/2$ is about 4.3°C, the average temperature of the liquid interface being 3.5°C. Therefore, $\Delta T_1 = 3.5$°C, $\Delta T_2 = 8.5$°C, $Ra_1 = \beta_1 g \, \Delta T_1 l_1^3/a_1 \nu_1 = 1.5 \times 10^4$, $Ra_2 = \beta_2 g \, \Delta T_2 l_2^3/a_2 \nu_2 = 1.04 \times 10^4$.

Let us examine the flow in the lower layer. As discussed earlier, the lower layer structure is similar to that of the flow in an independent liquid layer heated below and cooled above. For such a horizontal layer, the following patterns, applicable for the lower layer of a two-layer model, were established.

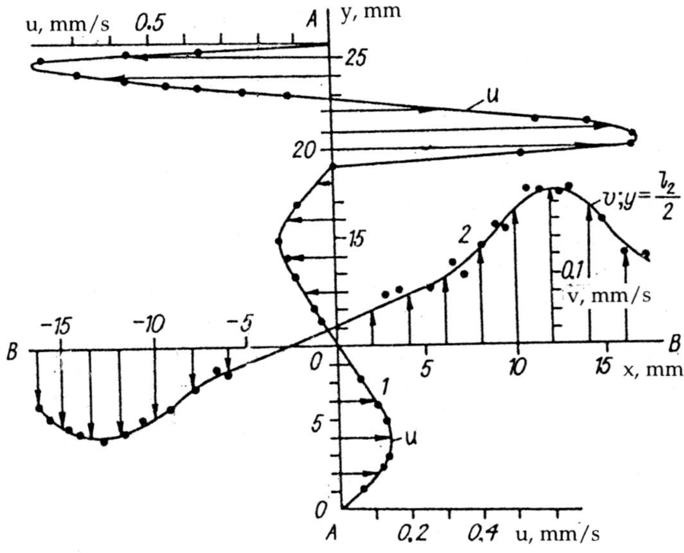

Fig. 4.8. Velocity profiles in a horizontal two-layer system of liquids: glycerine (lower layer) and hexadecane (upper layer). The velocity components are: u—horizontal and v—vertical; x and y axes and the position of sections A–A and B–B are shown in Fig. 4.7.

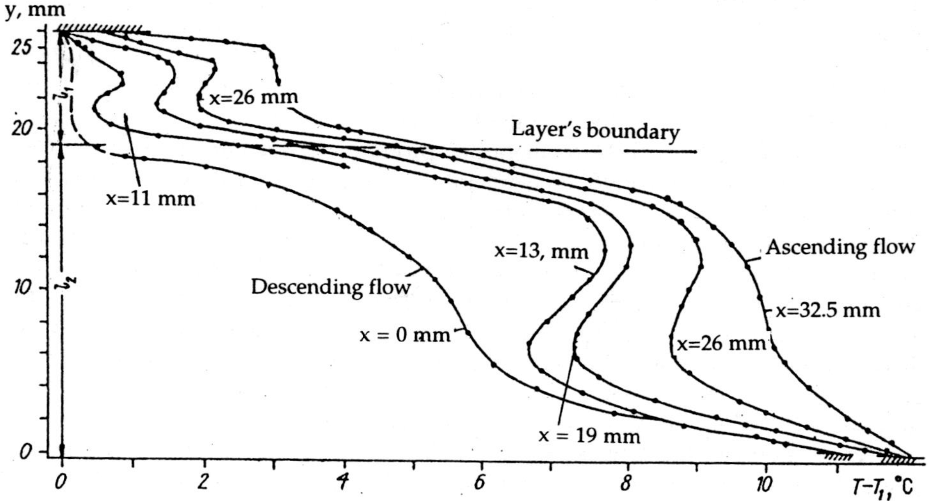

Fig. 4.9. Temperature profiles in a horizontal layer of glycerine-hexadecane determined in different vertical sections from the descending ($x = 0$) to the ascending flow ($x = 32.5$).

1) Leont'ev and Kirdyashkin (1968) experimentally showed that the relative difference between the temperatures of ascending T_a and descending T_d flows is independent of the Ra number:

$$(T_a - T_d)/\Delta T = 0.5, \tag{4.1}$$

where ΔT is the temperature drop between the lower and upper heat transfer surfaces. In fact, in the lower layer $T_a - T_d = 4.3°C$, $\Delta T = 8.5°C$ and their ratio 0.5.

2) The maximum horizontal velocity of free convection in the cell is determined from the following relation:

$$Pe_{max} = u_{max}\, l/a = 0.24\,(Ra - Ra_{cr})^{1/2}, \tag{4.2}$$

where $Pe = Re\, Pr$ is the Peclet criterion, $Re = u_{max}\, l/v$ is the Reynolds criterion and $Pr = v/a$ is the Prandtl criterion. This equation was derived experimentally (Kutateladze et al., 1974; Berdnikov and Kirdyashkin, 1978; Kirdyashkin, 1983) in the range $Ra_{cr} < Ra < 6 \times 10^4$ as well as theoretically by the finite amplitudes method for the vicinity of the point of stability loss $Ra > Ra_{cr}$ where Ra_{cr} is the critical value for two rigid surfaces and equal to 1700 (Chandrasekhar, 1961).

3) The following heat transfer laws were found from numerous experimental investigations:

$$Nu = 0.22\, Ra^{1/4} \text{ for } 3 \times 10^3 < Ra < 5 \times 10^4 \text{ and} \tag{4.3}$$

$$Nu = 0.1\ Ra^{1/3}\ \text{for}\ 5 \times 10^4 < Ra < 10^{10}, \tag{4.4}$$

where $Nu = \alpha l/\lambda$ is the Nusselt criterion, $\alpha = q/\Delta T$ the heat transfer coefficient, q the specific heat flux and λ the thermal conduction coefficient.

The values of specific heat flux (280 W/m) calculated using eqn (4.3) for the lower layer correspond to the experimental value of 300 W/m determined from the average temperature gradient at the interface for the upper as well as lower layers. This also shows that heat transfer in the lower cell corresponds to the principles of heat transfer for an independent horizontal layer. Let us determine velocity u_{max} from eqn (4.2) for an independent horizontal layer when the temperatures along the heat transfer surfaces are constant. In this case, $u_{max_1} = 0.46$ mm/s and $u_{max_2} = 0.11$ mm/s. It can be seen from Figure 4.8 that the experimental values are: $u_{max_1} = 0.9$ mm/s and $u_{max_2} = 0.15$ mm/s, i.e., the calculated values differ from the experimental values by two times for the upper layer and by 30% for the lower layer.

The differences in the values for the upper and lower layers are due to the presence of a horizontal temperature gradient at the interface in the two-layer system. The presence of a local heat flux in the form of a tubular cooler also promotes an increase in the horizontal temperature gradient. Equation (4.2) is valid for isothermal heat transfer surfaces. The superpositioning of the horizontal temperature gradient on the temperature field, characteristic of an independent layer with isothermal heat transfer surfaces, leads to an increase in flow velocity in the two-layer liquid system. Temperature profiles in the upper layer point to a longitudinal temperature gradient $A = dT/dx$ along the layer height with the exception of the area adjacent to the wall. The flow structure in the upper layer corresponds to the flow structure in a flat layer with a horizontal temperature gradient which has no stability threshold. In a horizontal layer with a horizontal temperature gradient, a steady flow is established; the velocity of this flow is directly proportional to the value of gradient A.

For the boundary conditions on horizontal walls, $y = \pm 1$, $u_{\pm 1} = 0$, we have the following equation for the flow velocity (Birikh, 1968):

$$u = (\beta g/6\nu)\ l_0^3\, A\ [(y/l_0)^3 - (y/l_0)]. \tag{4.5}$$

where $l_0 = l_1/2$.

At boundary conditions for the upper layer, $y = 1$, $\partial u/\partial y = 0$ (surface is free or the upper plate is displaced at velocity u) and at $y = -1$ and $u = 0$ (condition of adherence to the lower boundary of this layer), the velocity profile is determined as follows:

$$u = (\beta g/24\nu)\ l_0^3 A\ [4\,(y/l_0)^3 - (y/l_0)^2 - 6y/l_0 + 1]. \tag{4.6}$$

For hexadecane layer of height $l_1 = 7$ mm and extent $L = 27$ mm (see Fig. 4.9), the average longitudinal temperature gradient at the interface can be determined from eqn $(\partial T/\partial x)_0 = \Delta T_2/2L = 8.5/2.27 = 0.163°C/mm$. From eqn (4.5), for parameters of the upper layer, $u_{max_1} = 0.65$ mm/s; the calculated value differs from the experimental value by 28%. A higher value of flow velocity, as in the case of

130

the lower layer, may be associated with the local heat flux caused by the tubular cooler placed close to the cooling surface.

Let us determine the conditions at which the ascending and descending flows in the upper and lower layers are correlated and the horizontal size of convective cell in the upper layer is comparable to the horizontal size in the lower layer. Contrary to this condition are those at the interface of the two layers when convective cells of horizontal size comparable to the thickness of the upper layer are present in the upper layer. We determine the horizontal temperature gradients in the cells in the upper and lower layers for the latter case. For the unsteady lower layer the heat transfer law is in the form of eqn (4.4). In this case, the temperature drop in the lower layer at known heat flux q is found as follows:

$$\Delta T_2 = (10q/\lambda_2)^{3/4} (a_2 v_2)^{1/4}/(\beta_2 g)^{1/4}.$$

The temperature difference between the ascending and descending flows, as shown in Section 3.4, is equal to: $T_{x=1} - T_{x=0} = 0.5\ \Delta T$.

The average horizontal temperature gradient in the lower layer at the boundary with the upper layer $(\partial T/\partial x)$ will be as follows considering that the horizontal cell size is equal to the thickness of the layer:

$$(\partial T/\partial x)_2 = 0.5\ \Delta T_2/l_2 = l_2^{-1} (10q/\lambda_2)^{3/4} [(a_2 v_2)/(\beta_2 g)]^{1/4}, \tag{4.7}$$

and for the upper layer:

$$(\partial T/\partial x)_1 = 0.5\ \Delta T_1/l_1 = l_1^{-1} (10q/\lambda_1)^{3/4} [(a_1 v_1)/(\beta_1 g)]^{1/4}. \tag{4.8}$$

The flows in the upper and lower layers correlate when the ascending flow in the upper layer is located above the ascending flow in the lower layer and condition $(\partial T/\partial x)_1 < (\partial T/\partial x)_2$ prevails. It follows from eqns (4.7) and (4.8) that

$$(l_2/l_1)\ [\lambda_2^3 a_1 v_1\ \beta_2)/(\lambda_1^3 a_2 v_2\ \beta_1)]^{1/4} < 1. \tag{4.9}$$

At the present level of knowledge of the physical parameters of flow in the upper and lower mantle, they should be correlated according to the derived equation.

For the upper heat-insulated surface, $\partial T/\partial y = 0$ and for local discharge of heat in the upper layer, condition (4.9) is invariably satisfied since $(\partial T/\partial x)_1 \to 0$ in this case. This denotes that under conditions of free heat-insulated upper boundary, the horizontal sizes of cellular flows in the upper and lower layers will invariably be identical however thin the upper layer may be.

Experiments showed that when the viscosity differences between the lower and upper layers are high, condition (4.9) may be fulfilled. In this case, flows arise in the thin upper layer. A similar situation is possible even in the mantle with convective flows in the thin upper mantle of horizontal sizes comparable to the dimensions of the lower mantle flows, i.e., comparable to the mantle thickness. Therefore, the presence of large-scale horizontal movements on the surface of the upper mantle does not signify the presence of whole mantle flows. According to geophysical data, a relatively sharp rise in density with depth is observed in the mantle at depths of 400 and 670 km and the stratification of convective flows is possible at the boundaries of density variation. The presence of multilayer convection in the mantle depends on the nature of these sharp changes in density.

A correlation of above experiments with numerical modelling and observational data is discussed below.

4.3. Nature of Transition Layer (C) in the Upper Mantle

The degree of coherence between the lower and upper mantle convective cells is controlled by the nature of the transition layer C and by the possibility of the existence of independent convective flows in it. In our experiments (Kirdyashkin and Dobretsov, 1991; Dobretsov and Kirdyashkin, 1993b) and in numerical simulation (Cserepes and Rabinowics, 1985; Cserepes et al., 1988), an extreme case was formulated in which the boundary between the convecting layers was sharp and there was no transition zone. In Irvine's model (Irvine, 1988, 1991), another extreme case was examined wherein independent convection was present throughout a transition layer of constant thickness (400 to 670 km). Further, the boundary between the transition zone and asthenospheric cells in all cases was 'mechanically coupled' (i.e., the flow directions coincide but the layers differ in density and viscosity) and the boundary between the transition zone and lower mantle changed relative to the tectonic position: nearly one-half of the boundary area was also 'mechanically coupled' but the other half of the boundary coupled only 'thermally' (i.e., counterflow along the boundary occurs which ensures not only mechanical but also temperature gradient along the boundary (see Figs. 4.15 and 4.16)). Specifically, this is a three-layer stratified convection in the mantle and is the main hypothesis of T. Irvine who formulated it as a 'bilateral convergence model'.

Let us examine the phase transitions at the boundary around 400 and 670 km in order to study the rationality of these suggestions.

The experimental phase diagram of mantle peridotite (Fig. 4.10) serves as the limit of the possible temperature distribution in the ascending and descending flows in the transition layer and lower mantle as well as of pressure and temperature conditions of transition at the boundaries 400 and 670 km (Takahashi, 1986; Irvine, 1991).

The boundary around 400 km reveals a variation of density $\Delta\rho = 0.20$ g/cm^3, has a positive slope $dP/dT > 0$ and hence is permeable to convective flows and is 'smeared-out' in depth, i.e., it varies over 60 km in depth in the zone of ascending and descending flows; further, the boundary is displaced depending on the ferruginous level of olivine and redox conditions. Phase transition $\beta - \leftrightarrow \gamma$-phase at a depth of 500 to 650 km is similar and even more variable with depth (see Fig. 4.10). The boundary at depth 670 km corresponds to phase transition with $\Delta\rho = 0.4$ g/cm^3, has a negative slope $dP/dT < 0$ and hence is sharp, changes little with depth (less than 20 km), is impermeable to descending flows and permeable only to the most intense ascending flows (Ringwood et al., 1992).

These features are intensified also because the boundary at depth 670 km is not only a phase but also a chemical boundary. This is confirmed by calculations of density, which can be ensured by phase transitions only by 30% (Kuskov and Parfenov, 1992). A possible explanation is that restites accumulate at the bottom of the transition layer after melting of the subducted lithospheric plate. Therefore,

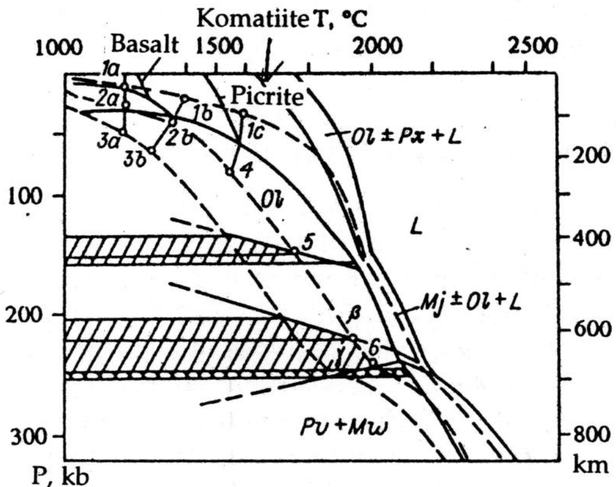

Fig. 4.10. Experimental phase diagram of mantle peridotites (Takahashi, 1986) and critical points (P and T) in the upper mantle (Sobolev, 1974; Dobretsov, 1981; Dobretsov and Ashchepkov, 1991): points: la to c; 2a, b; 3a, b; 4—estimates of pressure and temperature conditions for mantle xenoliths; 5, 6—position of 'wet' solidus on the boundary with the lower mantle. Transition intervals $Ol \leftrightarrow \beta$-phase, $\beta \leftrightarrow \gamma$-phase and γ-phase $\leftrightarrow$ Pv + Mw (perovskite + magnesial wüstite) for different geotherms are hatched. Broken lines depict the minimal, average and maximum temperatures in the mantle.

in the zones of descending flows, the $\Delta\rho \geq 0.4$ g/cm^3, boundary is even sharper, the subducting plate cannot penetrate the lower mantle (see Fig. 4.3) and a systematic flexure of the boundary of upper and lower mantle arises here. This pattern has been confirmed by the recent seismic data (Shearer and Masters, 1992) depicted in Figure 4.11.

The lowering of this boundary for Kuril-Kamchatka region is 40 km, with the adjoining part of the boundary raised by 10 to 15 km (see Fig. 4.11) but this rise is most distinctly noticed on correlating the depth with velocity corrected for temperature variation and calls for additional verification. The width of depression in the most caved part is about 10 to 15° or 1100 to 1700 km and the edges of the depression attain double the width (30° or 3300 km) while the length exceeds 5000 km, from Chukchi to the southern tip of Mariana arc. The axis of the depression is displaced westwards and north-westwards relative to trenches conforming to the dip of the subduction zone. The latter, determined on the basis of deep-focus earthquake data, falls exactly on this depression of the 660 km boundary (Fig. 4.11). This result coincides with the model according to which the subducting plates spread horizontally at a depth of about 660 km (see Fig. 4.3); it does not confirm penetration of the subducting plates into the lower mantle up to a depth of 900 km, as formerly assumed on the basis of results of seismic tomography. Similar anomalous depressions of the 660 km surface have been established west of the Tonga sea, in South America (displaced westwards under the east Atlantic conforming to the gentle dip of Peru-Chili subduction zone) and in the central part

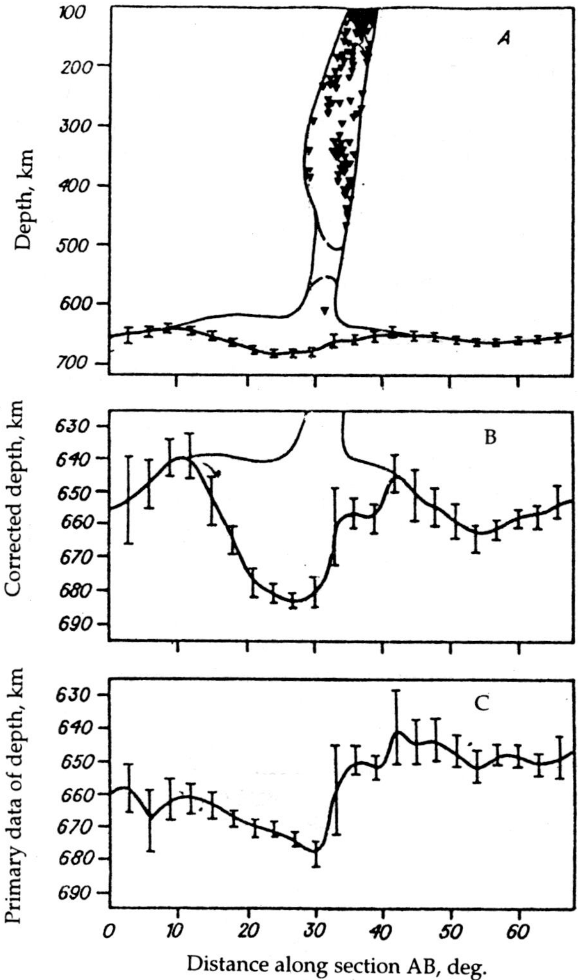

Fig. 4.11. Relief of the seismic boundary of the upper and lower mantle in the section through Kuril-Kamchatka trench:
position of line AB (see Fig. 4.12): A—position of subduction zone fixed from the epicentres of earthquakes (triangles), lines of relief of the boundary of upper and lower mantle (lower line with vertical bars depicts the limits of experimental error) and suggested lines restricting the accumulation of restites arising from the melting of the subducted plate (upper thin and dotted lines). Details of lenses of accumulated restites and lowering of the surface of lower mantle of corrected (B) and uncorrected (C) depths (after Shearer and Masters, 1992, modified).

of the Alpine-Himalayan belt; complex subduction zones are also known in the Caribbean basin (see Figs. 4.2 and 4.12). In other words, these depressions coincide with the active subduction zones prevailing at present or which prevailed in the recent past. However, exceptions do exist. Shallower depressions have been detected on this surface in the equatorial part of the Pacific Ocean (northern part of the Pacific Ocean, see Fig. 4.12) and in the equatorial Atlantic from South America to central Africa. Together with the Alpine-Himalayan anomalies, these form a broken equatorial belt only partly coinciding with the Mesozoic subduction zones. It may be suggested that descending flows prevail in the lower mantle and do not coincide with the movements of the lithospheric plates, as pointed out earlier.

The projections of the positive relief in the 660 km boundary (see Fig. 4.12) may suggest ascending flows and are most distinctly manifest under the north-western part of the Pacific Ocean (with 'weak' continuation up to the Hawaiian centre); north-western Atlantic (Iceland) and the adjacent part of North America and Antarctica; two major anomalies are also seen under the northern and southern parts of the Indian Ocean (the southern part extends through South Africa up to the Atlantic). Localisation of ascending flows is confirmed by the following facts: all the major mantle plumes originating in the lower mantle (compare Figs. 4.2 and 4.12) are situated here or hereabouts, projections of the core-mantle boundary are seen close to these anomalies (Hager, 1984; Hager et al., 1985), changes in velocities V_p and V_s in transition layer D_2 (Wysession et al., 1995; Wysession, 1995) and low-density anomalies in the lower mantle according to seismotomographic models (Inoue et al., 1990; Fukao et al., 1994). Using the seismotomographic model (Inoue et al., 1990), a model of circulation in the upper and lower mantle was constructed and, in particular, centres of major ascending and descending flows were detected (Irvine, 1988, 1991). Centres of ascending and descending flows and lines joining them are compared with the position of the main rises and subsidences of the relief of 670 km boundary in Figure 4.12 (Shearer and Masters, 1992). It can be seen that the centres of descending flows in Peru and Vietnam should be shifted towards the north-east and those of ascending flows in Balleny (Antarctica) and Okavango (South Africa) towards the north-west and north. A modified model of ascending and descending centres connecting the anomalies according to Shearer and Masters (1992) is shown in the inset map of Figure 4.12. The sublatitudinal line connecting the zones of descending streams, in particular, is displaced north-wards compared to the line plotted in the work of Irvine (1988).

The above analysis of the nature of sharp density variations at depths of 400 and 670 km and relief of the 670 km boundary points to stratification in large-scale convection flows at depth 670 km and less probably at depth 400 km and hence to the correctness of multilayer (two- or three-layer) model of convection flows in the mantle of the Earth. The increase in density and viscosity at the boundaries of transition layer C was pointed out in Chapter 1 (see Figs. 1.6 and 1.8). Identification of a low-viscosity low-density layer $(- \Delta V_p)$ below the C layer at depths of 700 to 1000 km in the seismotomographic models (Fukao et al., 1994; Su et al., 1994; Honda, 1995) shows the importance of the lower boundary of the layer at depth about 670 km. Along this boundary, restites of the subduction zone spread

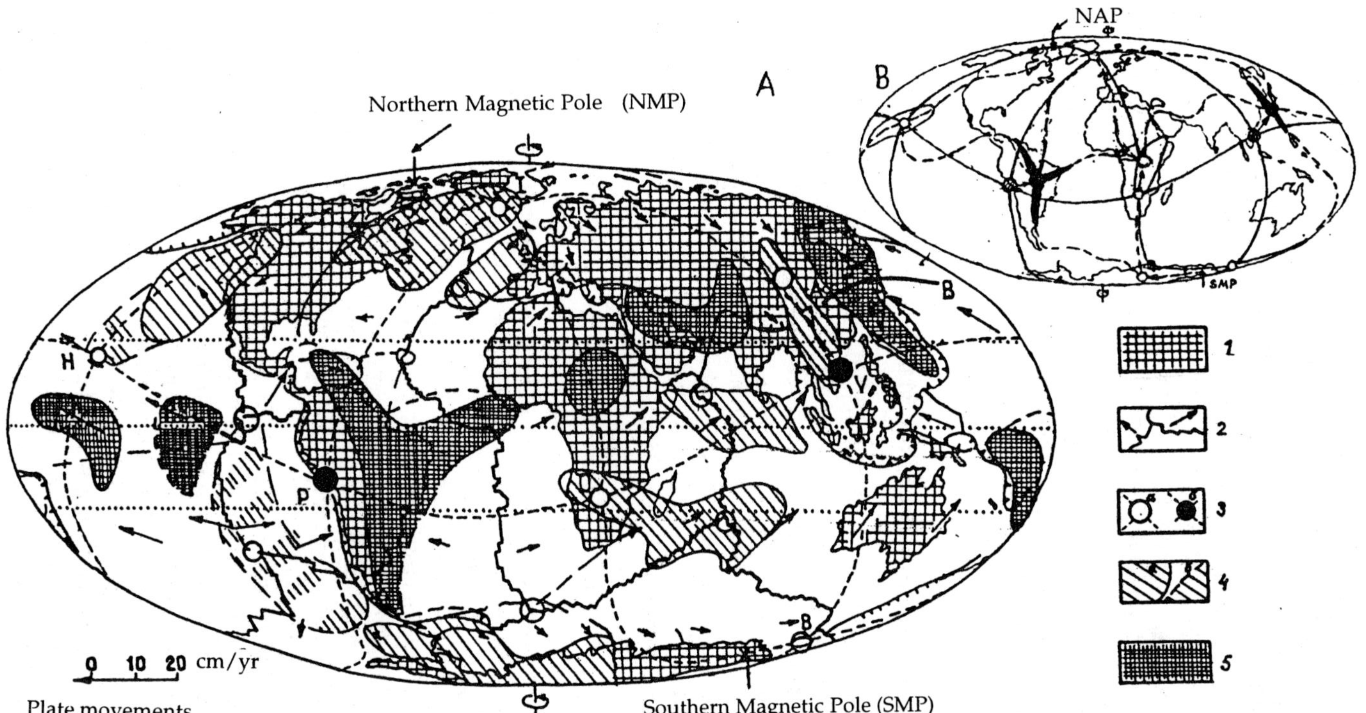

Fig. 4.12. A) Plate tectonics and flow structure in the mantle determined by the centres of ascending and descending flows of H (Hawaii), I (Iceland), O (Okavango) and B (Balleny) and two centres of descending flows of P (Peru) and V (Vietnam) as well as the relief of the 670 km boundary (Irvine, 1988; Inoue et al., 1990; Shearer and Masters, 1992). 1—Continents; 2—oceans and velocities of plate movements, proportional to the length of arrows; 3—centres of ascending (a) and descending (b) flows; 4—projections in the positive relief of the lower mantle boundary are distinct (a) and poorly manifest (b); 5—depressions of the 670 km boundary.
B) Comparison of the actual net connecting the tectonic centres of ascending and descending flows (continuous lines) and probable picture taking into consideration the relief of the 670 km boundary (broken line).

upwards (i.e., in the C layer) and accumulate while the lower plumes stop, spread and are transformed at the bottom (see Secs. 4.10 and 4.11).

The C layer was the main concentrator of water in the mantle in the early history of the Earth and possibly is so even at present. This is due to the high solubility (up to 3.5 wt, %) of H_2O in the spinel modification β-Mg_2SiO_4 and at very low temperature (below 1000°C), by the inflow of H_2O into water-bearing silicates such as phlogopite, clinohumite and chondrodite (Kawamoto et al., 1996). Another water-containing layer could have formed at the bottom of the continental lithosphere at depths of 100 to 200 km (K-richterite, phlogopite and pargasite, the latter from depth 120 km). The effect of these layers may have been decisive in the formation of deep magma-forming kimberlites, lamproites and other high-potassium magmas rich in volatile constituents.

4.4. Importance of Subduction Zones for Mantle Convection

In our model based on experiments (Kirdyashkin and Dobretsov, 1991; Dobretsov and Kirdyashkin, 1993), subduction zones play a leading role since they stabilise convection in the upper mantle and cause major downward flows in the lower mantle. The coherence of the downward flows in the upper and lower mantle in such zones can be interpreted as the penetration of subducting plates into the lower mantle. As pointed out before, in fact, such a proof of whole-mantle convection is only apparent and is not confirmed by detailed seismic tomography of the 670 km surface (Shearer and Masters, 1992).

Moreover, forces acting in the subduction, including the weight of the cold plate and its loading as a result of eclogitisation, form the main factors in the formation of systems of convection flows in the upper mantle and movements of lithospheric plates. According to the estimates of Lithgow-Bertelloni and Richards (1995) and Lithgow-Bertelloni et al. (1996), these mobile subduction forces ('slab pull') constitute up to 90% of all the mobile forces of plates. This is also confirmed by good correlation with the observed velocities of plate movements ($r = 0.90$). Exceptions are the eastern Pacific rise, Nazca and South American plates where ascending flows in the upper and lower mantle also play an important role. Moreover, the eastern Pacific Ocean rise and most other spreading zones can be regarded as the result of a passive reaction to the system of movements originating in the upper mantle by subduction forces.

Our experiments on two-layer convection indicate that the structure of not only asthenospheric, but also lower mantle flows is largely determined by subduction. Change in the position of subduction zones also induces changes in position of the descending flow in the lower mantle and hence the general structure of convection in it. Subduction zones are formed by a combination of several physical and chemical processes (compaction, solid phase transitions etc.) leading to the loading of the lithosphere; hence they are less dependent on thermal convection and represent themselves the boundary conditions for the lower and upper mantle convection flows. Therefore, when setting up numerical simulation, the organisational characteristics of subduction and the physical and chemical processes referred to above should be taken into consideration.

Continents represent heat insulators for mantle flows due to their large thickness, low temperature condition and high heat generation in their crust. This may lead to perceptible heating under them over periods, comparable to those equal to a single revolution of mantle material in the cell. This aspect in particular has been used in Trubitsyn's models (Trubitsyn and Nikolaichik, 1991; Rykov and Trubitsyn, 1995). However, large-scale flow initiation, including lithospheric displacements, is possible only in the formation of subduction zones.

Finally, the most important condition controlling the subduction process is the presence of a viscous accretion-subduction wedge and internal pressure arising in it. The subduction process and the role of accretionary wedge is described in Chapter 5 in greater detail. It is important to note here that the melting of subducting plates generates additional local convection cells and (or) plumes extending onto island-arc and back-arc regions. With them are associated the most important manifestations of magmatism, especially the 'fire ring' of andesite volcanoes around the Pacific Ocean.

However, there still remain two key issues which have not been adequately understood and which call for further study: 1) how do the subduction zones form and 2) what are the causes or mechanisms of reorganisation of asthenospheric flows which led to periodic (at intervals of 400 to 450 million years) piling up of continents into the Pangaea supercontinent and its splitting later and the continents separating, i.e., the factors responsible for the Wilson supercycle 'from Pangaea to Pangaea'? Apart from matching with the period of rotation of about 400 million years of convection cells in the lower mantle, demonstrated below, facts have to be invoked to explain the regular changes in the disposition of subduction zones and the development of great tensions under the supercontinent. These aspects are discussed in detail in Chapter 5.

4.5. General Structure of Convection in the Earth's Mantle

As pointed out earlier, convection flows in the lower layer of the mantle correspond in structure to convection in a single layer heated from below. However, due to internal heat sources, the situation differs significantly under real mantle conditions. As a first approximation, we can nevertheless hypothesise that the average heat flux through the lower mantle is equal to the corresponding flux on the surface of the ocean floor: $q = 0.059$ W/m^2. In this case the temperature drop and the velocity of convection flows in the lower mantle can be determined using the results of our experiments and the well-known eqns (4.4) and (4.7) for heat transfer and variation of flow velocity relative to the layer parameters using eqns (4.2) and (4.6).

For the lower mantle, Ra $> 5 \times 10^4$ and hence the temperature drop is determined from eqn (4.4):

$$\Delta T_2 = (10q/\lambda_2)^{3/4} \ (\alpha_2 v_2/\beta_2 g_0)^{1/4}. \tag{4.10}$$

It follows from the ahead equation that temperature drop is independent of the thickness of the convecting layer. The value of Ra_2, taking into consideration eqn (4.10), is determined from the following equation:

$$Ra_2 = \beta_2 g_0 \, \Delta T_2 l_2^3 / a_2 \nu_2 = l_2^3 \, (10q \, \beta_2 g_0 / \lambda_2 a_2 \, \nu_2)^{3/4}. \qquad (4.11)$$

The above analysis of experimental results shows that the velocity in the lower cell is 30% more than the value calculated using eqn (4.2).

Therefore, for our case at $Ra \gg Ra_{cr}$, let us adopt

$$Pe_{max} = 1.3 \times 0.24 \, Ra_2^{1/2} = 0.312 \, Ra_2^{1/2}. \qquad (4.12)$$

According to the estimates of Zharkov (1983), the physical properties for lower mantle of thickness $l_2 = 2.2 \times 10^6$ m are as follows:

$$g_0 = 10 \ \text{m/s}^2; \ \beta_2 = 1.5 \times 10^{-5} \ °C^{-1};$$

$$\rho_2 = 5000 \ \text{kg/m}^3; \ a_2 = 2 \times 10^{-6} \ \text{m}^2/\text{s}; \ \lambda_2 = 12 \ \text{W/m} \cdot °C.$$

As adopted earlier, the average heat flow $q = 0.059$ W/m^2. The results of calculation of ΔT_2, u_{max}, Ra_2 using eqns (4.10) to (4.12) for different values of ν are shown in Figure 4.13. According to Zharkov (1983), the coefficient of kinematic viscosity in the lower mantle is $\nu_2 = 10^{18}$ to 10^{19} m^2/s. For these values of ν_2, the velocity according to Figure 4.13, $u_{max_2} = 0.4$ to 0.8 cm/yr and temperature drop $\Delta T = 1150$ to 2000°C. If it is assumed that the global movements of continents and plates depend on the velocity of flows in the lower mantle, the more probable value will be $u_{max_2} = 0.8$ to 2 cm/yr.

It can be seen from Figure 4.13 that, in this case, the most probable parameters are: $\nu_2 = 10^{17}$ to 10^{18} m^2/s (corresponding to the above estimation of $\eta = 10^{20}$ to 10^{21} poises), $\Delta T_2 = 620$ to 1150°C and $Ra = 9 \times 10^5$ to 5×10^6. On this basis we can estimate the average temperature and temperature distribution in the descending and ascending flows in the lower part of the mantle shown in Figure 4.14 using the critical temperature points for the mantle established before (see Fig. 4.10) (Sobolev, 1974; Dobretsov, 1981; Zharkov, 1983; Dobretsov and Ashchepkov, 1991; Sorokhtin and Ushakov, 1991) and the experimental evidence on phase transitions at the boundaries of 480 to 650 km. The T and P values found for the upper mantle (see Fig. 4.14, points 1 to 5) are based on the stability of mantle xenoliths (see Fig. 4.10, points 1a, 1b, 1c, 2a, 2b, 3a, 3b and 4) and on the phase boundaries, including the curves of melting and hardening and fluidisation of peridotites in the mantle (see Fig. 4.10). The disposition of points 6 and 7 in Figure 4.14 is associated with the estimation of possible conditions of melting in the lower mantle and in the core and temperature drop in the core-mantle boundary (Dobretsov, 1980; Bochler, 1992). The temperature profiles for the ascending and descending flows have been calculated according to eqns (4.1), (4.4) and (4.10) and experimental results (see Fig. 4.10) while estimation of velocity in the lower mantle is based on Figures 4.8 and 4.13 and eqns (4.12) and (4.6) at $\nu = 2 \times 10^{17}$ m^2/s. In the lower mantle, $\Delta T_2 = 1350$°C but, on excluding adiabatic heating of 2150 km at 0.3°C/km (645°C), $\Delta T_2 = 700$°C. The maximum temperature difference between the ascending and descending flows is 350°C. For the upper mantle (B + C) (see Fig. 4.14), $\Delta T_1 = 1000$ to 1400°C and in the asthenosphere $\Delta T = 400$ to 500°C.

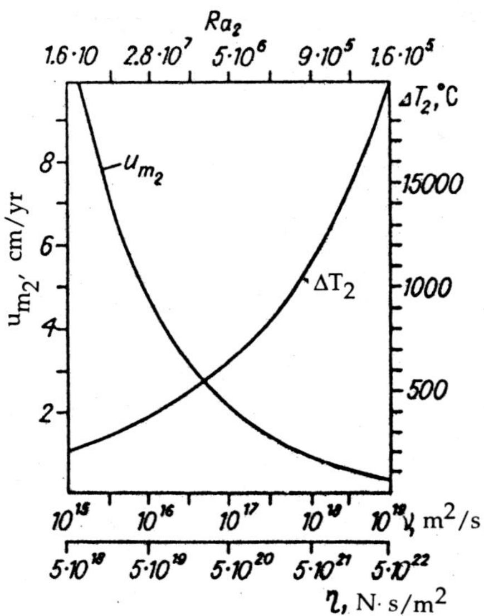

Fig. 4.13. Temperature drop ΔT_2 and maximum value of horizontal velocity $u_{\mathrm{max}2}$ in the lower mantle calculated using eqns (4.10) to (4.12) for different viscosities (ν and η) and Rayleigh numbers (Ra_2).

Without adiabatic temperature variation, these values will be 700 to 1000 and 300 to 350°C respectively.

The structure of mantle movements is shown in Figures 4.14 and 4.15. The boundary at 670 km depth, corresponding to the boundary of chemical and phase transitions, separates the lower and upper mantle convection flows. In the upper mantle, horizontal cells correspond generally to the large convection cells in the lower mantle (and to the sizes of large lithospheric plates). As pointed out, the main subduction zones correspond to the descending flows in the lower mantle. The phase transition at 400 km depth may be permeable to free convection flows. Assuming a viscosity difference above and below the 400 km boundary, slow convection flows were possible even between 400 and 670 km but may have vanished in the course of tectonic history.

The independent structure of 400 to 700 km layer can be established from the data of seismic tomography (Spakman, 1990; Spakman et al., 1993; Fukao et al., 1994). These anomalies of density in the transition layer are also responsible for the main anomalies of the geoid (Cazenave et al., 1988; Cazenave and Thoraval, 1994). From the petrological viewpoint, the young oceanic crust and oceanic type ophiolite formations are generated in the mid-oceanic ridges above the ascending flows of the upper mantle (see Figs. 4.12 and 4.15). A special type of ophiolites containing the boninite series is formed near the descending flows in the structures of island arcs and marginal seas (Dobretsov, 1981). It is possible that the

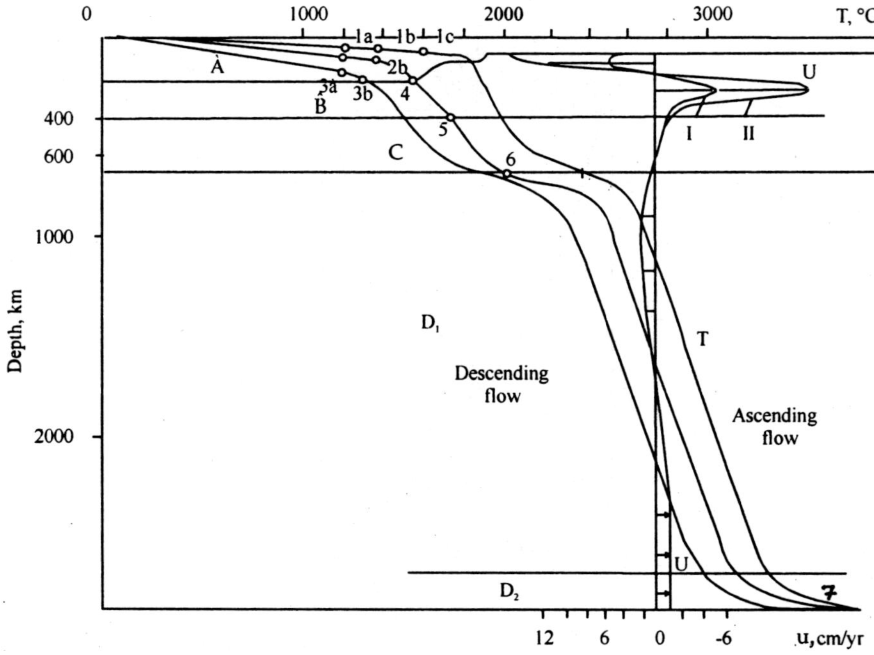

Fig. 4.14. Temperature distribution in the mantle of the Earth according to critical points 1 to 7 and experimental model (see Figs. 4.9 and 4.11): A—lithosphere; B—asthenosphere; C—transition layer of upper mantle; D₁—lower mantle; D₂—transition layer from core to lower mantle. Points 1 to 6 are according to Fig. 4.10. Velocity profiles in the lower and upper mantle for: I—upper steady surface according to eqn (4.5) and II—free upper surface according to eqn (4.6).

island arc was formed at places where the older and cold oceanic lithosphere was subducted while margins of the Andean type and associated basins were formed at places where younger plates were subducted (Wilson, 1990). Figure 4.15 depicts an additional convection cell under the continental margin which led to the formation of an island arc and marginal sea or volcanic cordillera and back rift.

However, the sections depicted in Figures 4.14 and 4.15 do not provide an idea of the spatial distribution of convection cells in the mantle. For this purpose, additional data are required and the seismographic image of the lower mantle may be used.

4.6. Structure of Convection in the Lower Mantle

One of the first attempts at determination of the actual structure of convection in the lower mantle was undertaken by Irvine (1988, 1991) on the basis of the hypothesis of six global centres and the so-called 'bilateral convergence model' (see Fig. 4.12). He suggested Hawaii and Iceland as centres of ascending flows in

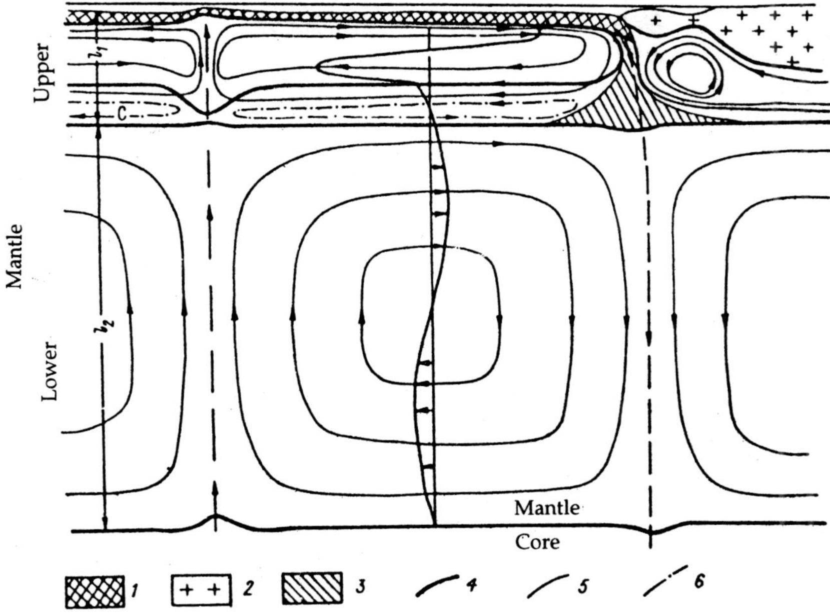

Fig. 4.15. Model of mantle flows (Dobretsov and Kirdyashkin, 1993a): 1—oceanic and 2—continental lithosphere; 3—subducted plate and possible restites; 4—boundaries between layers; 5—flows in asthenosphere and lower mantle; 6—possible flows in transition zone C.

the lower mantle. These centres have long been interpreted as the largest of lower mantle plumes. Two more centres have also been suggested: Okavango in South Africa and around Balleny island between Antarctica and Pacific Ocean-Atlantic ridge. The conditions around Balleny resemble those of Iceland centre including vigorous volcanic centres of Balleny island and Erebus in Antarctica (Waissel et al., 1977). Okavango represents a seismically active tip of the east African rift system (Fairheed and Henderson, 1977) and was a centre of massive trap effusions in the Jurassic and Cretaceous (Karroo and Parana) after which separation of South America from Africa commenced in the Late Cretaceous (Zonenshain and Kuz'min, 1993a). Okavango coincides also with the south-eastern extremity of the largest African lower mantle plume (see Fig. 4.2). Similarly, on the basis of palinspastic reconstructions, the Iceland centre in the Permian coincided with the centre of massive basalt effusions in the Tunguska syneclise (Morel and Irving, 1981; Zonenshain et al., 1990) accompanied along the periphery by kimberlite and alkaline-ultrabasic magmatism.

The descending flows are concentrated in the territories of Peru and Vietnam where the most rapid countercurrent movements of Nazca and South American plates (centre in Peru) and Indian, Chinese (Eurasian part), Philippine (Pacific Ocean part) plates (centre in Vietnam, see Figs. 1.3 and 4.12) are seen. It has further been suggested that movements around these centres in the lower mantle are symmetrical. This was demonstrated by matching the break-up period of about 54 to 56 million years around the Iceland centre (separation of North America, Greenland and Europe) and around the centre in Balleny (break up of Antarctica and Australia) (Sclater et al., 1977; Waissel et al., 1977). A little later, around 42 to 45 Ma ago, powerful eruptions commenced in the east African rift system (at that moment, above the Okavango centre) and in the Hawaiian chain of seamounts, which traced the course of movement of the Pacific Ocean plate above the Hawaiian plume for 43 million years before the present position (Sclater et al., 1977; Irvine, 1988) (see Fig. 4.18).

The importance of these centres is emphasised by the fact that in the geodynamic model of core-mantle boundary (CMB) constructed on the basis of seismic data (Hager, 1984; Hager et al., 1985), Iceland and Okavango centres are projected on one of the two large projections of core-mantle relief and Hawaii and Balleny centres on the other. The correlation with projections of 670 km boundary (see Fig. 4.12) and low-density zones in the lower mantle is similar (see Fig. 4.2). Moreover, Balleny centre almost coincides with the geomagnetic pole while the other pole lies on the arc of a large circle joining the Hawaiian and Iceland centres. Finally, the Peru and Vietnam descending centres lie on the geomagnetic equator (see Fig. 4.12). Since the geomagnetic field of the Earth according to the 'magnetic dynamo' model is explained by convection heterogeneity in the outer liquid core (see below), there is a good probability that both these patterns suggest an association of these centres with the largest of mantle plumes rising from the CMB, more precisely from boundary layer D_2.

These centres are inadequate, however, to effect large-scale convection in the lower mantle. Although the early tomographic images reflecting the density and thermal heterogeneities in the lower mantle (Dziewonski and Woodhouse, 1989; Inoue et al., 1990) do not contradict the disposition of the ascending centres (Hawaii, Okavango, Iceland and Balleny) in the hot mantle and descending centres (Peru and Vietnam) in the cold mantle, a more detailed tomography of the lower mantle (Fukao et al., 1994) (see Fig. 4.2) and especially the image of heterogeneities in the 670 km boundary (see Fig. 4.12) show a more complex pattern and close correlation between subduction zones and descending flows in the lower mantle as emerging from our model, but do not match with Irvine's model.

The sizes of the convection cells in Irvine's model are extremely large and contradict experimental and theoretical data (Cserepes and Rabinowicz, 1985; Cserepes et al., 1988; Dobretsov and Kirdyashkin, 1993b). It can be seen from the subequatorial section (Fig. 4.16A) constructed by Irvine (1988) through Hawaii-Peru-Okavango centres, that four convection cells have been identified in the lower mantle. In meridional sections, judging from the model of convection flows of the upper layer of the lower mantle (Irvine, 1988), four cells can be similarly separated. Thus, the minimum horizontal cell size in the lower mantle in Irvine's model is about 9000 km, this being four times the thickness of the lower mantle (2200 km). It follows from experiments and classical laws of convection that cells of cross-section only 1.8 times (or less) their thickness can be stable. In fact, in the latest tomographic model (Fukao et al., 1994) six cells have been reconstructed in the lower part of the lower mantle (see Fig. 1.3) and more cells only in the upper part of the lower mantle (see Figs. 4.2 and 4.12). This gives rise to the problem of accord between the theoretical structure of convection and tomographic data and also the representativeness of tomographic models (Irvine, 1991; Spakman et al., 1993; Fukao et al., 1994).

It can be hypothesised that the cells in the lower mantle as constructed by Irvine (1988, 1991) were unstable and that each of them ought to be divided into at least two cells. Taking into consideration the above correlation of the horizontal dimensions of convection cells in the lower mantle and in the asthenosphere as well as the data on rises and depressions in the relief of the 670 km boundary of the upper and lower mantle (Shearer and Masters, 1992), the above subequatorial section was reconstructed and eight cells of average horizontal size 5000 km seemed more probable in the lower mantle. This variant is depicted in Figure 4.16B. The large horizontal size of the lower mantle cell under the Pacific Ocean points to its instability: either it is split (shown by hatching in Fig. 4.16B) or the correlation between the ascending flows in the lower and upper mantles is disturbed. In particular, the gently sloping subduction zone under Peru comes into contact with the lower mantle under La Plata basin. It is here that the deflection of the 660 km boundary of the lower mantle is located and corresponds to the descending flow in it.

The instability of the lower mantle convection (at high Prandtl numbers) represents a special problem discussed below.

It is more complex to determine the horizontal cell sizes in the submeridional direction. Many cells in the asthenosphere are extended in the submeridional direction (under the Atlantic and Pacific Oceans, see Fig. 4.12, and also Dobretsov (1980)) due to the overall stress field in the Earth and origin of divergent plate boundaries in zones of minimal tension. At the same time, such an orientation of stress field in the lower mantle is less probable and cells that are isometric in plan, proximate to hexagonal (compare Figs. 1.3, 4.2 and 3.4), are wholly possible here. In some submeridional intersections, the number of major lithospheric plates is equal to 7 or 8 and it is easy to hypothesise the very same eight cells in the lower mantle as in the subequatorial section.

For example, in the section through Peru-Iceland-Vietnam-Balleny, six nearly equal plates are isolated (see Fig. 4.12) and only the seventh (Eurasian) is double in size; beneath this plate, two cells can be assumed in the lower mantle that are not manifest in the upper mantle since the asthenosphere here is mostly absent (see Fig. 4.2). In another section through Hawaii-Iceland-Okavango, only five plates are distinguished. The largest of them—Pacific Ocean plate—is divided transversely by depressions of the 670 km boundary in the lower mantle (Shearer and Masters, 1992), i.e., descending flow in the lower mantle shown in Figure 4.2 is possible here but is not reflected in the massive asthenospheric lens under the Pacific Ocean. It is situated in the continuation of the Alpine-Himalayan line of downward flows and divides the lower mantle under the Pacific Ocean plate into several cells (at least four, not counting Nazca as a plate and ignoring the small cells in the Caribbean basin, see Fig. 4.2). Thus, in thus section also, not less than eight cells are possible in the lower mantle.

If it is assumed that isometric polygonal cells of size $R_1 = 3000$ km at $R_2 = 2200$ km ($R_1/R_2 = 1.37$) predominate in the lower mantle, the area of one cell $S_1 = 2.83 \times 10^7$ km^2. The extent of boundary at depth 670 km

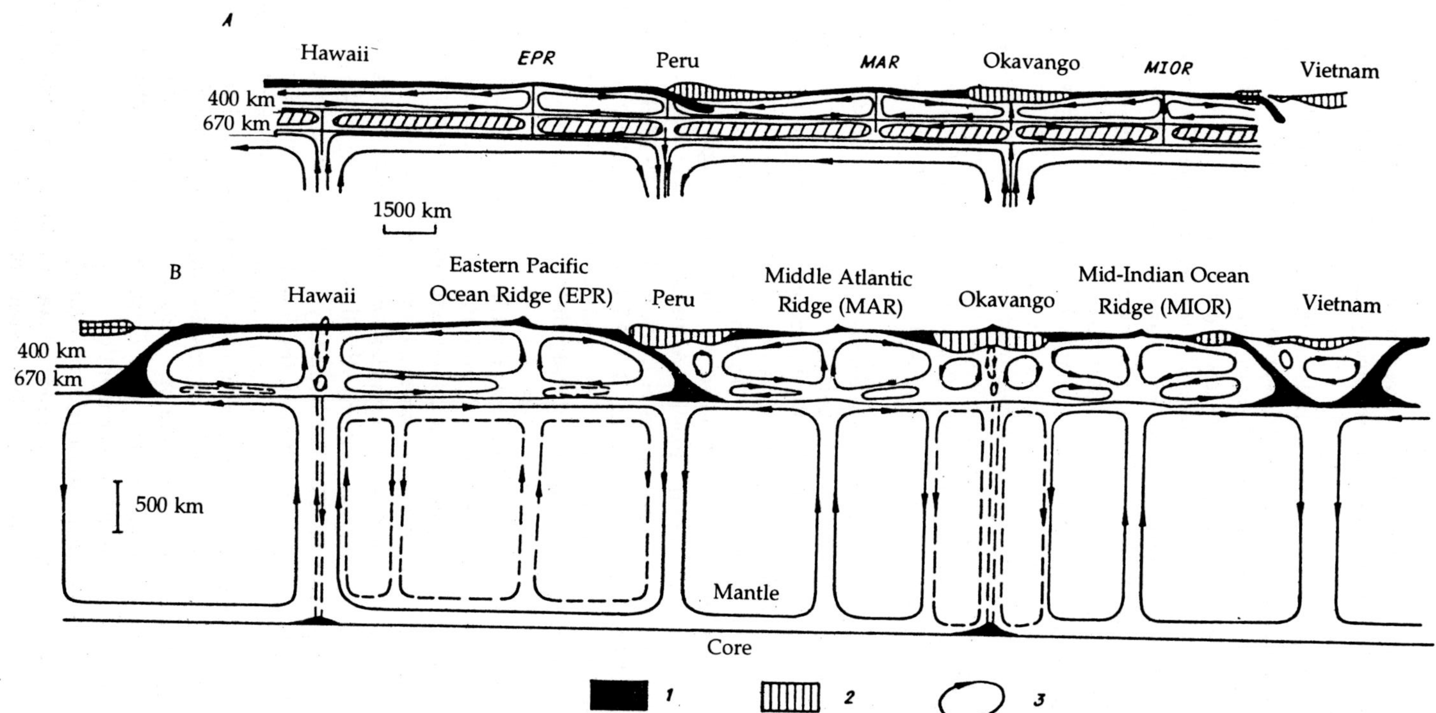

Fig. 4.16. Hypothetical sections through the Earth's mantle: position of sections and centres is shown in Fig. 4.12; A—according to Irvine (1988) and B—as reconstructed by the authors; 1—oceanic and 2—continental lithosphere; 3—convection flows in the asthenosphere and lower mantle.

(at $R_{bound} = 6300 - 670 = 5630$ km) corresponds to the area of a spherical surface $S_2 = 4\pi R^2_{bound} = 4 \times 10^8$ km^2 and the length of equator on this surface $2\pi R_3 = 35{,}400$ km. The number of polygonal cells n in area S_2 is equal to 14 ($n = S_2/S_1$) or at $R_1 = 3570$, $n = 10$. Such a number of cells can be arrived at also from the data of seismic tomography in the 2600 to 2900 km layer (see Fig. 4.2), five ascending and seven descending centres. The number and form of cells become complicated in the upper part of the lower mantle (700 to 1700 km) (see Figs. 4.2a and 4.17).

If it is assumed that the ascending flows are located at the centre of polygonal cells, as emerging from experimental data, ten large mantle plumes are obtained on the basis of seismic tomography in the upper part of the upper mantle (see Fig. 4.17). These plumes match with the upward lower mantle flows. In this case, in the above sections there should be a minimum of six (at $R_1 = 3000$ km) or five (at $R_1 = 3570$ km) polygonal double cells, i.e., 10 to 12 simple cells.

Thus, six to ten flows have to be added to the four large ascending flows of Irvine's model (1991). Some can be readily assumed in oceans at the points of triple junction where magmatism suggests the presence of intense lower mantle plumes and the positive relief of core-mantle surface and 670 km boundary may point to ascending flows in the lower mantle. Among such regions are Galapagos islands, Ascension and Mascareno plateau in the Pacific Ocean, Azores and St. Helena in the Atlantic, and Afar rift and Reunion east of Africa (see Fig. 1.3). Such points may also be present under the largest Eurasian continent, especially the Mongolian superplume, like Okavango in Africa (Windley and Allen, 1993).

On the whole, however, it is presently difficult to determine which of the 34 mantle plumes depicted in Fig. 4.17 (they can be counted up to a hundred) represent the ascending flows in the lower mantle and which are additional or emerge from the boundary of the upper and lower mantle. As shown below, some mantle plumes arise at the interface of the upper and lower mantles and are not associated with the lower mantle plumes. Contrarily, some plumes arising at the CMB do not reach the surface; they are lost below the 670 km boundary. Just how the 'rapid' mantle plumes are related to the slow ascending flows of independent lower mantle convection needs to be discussed.

Many authors have observed that the characteristics of lower mantle convection correlate poorly with present-day plate movements; they correlate better with the configuration of ancient subduction zones (Chase and Sprawl, 1983; Fukao et al., 1994) (see Fig. 1.3). The results (Fukao et al., 1994), pointing to correlation between the total volume of subduction and geophysical data characterising the lower mantle, confirm the assumption of the existence of direct correlation between large-scale thermal heterogeneity in the lower mantle and subduction during the Cenozoic and Mesozoic. The best correlation of seismic velocity and geoid parameters (harmonic coefficients of the second and sixth order) with volume of subducted material has been found in the interval 45 to 140 million years or in an integral form from 60 to 180 million years, almost corresponding to the maximum oceanic crust newly formed from the mantle and intensity of mantle plumes. Since the interval of averaging the volume of subducted material was 30 million years, only the large-scale periodicity of 90 million years for harmonic coefficients of the first and second orders can be recognised (Fukao et al., 1994).

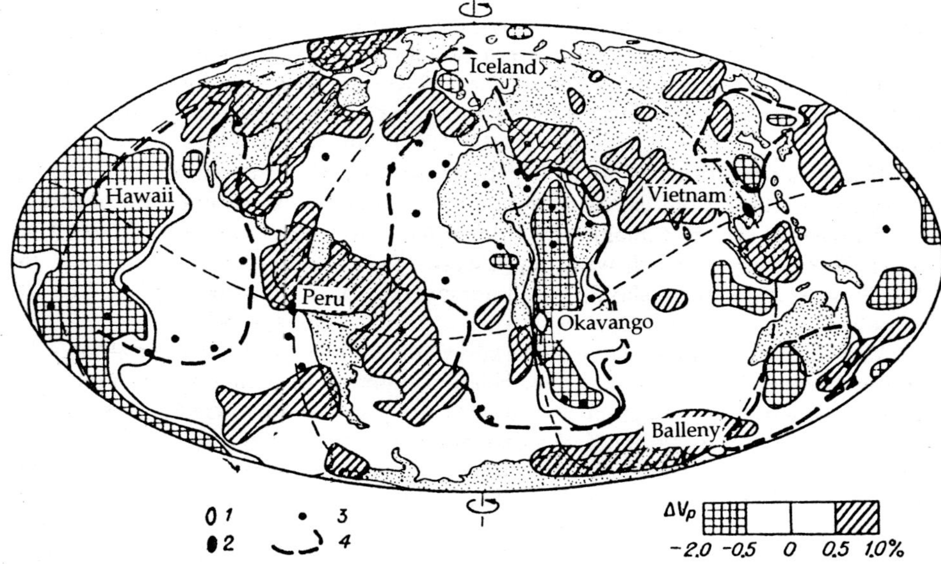

Fig. 4.17. Global seismic tomography of the lower mantle at depth 1203 to 1435 km (ΔV_p according to Inoue et al. (1990)) compared to the model of global convection in the lower mantle according to Irvine (1991) and surface manifestations of deep-level flows. Flow centres: 1—ascending, 2—descending; 3—volcanic 'hot spots'; 4—boundaries of hot fields according to Zonenshain and Kuz'min (1993a).

4.7. Mantle Plumes and Hot Spots

According to Zonenshain and Kuz'min (1983, 1993a, b) and Maruyama (1994), it is the tectonics of hot fields associated with mantle plumes that controls the global geodynamics of the Earth.

Let us examine the structure of mantle plumes. They may arise at three levels: in the upper mantle under the melting zone of subducted plates; at the interface of the upper and lower mantles at a depth of 670 km; and at the lower CMB. The nature of these plumes may be either purely thermal (during melting of the surrounding crystal material) or purely chemical (when there is a density difference between material of the plume and surrounding mass) or a combination of thermal and chemical (during partial melting of the surrounding material).

Mantle plumes emerge at the Earth's surface in the form of discharges (or intrusions) of molten magma of varying but usually basic and often even alkaline composition. The Hawaiian islands are an outstanding manifestation of such a hot spot. These islands, with their present-day volcanic eruptions, represent the concluding link of the Hawaii and Emperor chains in which volcanic eruptions range in age up to 42 million years in the Hawaiian chain and 43 to 70 million years in the Emperor chain (Fig. 4.18). These chains of volcanic islands with systematic variations in age are interpreted as the trace of movement of the Pacific Ocean plate above the Hawaiian hot spot that has existed for more than 70 million years. The geniculate bending of the submeridional Emperor range to the north-west strike of the Hawaiian chain reflects the variation in direction of movement of the Pacific Ocean plate, also fixed from the change of strike of anomalies commencing from No. 22.

The structure of the Hawaii-Emperor chain is distinctly superimpased, on the structure of the north-western part of the Pacific Ocean plate although evidently accompanied by non-volcanic (as at present) Shatsky and Hess rises. It is possible that these rises were formed during intensification of the magmatism of the Hawaiian plume 45 to 60 million years ago. Intensification of plume magmatism in the last 0 to 5 million years likewise resulted in the formation of a basalt plateau and thickening of the oceanic crust north and north-west of the Hawaiian islands. The possible plume rotations, as reflected in the semicircular arrangement of volcanoes, are shown by arrows in Figure 4.18. These changes in the configuration and sizes of plumes are discussed below.

The kimberlite fields represent another important example of the manifestation of mantle plumes. These are discussed in detail in Section 4.10.

The period of activity of present-day plumes and those that existed in the Mesozoic extends from 15 to 90 million years (Larson, 1991; Stothers, 1993). For example, the age of kimberlite fields in South Africa, reflecting probably the course of movement of the African plate above two hot spots, is dated 200 to 110 and 100 to 70 million years (Skinner et al., 1992). Definite patterns and analogies with the behaviour of Sun spots are noticed in the temporal and spatial disposition of hot spots during the Mesozoic (Simon and Weiss, 1991; Stothers, 1993): hot spots are localised in the middle latitudes $40 \pm 15°$ on the Earth and $30 \pm 10°$ on the Sun; new hot spots appear at high latitudes in both the hemispheres and migrate towards the equator, after which a new cycle commences (Fig. 4.19) and intensely affects the magnetic field (Stothers, 1993). On the Earth the period of such a cycle is 90 or 180 million years and on the Sun is 11 years. The short periods and more uniform distribution of spots on the Sun is not the only difference in the deep-level circulation on the Earth and on the Sun as determined by the difference in the matter of the gaseous Sun and highly viscous Earth's mantle, although several similarities are amazing (Simon and Weiss, 1991; Stothers, 1993). The origin of Sun spots and their migration towards the equator are mainly determined by Coriolis forces which are maximum in the high latitudes but disappear at the equator. On the basis of analogy with the Sun spots, manifestation of Coriolis forces is possible in the relatively low-viscous liquid core. This in turn points out that the core-mantle boundary represent the region of formation of hot spots.

From these and other facts, it becomes increasingly clear that plumes and multi- or whole-mantle convection are not alternative phenomena but combine to various extents in different life periods of the Earth, the intensity of mantle plumes probably emerging as a regulator of convection: in the period of their maximum intensity (e.g., in the period of the Cretaceous 'superplume') according to Larson (1991), general convection predominated and, in the period of their minimum intensity, multilayer convection is more distinctly manifest; on the whole, convection in the

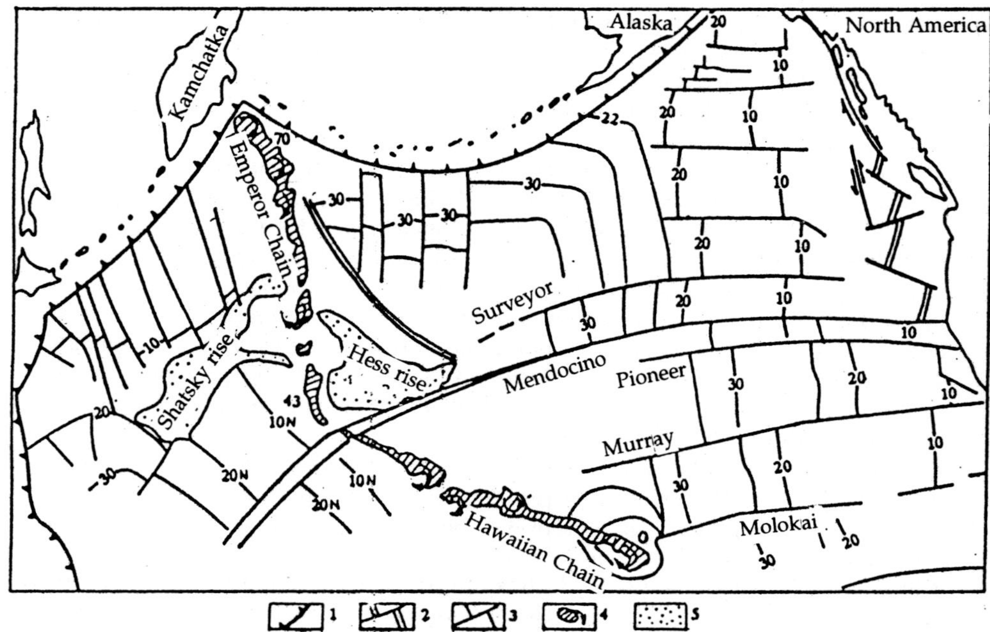

Fig. 4.18. Course of the hot spots of the Hawaii-Emperor chains (Kononov, 1989, modified). 1—subduction zone, Pacific Ocean plate (POP) boundary; 2—axis of spreading boundary of POP and North American plate; 3—magnetic anomalies, their numbers and transform faults; 4—Emperor-Hawaii chain of volcanoes; 5—Shatsky and Hess rises. Explanations in text.

Earth is unstable and unsteady, this being due to the high Rayleigh numbers (Ra) (Trubitsyn and Kharybin, 1988; Kellog and Turcotte, 1990; Trubitsyn and Nikolaichik, 1991; Kirdyashkin et al., 1994).

These relations could have undergone variation throughout the Earth's history. For example, McCulloch (1993) has suggested that in the Archaean (up to 2.6 billion years), when the Earth was hot, the subducted plates penetrated only to a depth of 200 km, the upper mantle had low viscosity and did not mix with the lower mantle and multilayer convection arose (Fig. 4.20). After the Archaean, the temperature of the upper mantle dropped, cold and heavy lithospheric plates began penetrating into the lower mantle, whole-mantle convection began to predominate and was disrupted by mantle plumes (see Fig. 4.20B). McCulloch's model does not take into consideration the fact that, around 1.8 billion years, the situation could have changed afresh when multilayer convection arose, which in the Cretaceous during 'superplume' could once again have changed into predominantly whole-mantle convection (Larson and Olson, 1991; Dobretsov and Kirdyashkin, 1993).

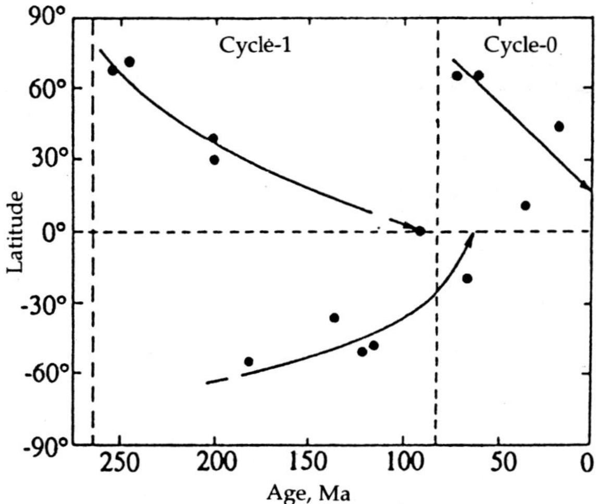

Fig. 4.19. Periodicity of manifestation of hot spots during the Mesozoic and their migration to the equator in one cycle. The boundary of two possible cycles is shown (Stothers, 1993) with minor modifications.

The foregoing review shows that probably the most important regulator of internal flows in the Earth, in any case in the preceding two billion years of its history, is the phenomenon of periodic mantle plumes arising at the core-lower mantle boundary. The conditions of their formation are not known for certain as yet. It is only known that the amplitude of the 'relief' at the CMB attains 10 to 20 km (one-half that at the 670 km boundary) and that volatile light constituents, primarily hydrogen, may have been dissolved in the liquid core (Trubitsyn and Nikolaichik, 1991; Schmalze and Hansen, 1993). Their separation from the core in the process of convection and accumulation in layer D may have given rise to gravitational instability, i.e., separation of plumes due to deficits in low-density matter enriched with fluid. Another purely thermal nature of plumes caused by local heating in the core salients is also possible.

In any case, mantle plumes arising at the CMB and probably enriched with hydrogen cease or are modified at the upper-lower mantle boundary around 670 km, as suggested in the model of Ringwood et al. (Ringwood and Irifune, 1988; Ringwood et al., 1992). We also hypothesise that the major plumes may have broken away from the lower part of the molten subducted plate and been away from the lower part of the molten subducted plate and been responsible for the formation of back-arc and interarc basins (Dobretsov et al., 1993).

To resolve these issues, experimental and theoretical study of the conditions of formation, movement, stability or disintegration of mantle plumes along with

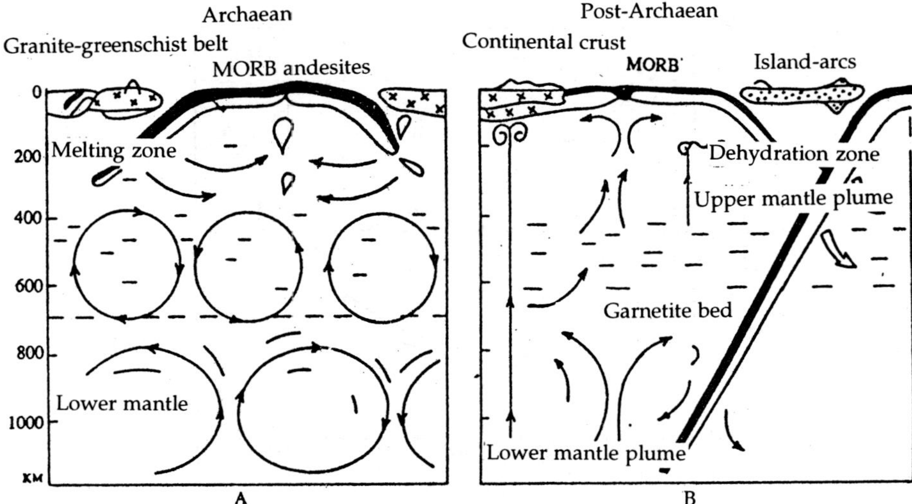

Fig. 4.20. The regimes of Archaean and post-Archaean mantle convection (McCulloch, 1993); limited (up to 200 km) subduction and three-layer convection are characteristic of the Archaean (A); the picture of subduction and convection in the post-Archaean period approximates the present one (B). (Reprinted with kind permission of Elsevier Science -NL, Sara Burgerhartstraat 25, 1055 KV Amsterdam, The Netherlands.)

studies on the general problem of stability (unsteady state) of mantle convection are essential. This aspect is discussed in successive sections of this chapter.

4.8. Experimental Study of Plumes (Melting within a Crystalline Body above a Local Heat Source)

As mentioned before, the nature of mantle plumes can be either purely thermal (with melting of the surrounding crystal material) or purely chemical (when there is a density difference between the material of the plume, as for example enriched with fluid, and the surrounding mass) or a combination of thermal and chemical nature (with partial melting of the surrounding material). It is difficult to judge the nature of these plumes from their manifestation at the Earth's surface (Dobretsov and Kirdyashkin, 1993).

Let us first examine the structure of thermal plumes arising as a result of the melting of a crystalline body in the lower part of which a local heat occurs. Published researches have been mainly devoted to studying the process of melting along an outer generatrix of a crystalline bulk in the form of a cylinder (Ho, 1984; Charoki and Sparrow, 1987) or a flat surface (Okada, 1984; Woon-Shing, 1989).

Kirdyashkin et al. (1987) reported the results of studying melting within a crystalline bulk above a local heat source placed within the bulk. It was found that during the melting of a channel of height far exceeding the horizontal size of the heat source, the molten volume had an axis of symmetry independent of the configuration of the local source. Melting within a crystalline bulk by a horizontal cylinder type heat source at small heights of melting comparable to the channel diameter was studied by Rieget et al. (1982). They indicated the need to take into account the free convective heat transfer even during small intervals from the commencement of melting. Wu et al. (1989) reported the results of studying melting around a vertical cylinder placed in a crystalline bulk of height equal to that of the heat source.

Our experiments in octadecane, eicosane and paraffin (mixture of paraffins from $C_{12}H_{26}$ to $C_{56}H_{114}$ and paraffin oil) showed that a melting centre in the form of a channel or jet is formed above the local heat source with temperature (T_s) higher than the melting temperature of the substance ($T_s > T_1$) (Fig. 4.21a).

The experimental unit is a rectangular stainless steel container with double walls between which water was pumped. The front wall, made of organic glass, is transparent. The heater is placed along the front wall at the base of the container, in the midsection along its breadth. The container ($120 \times 150 \times 370$ mm) was filled with the melt and cooled. When the heater was turned on, the process of melting and flow structure could be observed through the transparent wall.

In the initial period ($t = 120$ to 230 min, see Fig. 4.21a), a cylindrical channel formed. The diameter of the channel showed little variation and almost the whole of the heat supplied was expended in melting its upper portion. The process was unsteady since part of the heat was expended in heating the surrounding mass and part in melting and heating the liquid phase. In the limiting case, the plume reached the upper boundary and the amount of heat entering the melt from the heating source was wholly transmitted to the surrounding mass ($t = 420$ to 520 min).

The temperature profiles in the octadecane melt are shown in Fig. 4.21b. Its physical properties have been given by Vargaftik (1972). Temperatures were measured with thermocouples 0.06 mm in thickness at the free boundary of melt and air. Heat is dissipated into the surrounding medium along the cylindrical phase transition, expended in melting in the upper part of the channel, and transmitted into the air at the free surface.

The channel diameter remains unchanged as long as the melting of the channel proceeds in its upper portion. Further, a boundary layer regime is observed in the channel and an ascending jet-like flow is formed along the channel axis and descending flows in the boundary layer at the phase boundary (see Figs 4.21, 4.23A).

Temperature measurements (see Fig. 4.21b) show that in the region of the jet flow, whose diameter is much smaller than that of the channel diameter, a temperature reduction can be seen along the height. An unstable stratification ($\partial T/\partial x < 0$) is evidenced in the jet and stable stratification in the region between the jet flow and the boundary layer where velocity u is close to zero. The stable stratification results from the fact that the temperature in the upper part of the layer is higher than that in the lower ($\partial T/\partial x > 0$). Along the channel walls, local melting as well as crystallisation are possible depending on the variation of the flow regime in the boundary layer at the interface with the formation of secondary cellular flows (Fig. 4.25).

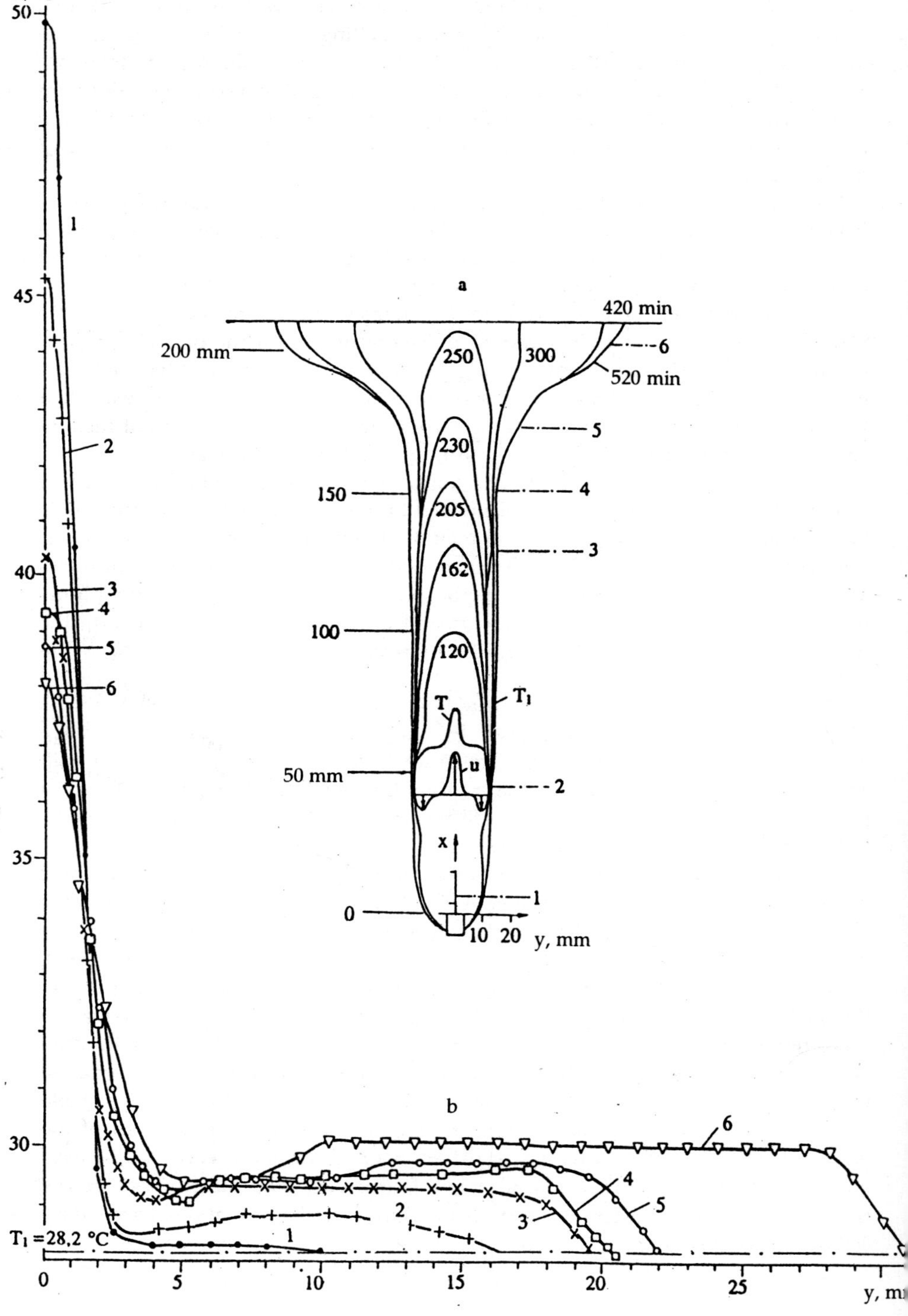
T, °C
50
45
40
35
30
T₁ = 28,2 °C
1
2
3
4
5
6
a
420 min
200 mm
250
300
6
520 min
5
230
4
150
205
3
100
162
120
T
T₁
50 mm
u
2
x
1
0
10 20 y, mm
b
6
30
4
5
3
2
1
0 5 10 15 20 25 y, mm

The boundary layer regime for a flat vertical layer exists for Rayleigh numbers (Kirdyashkin et al., 1971)

$$Ra = \beta g \, \Delta T \, l^3/av > 10^4.$$

For an axisymmetrical flow, a boundary layer regime can be expected for Rayleigh numbers:

$$Ra_r + (\beta g \Delta T/av) \, (D(x)/2)^3 > 10^4, \tag{4.13}$$

where $D(x)$ is the channel diameter at height x. It can be seen from Figure 4.21 that at $t = 300$ to 520 min, $Ra_r = 7.7 \times 10^5$.

Channel variations in case of melting in a paraffin body ($T_1 = 52°C$) at constant heater capacity ($N = 9$ W) and its diameter $D_h = 5$ mm are shown in Figure 4.22. During melting of the channel, as shown in Figure 4.21, a free convection flow is observed in the melt in the boundary layer regime. As the upper boundary of the crystalline layer is reached, the melting zone in the upper portion widens to dimensions when the heat is transferred to the surrounding mass and to air ($t = 150$ to 250 min). In this case, a thermal conduction regime is observed at the bottom of the channel (see Figs. 4.22 and 4.23B) when the heat transmission from the jet to the wall proceeds by thermal conduction. In the thermal conduction regime the temperature profile in the cylindrical co-ordinates corresponds to heat conduction, but the velocity profile in the channel is convective. In the upper part of the channel, however, the boundary layer regime may persist. Figure 4.23 depicts these regimes schematically. The situations depicted in Figures 4.20 to 4.23 may be relevant to mantle plumes and to magma channels in the upper part of the mantle and in the crust. The interface shown in Figures 4.21 and 4.22 may originate in particular in the mantle or crust, on reaching a layer with a high melting point or when a fluid layer occurs in the upper part of the channel. In this case, as shown in Figures 4.21 and 4.22, the size of the upper part of the channel increases and the melting zone adjacent to the phase boundary enlarges due to the appearance of mushroom-like forms. Such a situation may prevail at the upper-lower mantle interface or at the contact with carbonate rocks in the crust. A similar situation is possible at the magma chamber of a volcano.

Profiles and sections of channels obtained by melting paraffin using a heat source of $N = 3$ to 7 W capacity and constant temperature at the outer boundary of the paraffin mass are shown in Figure 4.24. Experimental investigations were carried out in a cylindrical body of 200 mm diameter with thermally controlled wall. The interface of the liquid and solid phases (i.e., channel form) was determined in the following manner; the cylinder was first filled with liquid paraffin and cooled for two days; the heater was subsequently turned on the moment the channel height underwent no further change. The melt discharged rapidly through

---

Fig. 4.21. Melting in an octadecane bulk above a local heat source. a—change of phase boundary with time at heating source capacity $N = 5$ W, heater diameter $D_h = 6$ mm, container temperature $T_w = 24°C$; b—temperature profile in the melt for different heights x (mm): 1—6.3; 2—43.8; 3—128.3; 4—151.2; 5—172.6; 6—201 (sections 1 to 6 are shown in Fig. 4.21a).

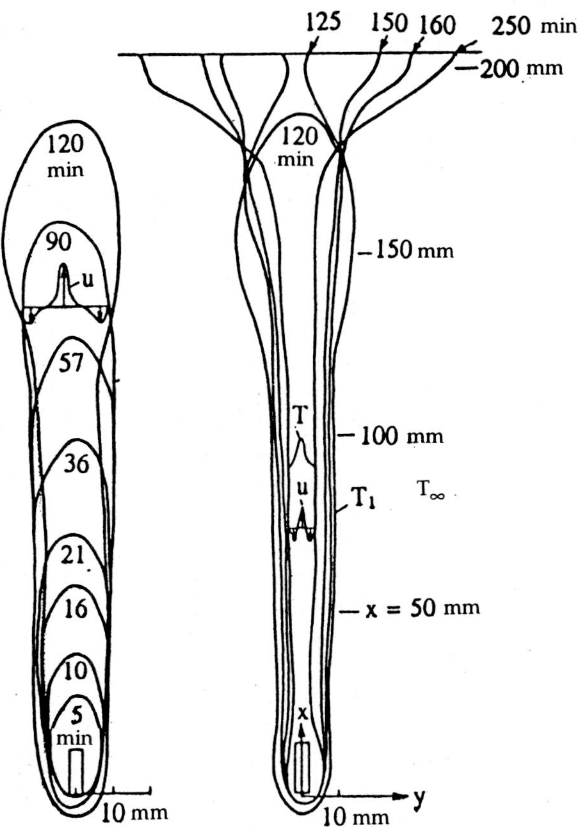

Fig. 4.22. Interface variation in time within the melt-polycrystalline paraffin mass above a local heat source with D_h = 5 mm at N = 9 W and T_w = 22°C and change of free convection regimes in the channel.

the lower orifice and the paraffin cylinder was sectioned in the longitudinal and lateral sections to determine the form of the molten channel. Depending on the heat source power the form of the molten channel varied, this being associated with the different regimes of free convection in the melt. At heat source power N = 3 W (see Fig. 4.24a), a thermal conduction regime was essentially observed. At N = 4 W (see Fig. 4.24b), the thermal conduction regime persisted in the lower part (zone B). In the region of the thermal conduction regime, the axisymmetrical flow was often not sustained, as can be clearly seen in Figure 4.24c at the same value N = 4 W (zone C): ascending flow was observed in one wall and descending on another. The cross-section of the channel had an asymmetrical ellipsoidal form (see Fig. 4.24c, d, e). This denotes the absence of a steady heat transfer regime in the channel and melting on one side of the channel and crystallisation on the other.

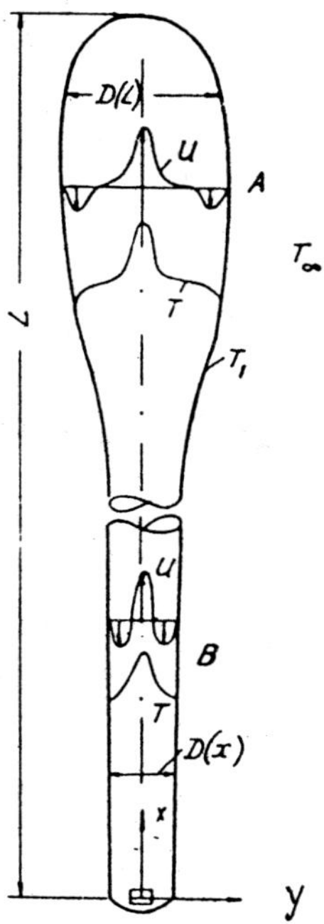

Fig. 4.23. Scheme of free convection flow regimes in the melt. A—boundary layer regime; B—thermal conduction regime.

In the limiting regime, when the channel height no longer changed, the regime became quasi-steady: the amounts of molten and crystallised material were the same but the shape and location of the channel changed continuously near the 'median' axis parallel to the gravity vector (see Figs. 4.24 and 4.25).

On increasing the capacity of the heat source (see Fig. 4.24d), the complex unsteady flow structure in the channel became even more distinct: a boundary layer regime was observed in the upper part and a thermal conduction regime with asymmetric and periodically varying flow structure in the lower. Jumping of the ascending flow from one side to the other was evidenced at different levels of the channel. At maximum capacity of the heat source (see Fig. 4.24e), the maximum channel height L also increased.

Since instability and periodic displacement of the channel in the horizontal plane were established for the first time and could be very significant in the geological context, special investigations were carried out on such instability, as shown in Figure 4.25. Additional experiments were carried out in eicosane at constant heating power (7 W) and diameter of heater (15 mm) in the unit with a transparent front wall (design explained earlier). The physical properties of eicosane were taken from Vargaftik (1972). Melting was initially carried out for 8 h;

156

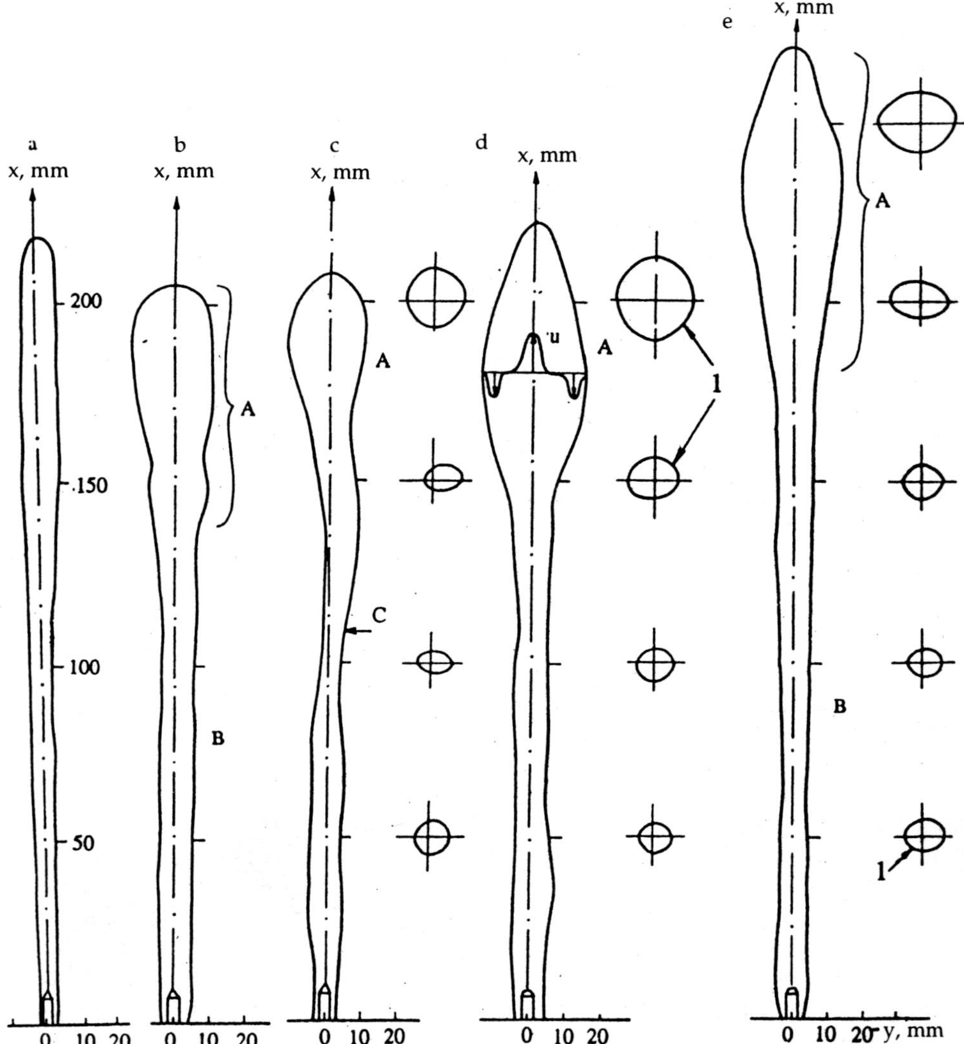

Fig. 4.24. Boundary of the separation of melt-polycrystalline paraffin mass on melting above a local heat source with $D_h = 4$ mm when all the heat enters the paraffin body.
Zones: A—boundary layer regime; B—thermal conduction regime; C—asymmetrical flow: a) $N = 3$ W, $T_w = 23°$C; b) $N = 4$ W, $T_w = 22°$C; c) $N = 4$ W, $T_w = 21°$C; d) $N = 5.6$ W, $T_w = 24.5°$C; e) $N = 7$ W, $T_w = 21°$C, where T_w is the temperature of the cylindrical container ($D = 200$ mm). 1—Channel cross-section for different heights (x).

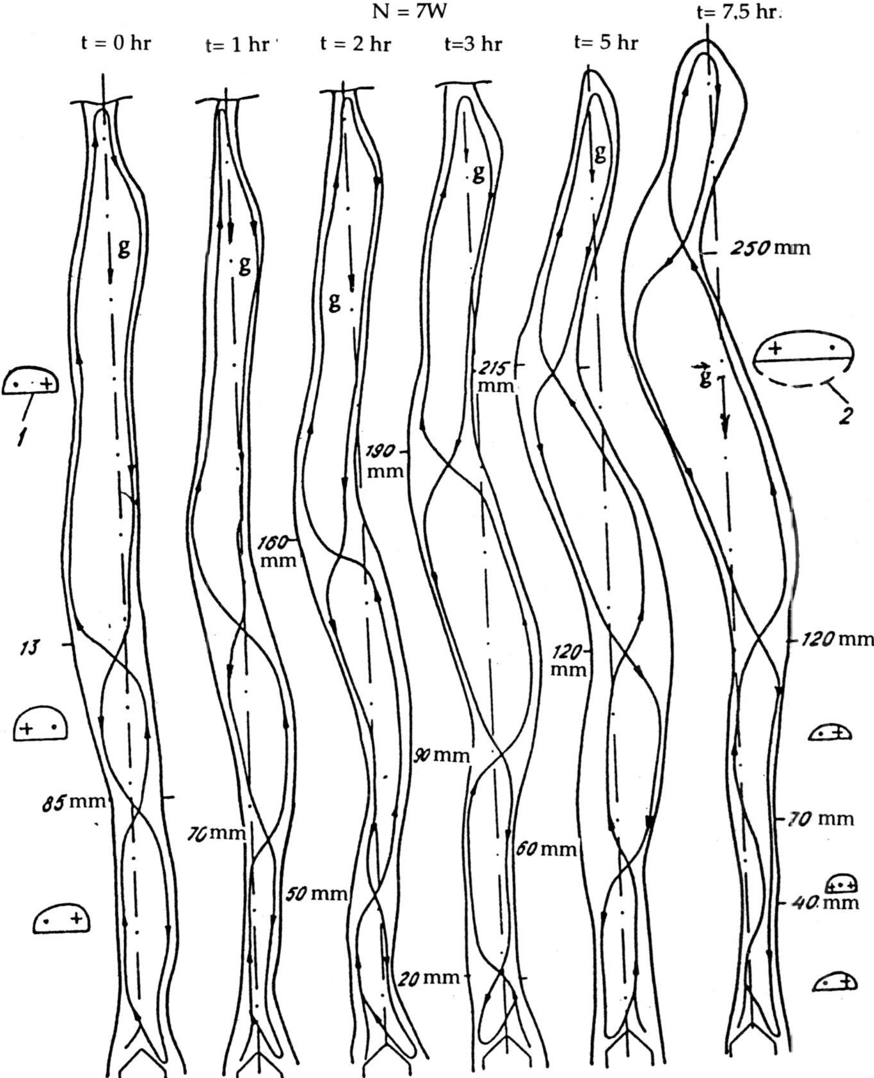

Fig. 4.25. Nature of interface variation with time in melt-polycrystalline mass (eicosane) on melting over a local heat source: D_h = 15 mm, N = 7 W and T_w = 21°C; 1—cross-section of the channel for different heights (x); g—gravity acceleration. Other symbols explained in text.

the liquid was discharged from the channel, the channel filled again the next day with eicosane and the experiment continued under the same heating conditions for another 7.5 h. Thus the duration of the experiment could be extended as desired. The channel section and flow direction in it recorded for different periods of additional heating are shown in Figure 4.25.

In all cases (see Fig. 4.25), unsteady flow is seen and periodic fluctuations are observed in space and time. The flow is divided along its height into cells of length 2.5 to 4 times more than the channel diameter. At the cell boundaries displacement of the flow from one wall to another and their partial mixing are observed. On the basis of experimental results, the channel can be regarded as a travelling wave with helical rotation around the vertical passing through the source. The amplitude of deviations of the phase boundary from the axis ranges from 1 to 2 diameters and increases upwards along the channel. Figure 4.25 shows the cross-sections of the channel (1) where the straight line represents the boundary of the glass plate (glass wall in front) since the heater was close to it. When the heater was situated far from the plate, the channel was symmetrical and had an elliptical form (2); its small axis was one-half of the large one. Figure 4.25 depicts the following conditions: Ra = 8.6 $\times$ 10^4 and ΔT_1 = 28.6°C (T_s = 65.5°C and melting point T_1 = 36.9°C) and channel radius r = 12/2 = 6 mm. At such Ra values and Pr = 45.6, the boundary layer regime of free convection flow was established in the melt channel.

Similar patterns are characteristic of surface manifestations of mantle plumes when the lithospheric plate above them comes to a stop or is relatively less displaced (see Figs. 4.18 and Fig. 4.33).

4.9. Theoretical Estimations of Mantle Plume Parameters

Let us estimate the stability of plumes (channels) in the mantle, their possible parameters and formation conditions utilising the above experiments and known relations for convection.

First, we analyse quantitatively the limiting steady regime in a plume channel when heat is transmitted to the surrounding crystalline bulk and the height of the channel is the maximum possible for a given heat power. The interface of the melt-solid body can be regarded as a cylinder of rotation of diameter $D(x)$ depending on vertical co-ordinate x. The melting point of the crystalline matter is adopted as T_1 at pressure P = const. Let us examine the condition when the cylinder diameter $D(x)$ is much smaller than the height of the molten body $D(x)$ /L << 1 (see Figs. 4.21 to 4.23). In this case, heat transfer results in heat dissipation from the cylindrical surface at $T = T_1$ into an infinite solid volume whose temperature $T = T_\infty$ at y >> $D(x)$. It is known (Mikheev, 1947; Kutateladze and Borishanskii, 1959) that in this case, the dimensionless value of the heat transfer coefficient (Nusselt number, Nu) has a constant value:

$$\text{Nu} = \alpha D(x)/\lambda_t = qD(x)/\Delta T_1\lambda_t = 0.5, \tag{4.14}$$

where $\alpha = q/\Delta T_1$ is the heat transfer coefficient, q the specific heat flux from cylindrical surface of diameter D, $\Delta T_1 = T_1 - T_\infty$ the temperature drop between the surface and the crystal material and λ_t the thermal conduction coefficient of the crystalline body. Equation (4.14) is valid for values of Fourier criterion body (Fo = at/r^2 > 0.04 to 0.05), where a is the thermal diffusion coefficient, t the time reckoned from the moment of heat supply and $r = D/2$. For example at a = 10^{-6} m^2/s and r = 10^3 m, t = 0.05 r^2/a = 1.6 $\times$ 10^3 years.

It follows from eqn (4.14) that

$$q(x) = 0.5\ \lambda_t\ \Delta T_1/D(x).\tag{4.15}$$

In this case a smooth change of $D(x)$ along the channel height has been assumed. At $\Delta T_1 = \text{const}$ and $\lambda_t = \text{const}$,

$$q(x)\ D(x) = 0.5\ \lambda_t\ \Delta T_1 = \text{const}.\tag{4.16}$$

For an asymptotic variation of the melting process of the channel with time at fixed capacity N of the heat source, let us determine the heat supplied to the solid mass and the maximum height of the molten channel L considering that $D(x) \ll L$,

$$N = \int_0^L q(x)\ \pi\ D(x)\ dx.\tag{4.17}$$

Taking into consideration eqn (4.16), we derive from eqn (4.17):

$$N = 0.5\ \pi\ \lambda_t \Delta T_1 L;\ \ L = N/0.5\ \pi\ \lambda_t\ \Delta T_1.\tag{4.18}$$

It can be seen from eqn (4.18) that the height of the molten channel at $\lambda_t = $ constant depends on the capacity of the heat source and temperature drop ΔT_1 between the interface and the body, but is independent of $D(x)$. The diameter of the channel being melted influences the melting time.

It must be remembered that at maximum height L, the form of the interface at the upper end of the channel corresponds to a hemisphere from which heat transmission into an infinite volume is determined from the relationship $\text{Nu} = 2$ (Kutateladze and Borishanskii, 1959). Therefore, eqn (4.18) is relevant at $L \gg D(L)$. The results of calculations using eqn (4.18) for different parameters are shown in Figure 4.26. In particular, at $L = 670$ km, $N \approx 5 \times 10^6$ kW, and at $L = 2900$ km, $N \approx 5 \times 10^7$ kW, suggesting $\Delta T \sim 10^3\ ^\circ$C and λt is different for L_1 and L_2 (lines 3 and 1 in Fig. 4.26).

Let us estimate the heat source power depending on its size and temperature. For melting in a crystalline body, the temperature of the source (T_s) is more than the melting point ($T_s > T_1$). Without going into details of the possible configuration of the heat source, we represent it as a horizontal surface with temperature T_s of the surface in contact with the melt at increased average temperature $\overline{T}_0$ above the surface (Fig. 4.27). It must be pointed out that the temperature of the descending flow acting on the heat dissipating surface is proximate to that of melting point T_1. However, the average temperature will be high at height l_h of turbulent flow adjacent to the wall (see Fig. 4.27). Therefore, by adopting the temperature of the liquid above the surface T_1, we raise the heat flux from the source, i.e., the temperature drop at the height of the turbulent layer $\Delta T_0 = n(T_s - T_1)$, where $n < 1$. Let us examine the heat transfer between the horizontal surface and melt in natural convection in the case of heating (from below) a large volume of melt with a horizontal surface.

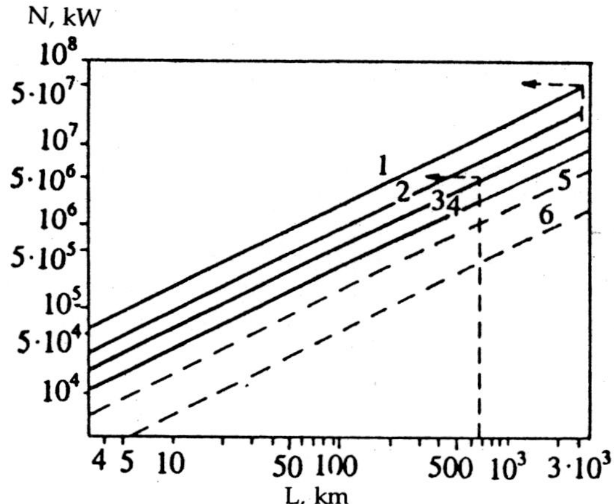

Fig. 4.26. Maximum height of the melted channel L against the capacity of the heat source calculated using eqn (4.18):
1—$\lambda_t = 12$ W/m · °C, $\Delta T_1 = 10^3$ °C; 2—12 W/m · °C, 500°C; 3—4 W/m · °C, 10^3 °C; 4— 4 W/m · °C, 500°C; 5—12 W/m · °C, 100°C; 6—4 W/m · °C, 100°C.

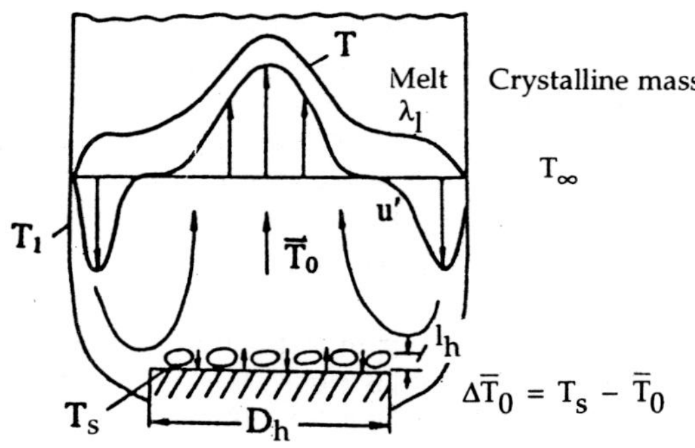

Fig. 4.27. Scheme of convection free flows above the heat source.

The heat transfer law in this case will be as follows (Leont'ev and Kirdyashkin, 1965):

$$Nu = 0.1\ Ra^{1/3}, \tag{4.19}$$

where Ra is the Rayleigh number determined from the temperature drop between the heat transfer surface and liquid away from the surface, $Ra = \beta g\ \Delta\overline{T}_0 D_h^3/a\nu$ and Nu number is equal to $qD_h/\Delta\overline{T}_0\lambda_l$, where λ_l is the thermal conduction of the melt.

According to eqn (4.19), the specific heat flux from the surface and power of heat source are respectively equal to:

$$q = 0.1\ \lambda_l\ \Delta\overline{T}_0{}^{4/3}\ (\beta g/av)^{1/3},$$

$$N = q\pi(D_h^2/4) = (0.1\pi/4)\ \lambda_l\ D_h^2\ \Delta\overline{T}_0{}^{4/3}\ (\beta g/av)^{1/3}. \tag{4.20}$$

Let us determine the heat source power relative to its sizes and melt parameters. The physical properties of the melt are accepted as follows (Zharkov, 1983): the assumed melt density at great depths $\rho = 4000$ kg/m^3, coefficient of volume expansion $\beta = 3 \times 10^{-5}$ °C^{-1}, $\lambda_l = 4.4$ W/m · °C and $a = 10^{-6}$ m^2/s. The dynamic viscosity of volcanic magma in field measurements $\eta = (1.40 \times 10^2$ to $3 \times 10^4)$ poises and thus the kinematic viscosity $v = \eta/\rho = (3.5 \times 10^{-3}$ to $0.75)$ m^2/s.

At small temperature difference $T_s - T_1$, even greater viscosity values are possible. Let us therefore estimate the variants at $v = 1$, 10 and 10^2 m^2/s. For the adopted values of the physical properties of the melt, the heat source power was calculated relative to D_h and $\Delta\overline{T}_0$. The results of calculations are depicted in Figure 4.28. The minimum power for melting the required height and the minimum diameter of the heater required for generating this heat energy can be evaluated by comparing Figures 4.26 and 4.28. From Figure 4.26, we find $N \approx 5 \times 10^7$ kW for the core-mantle boundary. This estimate can be verified from the amount of heat carried by the plume based on geological and geophysical data given by Larson and Olson (1991). According to these estimates, the amount of heat transmitted the surface with plumes and the heat required for melting this magma is 1.7×10^6 to 1.7×10^7 kW. To this power should be added the heat dissipated into the surrounding volume at $L = 2900$ km (CMB): 2×10^6 to 5×10^7 kW or from depth 670 km: 5×10^5 to 5×10^6 kW (Fig. 4.26). Thus, during the formation of plumes at the CMB, the power of the heat source is 4×10^6 to 7×10^7 kW. In this case, according to Figure 4.28, at $\Delta\overline{T}_0 = 25$ to 100°C and $v = 1$ to 10^2 m^2/s (lines 3 to 6), the diameter of the heat source D_h will be 2 to 25 km. This is close to the above-stated difference of about 10 km in the relief at the CMB where the relief has a hemispherical form.

It can be seen from Figure 4.28 that temperature differences between the heat source and the melt above it (melting temperature of the rock) are not great. At $\Delta\overline{T}_0 = 400$°C, the required size of the heat source is small and measures 2.5 km. The Rayleigh number value for $D_h = 10$ km, $\Delta\overline{T}_0 = 50$°C and $v = 100$ m^2/s, according to eqn (4.13), will be:

$$Ra_r = \rho\ \Delta\overline{T}_0 g r^3/8av = 10^{14},\ \text{where}\ r = D_h/2.$$

This points to the existence of a boundary layer regime. The flow pattern of this regime was depicted in Figure 4.23. In the central part is seen an ascending jet flow and along the generatrix of the channel, a descending flow (in the boundary layer at the interface, much smaller channel diameter). The channel melting time depends on the heat source power and depth of disposition of this source.

Let us determine the duration of channel melting relative to the parameters of the heat source and the surrounding volume. We assume that an insignificant variation of channel diameter along the height or the average value of diameter D can be adopted for a given height interval. Let us study a case wherein $x \gg D$ (Fig. 4.29).

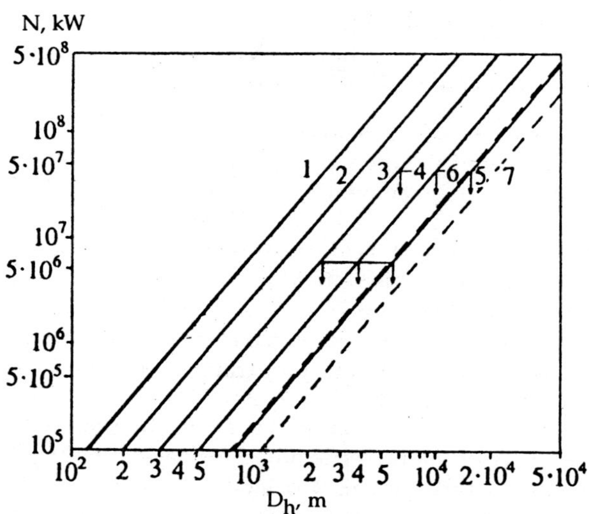

Fig. 4.28. Dependence of the heat source power capacity on its diameter D_h and temperature drop $\Delta\overline{T}_0 = \overline{T}_0 - T_s$ at $\lambda_l = 4$ W/m · °C according to eqn (4.20):
1—$\Delta\overline{T}_0 = 400$°C, $v = 1$ m²/s; 2—200°C, 1 m²/s; 3—100°C, 1 m²/s; 4—50°C, 1 m²/s; 5—25°C, 1 m²/s; 6—50°C, 10 m²/s; 7—50°C, 100 m²/s.

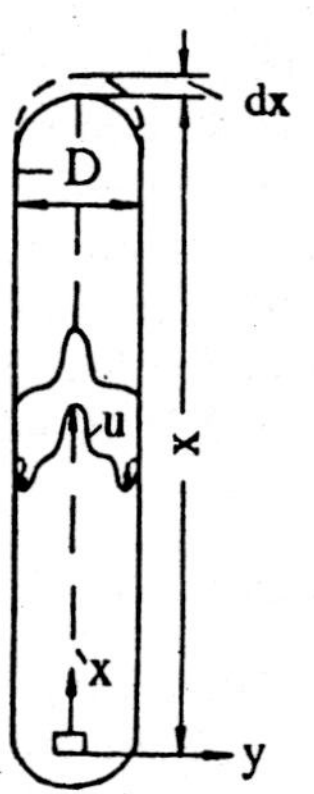

Fig. 4.29. Sketch showing unsteady melting of channel above a local source of heat.

The power of the heat source expended in melting N_n will be determined as:

$$N_n = N - N(x),\qquad(4.21)$$

where N is the capacity of the heat source and $N(x)$ the heat energy transmitted to the crystalline body at channel height x, i.e.,

$$N(x) = q\,\pi L(x).\qquad(4.22)$$

Taking into account eqn (4.18), value $N(x)$ is determined using this equation:

$$N(x) = 0.5\,\pi\,\lambda_t\,\Delta T_1\,x.\qquad(4.23)$$

By substituting eqn (4.23) into (4.21), we derive:

$$N_n = N - 0.5\,\pi\lambda_t\,\Delta\overline{T}_1 x.\qquad(4.24)$$

Let us derive elementary balance between heat energy expended during period $dt(N_n dt)$ and the amount of heat required for melting the channel by height dx (see Fig. 4.29):

$$N_n dt = (\pi D^2/4)\,\rho\,(B + c\,\Delta\overline{T}_0)\,dx,\qquad(4.25)$$

where $\pi D^2\,\rho B\,dx/4$ is the amount of heat expended in melting and $(\pi D^2/4)\,(\rho c\,\Delta T_0\,dx)$ the amount of heat required for heating the melt by value $\Delta\overline{T}_0$; B the heat of melting; c the heat power and ρ density of the melt. The rate of melting of the channel can be determined using eqns (4.24) and (4.25):

$$u_n = dx/dt = 4\,(N - 0.5\,\pi\lambda_t\,\Delta T_1 x)/\pi D^2\rho\,(B + c\,\Delta\overline{T}_0).\qquad(4.26)$$

In eqn (4.26):

$$dt = \frac{D^2\rho\,(B + c\,\Delta\overline{T}_0)\,dx}{2\,\lambda_t\,\Delta T_1\,(N/0.5\,\pi\,\lambda_t\,\Delta T_1 - x)}\qquad(4.27)$$

In eqn (4.27), member $N/0.5\,\pi\lambda_t\,\Delta T_1$ is similar to eqn (4.18) and has the dimensionality of length. It indicates the possible height of melting L.

Equation (4.27) can be rewritten as:

$$dt = D^2\rho\,(B + c\,\Delta\overline{T}_0)\,d\overline{x}/2\,\lambda_t\,\Delta T_1\,(1 - \overline{x}),\qquad(4.28)$$

where $\overline{x} = x/L$ and $L = N/0.5\,\pi\,\lambda_t\,\Delta T_1$.

Analysing eqn (4.28) we find the time required for melting the body to height x for known diameter D,

$$t = [-D^2\rho\,(B + c\,\Delta\overline{T}_0)/2\,\lambda_t\,\Delta T_1]\,\ln(1 - \overline{x}).\qquad(4.29)$$

The melting time of a channel of height x depends on the channel diameter and on the capacity of the heat source N. As N increases, according to eqn (4.18), the linear dimension L increases and the absolute value $\ln(1 - \overline{x})$ decreases for a fixed value of x.

Let us represent eqn (4.29) in dimensional form. For this purpose, we express L in terms of the heat source power according to eqns (4.18) and (4.20) and substitute this value of L into eqn (4.29), where $\bar{x} = x/L$. In this case the time required for melting the channel will be determined from the following equation:

$$t = [-D^2 \rho \ (B + c\Delta\bar{T}_0)/2 \ \lambda_t \ \Delta T_1] \ \ln$$

$$\times \ (1 - 20 \ \Delta T_1 \ x \ \lambda_t/D_h^2 \lambda_l \ \Delta \bar{T}_0^{4/3} \ (\beta g/av)^{1/3}). \tag{4.30}$$

We consider that channel diameter $D = D_h$. It follows from eqn (4.30) that on equating the logarithmic term to zero, $t = \infty$. Therefore, the minimum possible diameter of the source and the channel diameter close to it:

$$D_{\lim} = [20 \ \lambda_t \ \Delta T_1 x/\lambda_l \ \Delta \bar{T}_0^{4/3} \ (\beta g/av)^{1/3}]^{1/2}. \tag{4.31}$$

The physical properties of the mantle are similar to those used in Figure 4.28, namely,

$$\beta = 3 \times 10^{-5} \ 1/°C; \lambda_t = \lambda_l = 4 \ W/m \cdot °C; B = 210 \ kJ/kg \cdot °C;$$

$$c = 1.2 \ kJ/kg \cdot °C; \rho = 4000 \ kg/m^3; \Delta\bar{T}_0 = 50°C; \Delta T_1 = 500°C;$$

$$x = 2.9 \times 10^6 \ m; a = 10^{-6} \ m^2/s.$$

Figure 4.30 shows the dependence $D_{\lim} = f(v)$ for these physical properties. At $D < D_{\lim}$ (below the curve), $t = \infty$ and the plumes would never emerge at the surface if they had originated at the core-mantle boundary. For example, at $v = 10$ to $10^3 \ m^2/s$, $D_{\lim} = 5$ to 15 km.

The time of ascent of the plume (channel melting) relative to the diameter of the source D_h for two plume viscosities $v_1 = 1 \ m^2/s$ and $v_2 = 10^3 \ m^2/s$ is shown in Fig. 4.31.

At $D_h > D_{\lim}$, the value of t tends rapidly to its asymptotic value but depends strongly on viscosity. At $v_2 = 10^3 \ m^2/s$ and at $D = 100$ km, the melting time is 2 million years; at $D = 20$ to 100 km, $t = 2$ to 4 million years and at $D \to D_{\lim} = 15.4$ km, $t \to \infty$.

The rate of channel melting relative to the diameter of the source is determined by substituting eqn (4.20) in (4.26):

$$u = 0.1 \ \lambda_l \ \Delta\bar{T}_0^{4/3} \ (\beta g/av)^{1/3}/\rho \ (B + c \ \Delta \bar{T}_0)$$

$$- 2 \ \lambda_t \ \Delta T_1 \ x/\rho D^2 \ (B + c \ \Delta\bar{T}_0). \tag{4.32}$$

At $D >> D_{\lim}$, the second term of eqn (4.32) is small and

$$u \approx 0.1 \ \lambda_l \ \Delta\bar{T}_0^{4/3} \ (\beta g/av)^{1/3}/\rho \ (B + c \ \Delta \bar{T}_0). \tag{4.33}$$

The channel melting time is determined as

$$t = x/u. \tag{4.34}$$

The rate of channel melting and time of plume ascent for $D >> D_{\lim}$ relative to D calculated using eqns (4.33) and (4.34) are plotted in Figure 4.30. At $v = 10$ to $10^3 \ m^2/s$, $t = 440,000$ to $2,030,000$ years and at $v = 10^5 m^2/s$, $t = 10$ million years, the rate of ascent being 0.3 m/yr.

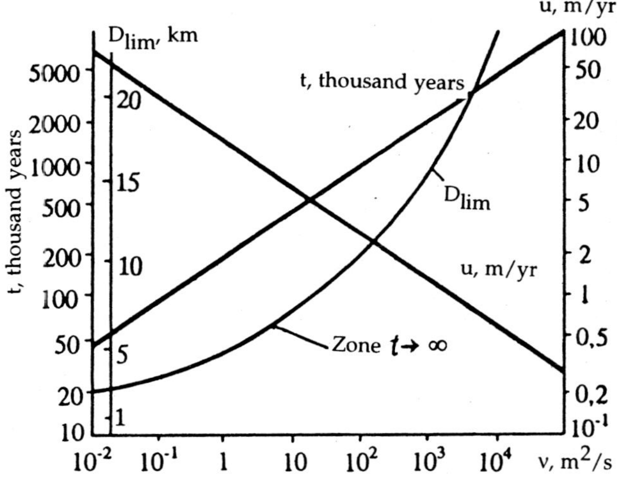

Fig. 4.30. Melting rate of channel (u) and melting time (ascent) of plume at $D >> D_{lim}$ and kinematic viscosity v according to eqns (4.33) and (4.34) and D_{lim} according to eqn (4.31).

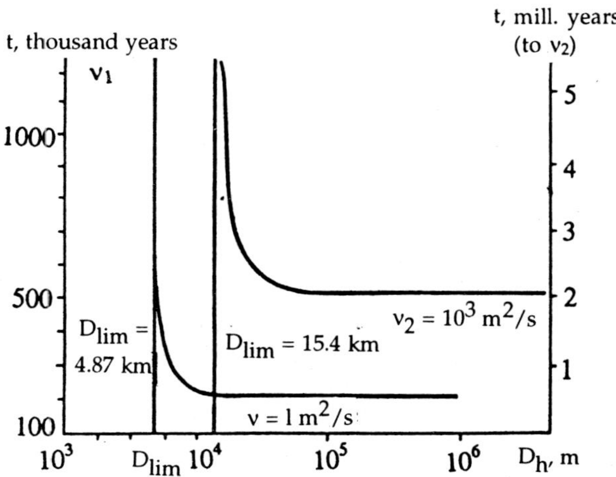

Fig. 4.31. Time of plume melting (channel melting) relative to the diameter of source for two different viscosities v_1 and v_2.

It is difficult to determine viscosity on channel melting. Direct measurements of the viscosity of igneous lavas show that $v \approx 1\ m^2/s$. It can be assumed that the composition of the plume has many constituents and, considering the possibility of partial melting, many phases, and is more viscous than a pure silicate fluid. If viscosity exceeds that of the lavas by a few orders ($v \approx 10^2$ to 10^3 m^2/s) the ascent time of this plume comprises several million years. This shows that the melting time for the mantle, on a geological scale, is insignificant and does not cause great delay from the moment of plume formation at the CMB to its manifestation at the surface. The periods of plume survival are determined mainly by processes responsible for the formation of heat sources—be they at the CMB (low mantle boundary, at a depth of 670 km) or in the subduction zone.

4.10. Geological Manifestations of Mantle Plumes

The experiments and calculations described above pertain to purely thermal plumes but the extent to which they are applicable, e.g., fluid or fluid-thermal plumes in the mantle, is the first problem yet to be solved. The investigations mentioned above show that, from the energy point of view, the existence of heat plumes is entirely real. In any case, heat plumes represent the limiting case of plumes of complex nature. The main problem for plumes of predominantly thermal character lies in elucidating the possibility of intense local heat sources with temperature exceeding the melting point of mantle rocks at a given depth by 25 to 100°C. These conditions are not exceptional. Salients at the core-mantle boundary (height 20 km) or at the boundary of the upper-lower mantle (50 km) may serve as local sources of heat. Their diameters, comparable to the height, have quite permissible dimensions according to the estimates made above. The minimum diameters of the heat sources are depicted in Figure 4.30 and are equal to 5 to 20 km.

There are three possible types of plumes: ascending from the CMB at a depth of 2900 km, from the upper-lower mantle boundary at a depth of 660 km and from the boundary of the plate subducting in the rear portion of the subduction zone from a depth of 100 to 300 km. These plumes have different scales of the process, primarily the length and time of its ascent. Estimates were made above for the lower and upper mantle heat plumes. Their ascent time is 0.5 to 5 million years. Such a small duration of ascent possibly represents the limiting value; it can also be longer under more complex conditions of melting. The small duration can, however, explain the apparent immobility of the hot spot and independent movement of plumes relative to the moving lithospheric plates. The periodicity of magnetic reversals and geological processes associated with mantle plumes (Larson and Olson, 1991; Mazaud and Laj, 1991) of the order of 30 or 15 million years also attests to a brief duration (less than 5 million years) of the ascent of mantle plumes. An independent analysis of the manifestation of the most important geological episodes over the preceding 250 million years revealed the main periodicity of 26.6 million years (Rampino and Caldeira, 1993) which, with some error, can be rounded to 30 million years. It will be demonstrated below that the periodicity of mantle plumes is 30 million years and that their minimum period is about 10 to 15 million years. It is important to note here that the periodicity of plumes of about 30 million years is possible only when the time of ascent $t \leq 5$ million years.

As in the case of most mantle processes, geological (mainly magmatic), geophysical and geochemical data are used for proving and describing mantle plumes. The most important hot spots characterised by specific alkaline-basalt magmatism and representing the probable manifestation of mantle plumes have been depicted in Figure 4.17. They are joined into hot fields (Zonenshain and Kuz'min, 1993a) which coincide with the important 'lighter' anomalies in the lower mantle ($\Delta V_p = -2.0$ to 0.5%) which are interpreted as ascending flows in the lower mantle. The relations between power of mantle plumes and magnetic field reversal (Larson, 1991; Larson and Olson, 1991) and salients of the 670 km boundary (Shearer and Masters, 1992; Larson and Kinkaid, 1996) as well as the analogy in formation and periodicity of migration towards the equator of Sun spots and hot spots of the Earth (Stothers, 1993), suggest the genesis of most large hot spots at the CMB.

Let us examine the actual data on the most important manifestations of mantle plumes, including alkaline-basaltic oceanic islands, kimberlite and plateau-basalt regions. Of the prevailing hot spots, only the Hawaiian (Fig. 4.18) possibly reveals the geochemical characteristics, especially the maximum anomalies with respect to helium isotopes (Irvine, 1988; O'Nions and Tolstikhin, 1995; Hawaii, 1996) which may prove their origin from the CMB. However, such geochemical anomalies are more often observed in the ancient manifestations of plumes (Fig. 4.37). The Hawaiian plume has been in existence for more than 70 million years (Fig. 4.18). The intensity of the Hawaiian plume, one of the most powerful among those extant, has been estimated at about 3×10^8 kW (Davies and Richards, 1992) which is close to the maximum value obtained by us (Fig. 4.28). This estimation is exaggerated, however, since it was made from a calculation of the work of the plume based on deformations of the moving lithosphere with time, which is less than the time of formation of the plume itself.

According to our calculations the average power of plumes may vary in the range 4×10^6 to 7×10^7 kW. According to Molnar and Atwater (1973), Nataf et al. (1981) and Jurdi and Stefanic (1990), the total power of all the present-day plumes is 2.5×10^8 kW, which for number of plumes $n = 36$ (Fig. 4.17) gives an average power of 7×10^7 kW and for $n = 100$, 2.5×10^6 kW per plume.

Another example of a stable plume that has existed for over a hundred million years is the Iceland plume. The trace of movement of the North American plate above the Iceland plume in the preceding 50 million years as well as the effusions of flood basalts (fb) in eastern Greenland from this plume about 40 million years ago are shown in Figure 4.32a. The earlier history of this plume and its track from Alpha ridge through Greenland or Baffin Bay have been discussed (Forsyth et al., 1986; Lawler and Müller, 1994). Figure 4.32a shows yet one more plume around the axis of the mid-Atlantic ridge, the Azores plume; other plumes of Circe, St. Helena, Tristan de Cunha, Bouvet and, close to eastern Pacific Ocean rise, Easter and Galapagos plumes occupy a similar position (Fig. 1.3). In all, 11 of the 36 plumes depicted in Figure 4.17 are localised around the mid-oceanic ridges alone.

All these ridge-centred plumes have their own specific features caused by interaction of plumes with ascending flows of the mid-oceanic ridge (Feighner et al., 1995; Ito and Lin, 1995; Rible et al., 1995). The schematic model and parameters characterising this interaction are shown in Figure 4.32b, c (Ito and Lin, 1995). The flow of mantle material incoming to the hot spot is displaced towards the ridge axis and spreads along it for distance W and the temperature under the ridge

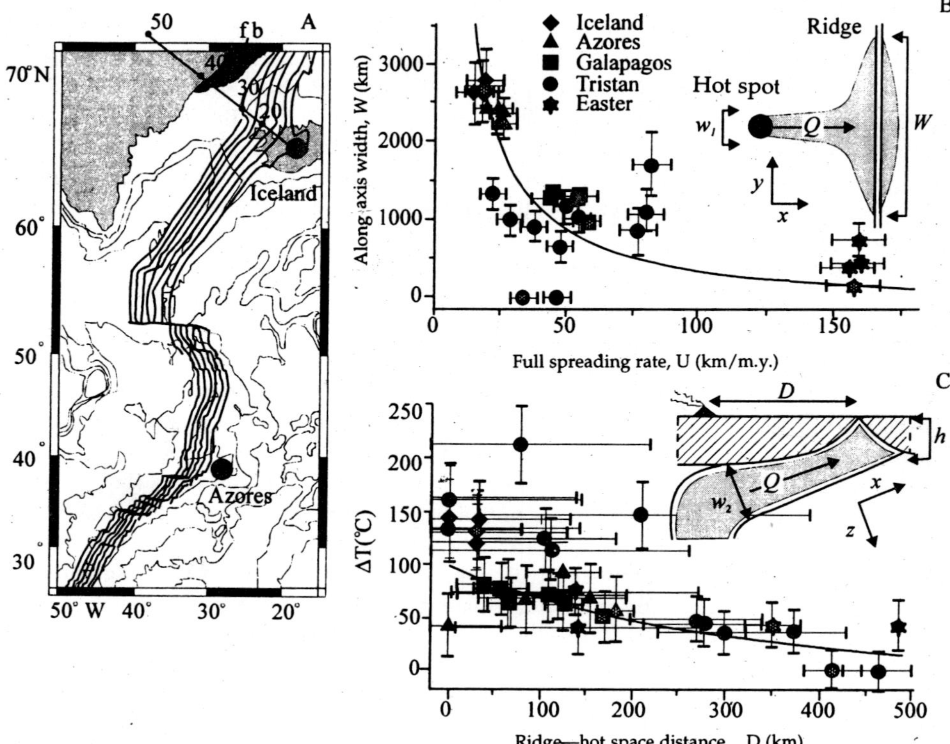

Fig. 4.32. A. Bathymetric Map of the Mid-Atlantic ridge with the Iceland and Azores hot spots: 0 to 30 Ma age isochrones and location at 30, 40 and 50 Ma for Iceland and at 0 to 25 Ma for Azores on the North American plate are shown; fb areas in eastern Greenland show Iceland hot spot flood basalts at 40 Ma.
B. Width along the axis of residual bathymetric anomalies versus full spreading rates during the time corresponding to the isochrone age solid curve in A; theoretical relation curve discussed in the text. Inset illustrates a view of plume-ridge flow pattern; dot pattern marks the plume material.
C. Temperature anomalies along the isochrone derived from residual bathymetry and gravity anomalies and plotted against D. Solid curve predicted by conductive cooling. Inset illustrates the deep cross-section of plume conduit between a hot spot and ridge axis (modified from Ito and Lin, 1995).

increases by value ΔT. It can be seen from this model that the flow of mantle material Q through a channel of section $w_1 w_2$ is equal to:

$$Q = w_1 \, w_2 \, (V - U/4), \tag{4.35}$$

where V is the average velocity of flow towards the ridge and $w_1 w_2 V$ the total flow towards the ridge; $w_1 w_2 U/4$ is the outflow from the ridge for spreading rate

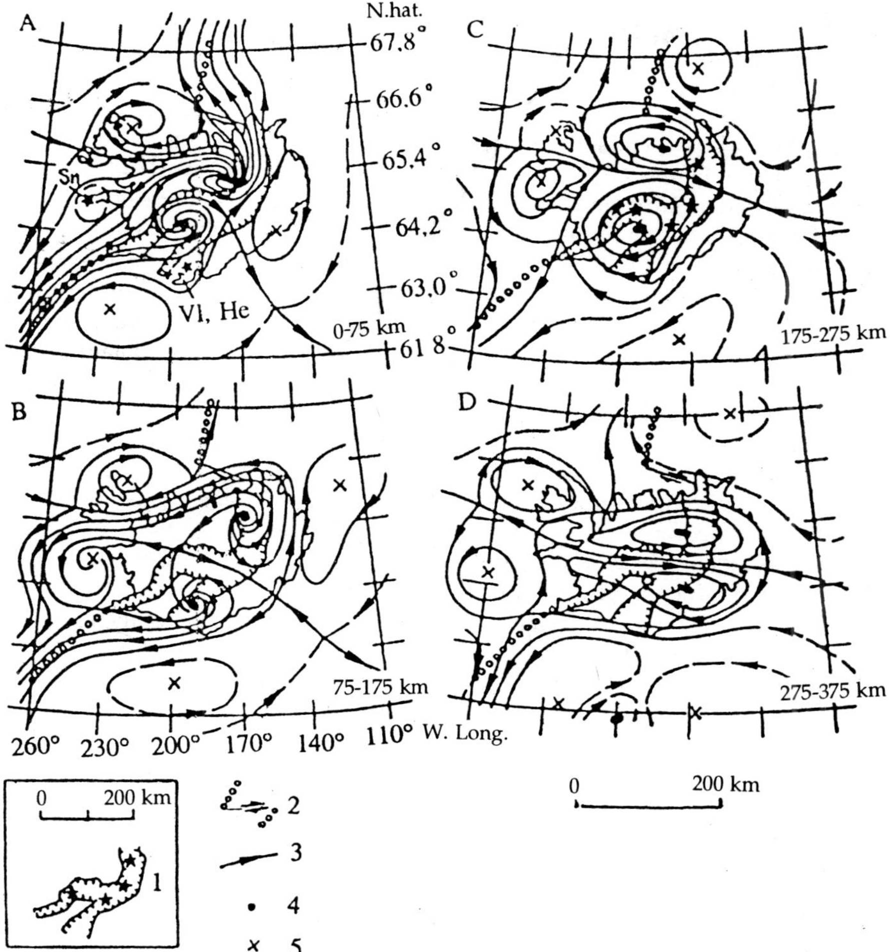

Fig. 4.33. Interpretation of seismic data above Iceland within the framework of the model of hot streams.
1—Iceland rift zone with major volcanic centres; 2—axis of mid-oceanic ridge outside Iceland; 3—flow lines; 4 and 5—centres of ascending and descending flows respectively (Irvine, 1991, modified).

U. The same flow Q concomitantly ensures spreading under the ridge for a distance W, i.e.,

$$Q = hUW/2; \quad W = 2\,Q/hU, \tag{4.36}$$

where h is the thickness of the lithosphere beyond the limits of the ridge. Hence,

$$W = 2\ w_1w_2/hU\,(V - U/4) = 2\ w_1\ w_2/h\,(V/U) - w_1w_2/2h. \tag{4.37}$$

The theoretical line best fitting the observed values of W and U (including the present-day positions) is calculated as $w_1w_2 = 3 \times 10^4$ km^2 and $V = 70$ km/million years, which corresponds to an average plume flow of $Q \sim 2.2 \times 10^6$ km^3/million years. The value of W is determined fairly reliably from residual anomalies of topography and independently from geochemical anomalies and thus the dependence depicted in Figure 4.32C as well as the value $Q \sim 2.2 \times 10^6$ km^3/million years can be regarded as fairly accurate. For the same area of cross-section of the plume channel at depth, the rate of its ascent $\overline{V}$ will be the same, ~ 70 km/million years, which corresponds to the time of ascent of the plume of 10 million years from the boundary with the lower mantle and ~ 40 million years from the boundary with the core. If, however, the channel cross-section is assumed as one order less, for example, for a circular cross-section of radius 35 km (see below), we obtain a value of $\overline{V} = 700$ km/million years and ascent durations of 1 and 4 million years, which are close to our estimates for a purely thermal model (see Fig. 4.30). Other more complex numerical models are also available. In the model of Feighner et al. (1995) in particular, W is proportional to $(Q/U)^{1/2}$ and not (Q/U) as in eqn. (4.36).

The values of ΔT determined on the basis of empirical dependences of ΔT and anomalies of topography as well as using residual anomalies of the gravitational field are less reliable. Most of the points in Figure 4.32C correspond to the theoretical curve of conductive cooling for an initial temperature $\Delta T_0 = 100°$C (Ito and Lin, 1995) but the points for Iceland and some points characterising the early history of Tristan de Cunha plume are close to $\Delta T_0 = 150°$C. Another independent modelling carried out for reconstructing the dynamic relief above the Iceland plume (Ribe et al., 1995) provides excellent correlation between the observed and theoretical relief for $\Delta T_0 = 93°$C $\sim 100°$C. It is possible that the Iceland and Tristan plumes are hotter than other plumes but lost their heat due to additional melting during rapid decompression (Ito and Lin, 1995). An independent test calculation showed that for $\Delta T_0 = 100°$C and $U = 20$ to 100 km/million years, the thickness of the crust above the spreading axis comprises from 4.5 to 9 km and this value corresponds to the observed values (Su et al., 1994; Ito and Lin, 1995).

The dynamic pattern of the interaction of ascending plume with rapid convection flows is quite complex. According to Irvine (1991), it is similar to a turbulent tornado in an atmospheric cyclone. Such a scheme in principle is plotted in Figure 4.33 on the basis of the actual seismic data under Iceland (Triggvason et al., 1983). In the surface layers (zero to 75 km (A) and 75 to 175 km (B)), two central ascending flows are surrounded by four descending ones. In layer A (zero to 75 km) these flows are confined to the linear structure of the Mid-Atlantic ridge and their sizes are 800×200 km. In layer B (75 to 175 km) the pattern is more symmetrical and the cross-section of flows 500×500 km. In the deeper layers C and D, the flow distribution is asymmetric: ascending flows are concentrated in the east and descending ones on the western side of the vortex; further the cross-section of layer C is $250 \times 250 = 6.25 \times 10^4$ km^2. This is double the aforesaid value of 3×10^4 km^2, which is close to the theoretical estimate of Sobolev (1979) for kimberlite pipes: for $r = 20$ to 100 km, $s = (0.75$ to $3.0) \times 10^4$ km^2. Nevertheless, all these estimates are within the limits of calculation error.

In our experiments (Kirdyashkin et al., 1987; Dobretsov et al., 1993), the steady-state plume had a boundary layer regime in the upper part; the ascending flow occurred at the centre and the descending one along the margins. In the bottom part this regime changed into one of thermal conduction where the ascending flow was situated on one of the margins of the elliptical section of the channel, and the descending one along the opposite edge. Cells with a different regime moved upwards in the form of a helix (see Figs, 4.23 and 4.25). At the surface, this feature can be seen as s-shaped or loop-shaped plume traces reflecting its action period.

Such movements have been observed in the Hawaiian (Fig. 4.18) and Iceland plumes (Fig. 4.33) but have been plotted more accurately for the Udokan plume (Rasskasov et al., 1996). The Udokan plume is located on the north-eastern flange of the Baikal rift zone and the margin of the Mongolian superplume (see Fig. 4.34). This superplume, about 2000 km in diameter is divided into several plumes, giving rise to basalt fields of different sizes: Udokan, Vitim, Khamar-Daban, Khangai and other fields (Ashchepkov, 1991; Dobretsov and Ashchepkov, 1991; Ashchepkov et al., 1994). In the Udokan plume (Fig. 4.34, bottom), six different fields ranging from 10 to 25 km in cross-section and formed over 0.2 to 1.4 million years each constitute two loops of about 40 km in diameter, i.e., 2 to 2.5 diameters of the plume; this value agrees with theoretical and experimental concepts (see Fig. 4.25).

Kimberlite and alkaline-basalt fields of plateau basalts represent the most important manifestations of mantle plumes. Kimberlite magmas, according to the latest experimental and mineralogical data (Ringwood and Irifune, 1988; Ringwood et al., 1992; Haggerty, 1994), originate close to the upper-lower mantle boundary. In this case a lower mantle plume is presumed under them. The plume spreads like a mushroom close to the boundary of the refractive garnetite material of layer C (see Fig. 4.22). This broad spot could have given rise to some local plumes in the upper mantle corresponding to the concentration of kimberlite pipes on the surface. The other, alkaline-basalt fields may represent a similar mushroom-shaped lower mantle plume, for example the Mongolian superplume mentioned above (Windley and Allen, 1993) over which some local fields arise, i.e., local upper mantle plumes (Khangai, Vitim, Udokan etc.). Most other 'hot fields', according to the model of L.P. Zonenshain and M.I. Kuz'min, consisting of some interconnected upper mantle jets, may have originated under the influence of one or two lower mantle plumes. Finally, a mantle plume spreading at the base of the thick continental lithosphere is presumed under the plateau basalt fields (Bercovici and Mahoney, 1994; Kerr et al., 1995).

Kimberlite plumes, according to the model of Haggerty (1994), form in periods of a quiescent magnetic field (Fig. 4.35), which may signify quiescent convection in the outer liquid core as a result of intense heat rejection by mantle plumes (Larson and Olson, 1991). A maijerite garnet 'stable' in transition zone C of the upper mantle at depth 400 to 650 km as well as stishovite, ferriperovskite and magnesiowüstite found in the form of inclusions in diamonds serve as proof of the lower mantle origin of plumes giving birth to kimberlite magmas (Haggerty, 1994). The presence of maijerite was established not only in the Mesozoic kimberlites of South Africa, but also in the Palaeozoic kimberlites of Arkhangel'sk and Siberia (Sobolev, 1994, pers. comm.).

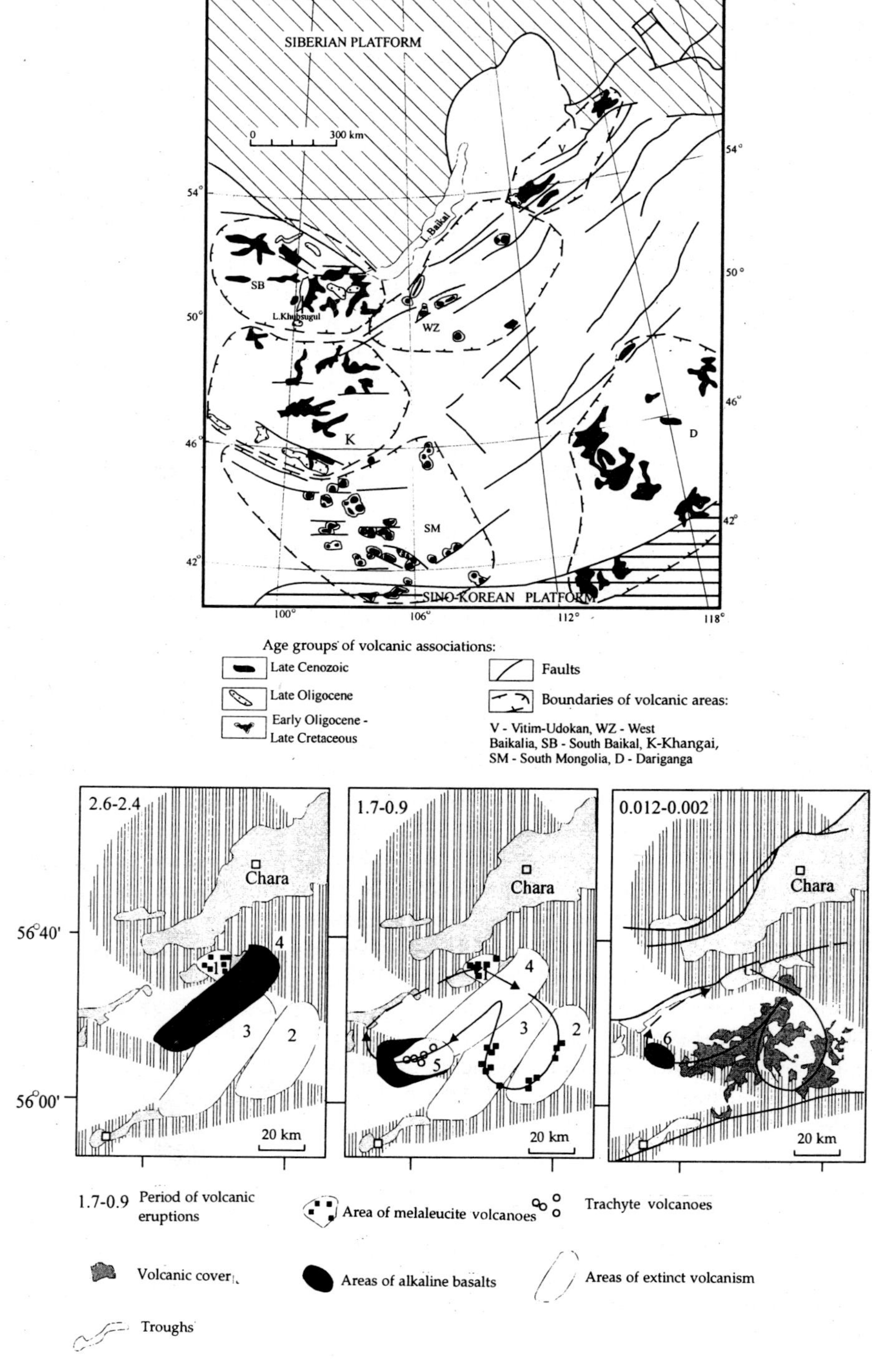
100°
106°
112°
118°
124°
SIBERIAN PLATFORM
0
300 km
54°
L. Baikal
54°
50°
SB
L.Khubsugul
50°
WZ
46°
K
D
46°
42°
SM
42°
100°
106°
112°
118°
SINO-KOREAN PLATFORM
Age groups of volcanic associations:
Late Cenozoic
Late Oligocene
Early Oligocene -
Late Cretaceous
Faults
Boundaries of volcanic areas:
V - Vitim-Udokan, WZ - West
Baikalia, SB - South Baikal, K-Khangai,
SM - South Mongolia, D - Dariganga
2.6-2.4
Chara
56°40'
4
3
2
56°00'
20 km
1.7-0.9
Chara
4
3
2
5
20 km
0.012-0.002
Chara
6
20 km
1.7-0.9
Period of volcanic
eruptions
Area of melaleucite volcanoes
Trachyte volcanoes
Volcanic cover
Areas of alkaline basalts
Areas of extinct volcanism
Troughs

Kimberlites aged 1.8 billion years (Premier Pipe, South Africa) and 1.1 billion years (Africa, Brazil, Australia, Siberia, India and Greenland) are known. Diamond is even older (in many cases, 3.2 to 3.3 billion years; Richardson et al., 1993) although the latest data invite a review of concepts on the antiquity of diamond in kimberlites (Shimizu and Sobolev, 1995).

Nevertheless, the major maximum ages recorded for kimberlites—Palaeozoic and Mesozoic—are: 460 to 440, 410 to 400, 380 to 370 and 320 to 250 million years (possibly two independent stages of 320 to 300 and 280 to 250 million years), 220 to 180, 160, to 140 and 120 to 90 million years which, as pointed out before, agree with the super chrones of the permanent magnetic field (Fig. 4.35). It is significant that the interval between the centres of short or boundaries of long intervals measure 60 or 30 million years. This interval corresponds to the 'main' periodicity (460, 405, 375, 310, 280, 250, 220, 180, 150, 120 and 90 million years).

The geometry of plumes shown in Figure 4.35 conforms to the model experiments on chemical-dense plumes (Griffiths, 1986a, c; Campbell and Griffiths, 1990; Kellog and Turcotte, 1990; Ringwood et al., 1992) while the relative height reflects the intensity of kimberlite magmatism for the corresponding epochs. The maximum activity of kimberlite magmatism was manifest in the Mesozoic stages. All the stages of kimberlite magmatism are manifest in Siberia with proper periodicity (Pearson et al., 1995).

The periods of kimberlite magmatism and the epochs of the effusion of the most important fields of plateau basalts shown in the upper portion of Figure 4.35 do not match and suggest the absence of association between kimberlite and plateau basalt magmatism. The most important fields of plateau basalts (also called flood basalts due to their bulk and catastrophic effusions over a short period) are shown in Figure 4.36. It is significant that most of them are associated with the currently active hot spots but these plumes formed vast provinces of plateau basalts only once (in the first period of formation) or twice (Bercovici and Mahoney, 1994). The largest province of plateau basalts is represented by Siberian traps (Fig. 4.37) aged 253 to 248 million years (Basu et al., 1995). Through Taimyr and the eastern part of Kara Sea, these traps extend into Franz Josef Land and the northern Barents Sea basin (Bogdanov and Khain, 1996). The age of plateau basalts in this direction probably becomes younger; 242 to 240 million years in Taimyr (Vernikovskii et al., 1995) and to the Early Jurassic 220 to 200 million years in the Barents Sea (Bogdanov and Khain, 1996). These thick (up to 4 km in thickness, Fig. 4.37) effusions are associated with formations of the Ob' palaeo-ocean (Aplonov, 1987), Pechora-Barents Sea and western Siberian sedimentary basins (see Chapter 5). The total volume of traps of Tunguska syneclise amounts to 33×10^6 km^3 (Al'mukhamedov et al., 1996) and the total volume including Western Siberia, Kara and Barents Sea syneclise five times more, i.e., 130×10^6 km^3.

Fig. 4.34. *Top*: Location of volcanic fields within the Mongolian superplume and location of the Vitim-Udokan field (modified from Yarmolyuk and Kovalenko, 1995).
Bottom: Evolution of field volcanism and migration of Udokan plume from 10 to 12 Ma (area 1), 9.6 to 8.2 Ma (area 2), 4 to 3 Ma (area 3) to 2.4 to 2.6 Ma (area 4) up to the youngest volcanoes (areas 5 and 6). Line of general migration of volcanic area 1 to 6 shows loop-shaped plume movement (Rasskasov et al., 1995).

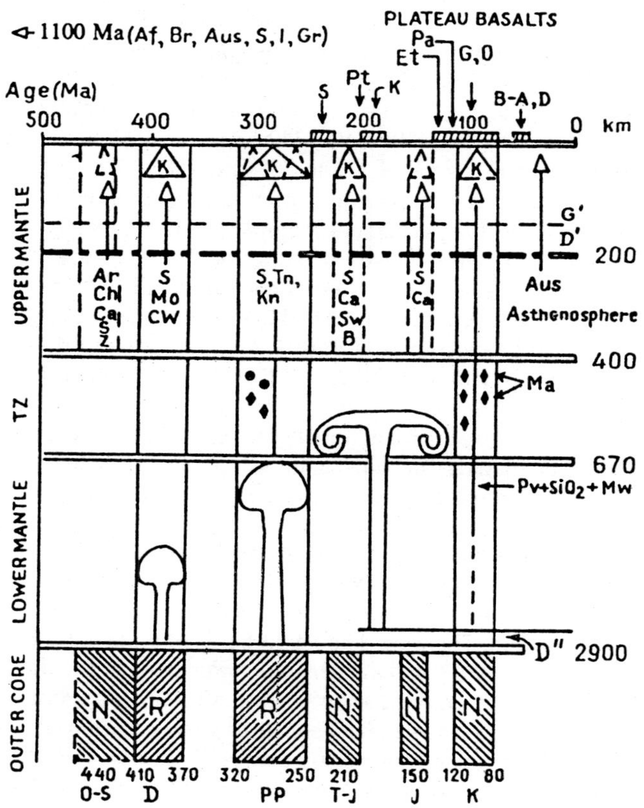

Fig. 4.35. Sketch showing melt of the 'superkimberlite plumes' (Haggerty, 1994). The time axis (zero to 500 million years) shows (bottom) the epoch of normal (N) and reverse (R) permanent magnetic field (superchrones) and (above) the epochs ('pulses') of kimberlite magmatism (including a pulse of 1.1 Ga) and plateau basalts.
Along the vertical are shown the structure of the lower mantle (2900 to 670 km), transition zone (TZ 670 to 400 km), upper mantle including the asthenosphere (400 to 200 km) and lithosphere (< 200 km), most important mineral reactions (perovskite (Pv) + SiO$_2$ + magnesiowüstite (Mw); maijerite garnet (Ma) and transition of diamond into graphite (D' and G'). Kimberlites: Af—Africa, Ar—Arkhangel'sk, Aus—Australia, B—Botswana, Ca—Canada, Ch—China, CW—Colorado, Gr—Greenland, I—India, Kn—Kentucky, Mo—Missouri, S—Siberia and Z—Zimbabwe. Plateau basalts: B-A—Brito-Arctic province, D—Deccan, Et—Etendeca, G—Gondwana, K—Karroo, O—Ontong plateau, Pa—Parana, Pt—Patagonia and S—Siberia.

Extrusions of alkaline basalts in Mendeleev ridge and Alpha ridge aged 120 to 100 million years, positively associated with the Iceland hot spot, represent the possible continuation of these processes (see Fig. 4.32). The magnitudes of these extrusions are, significantly smaller, however. Another example is the extrusion of

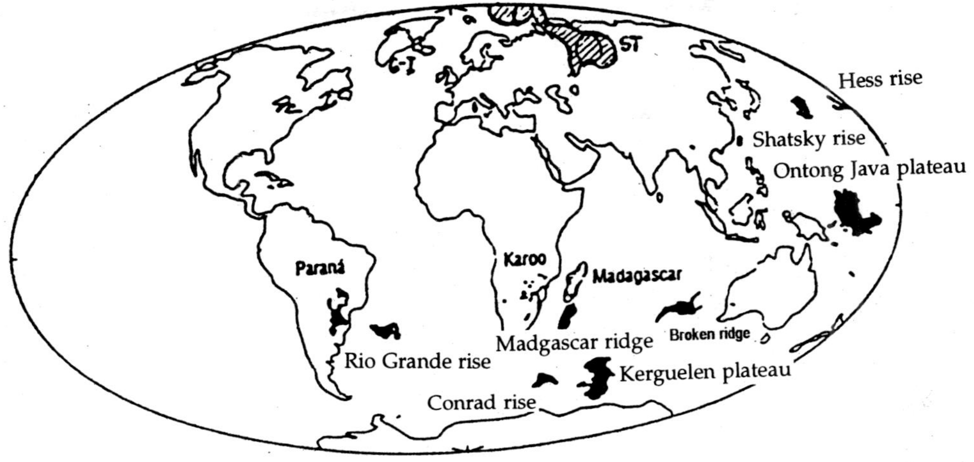

Fig. 4.36. World map showing location of the largest flood basalt provinces mentioned in the text. ST—Siberian (and Barents Sea) traps (modified from Bercovici and Mahoney, 1994).

Karroo dolerites and other plateau basalts associated with the break-up of Gondwana (Fig. 4.38). Formation of the Karroo dolerites in South Africa and eastern Antarctica associated with the Bouvet hot spot and plateau basalts of Ferrar province in Antarctica, Tasmania and Southern Australia, associated probably with the future hot spots Conrad and Kerguelen (Fig. 4.38a), occurred almost simultaneously, as suggested by the new U-Pb age determinations on zircon and baddeleyite in Ferrar province (183.6 ± 1.0 million years) and Karroo (183.7 ± 0.6 million years) (Encarnacion et al., 1996). These profound extrusions preceded rifting and break-up of eastern Gondwana and were accompanied by bimodal volcanism (Chon-Aike + Rohas Verdes) south of Gastre Fault System (Storey, 1995) (see Fig. 4.38a, c).

The next profound extrusions occurred around 130 million years ago and are associated with the hot spots of St. Helena and Tristan (Fig. 4.38b). Basalt provinces of Parana (see Fig. 4.36) aged 127 to 132 million years and Etendeca in southwestern Africa aged 129 to 132 million years (Renne et al., 1996) were formed above the early Tristan plume. Somewhat later, 115 to 110 million years ago, another outbreak resulted in the Kerguelen plume with formation of the Kerguelen-Rajmahal lava province (Fig. 4.38c). After formation of small volcanic islands, another outbreak of the Tristan plume led to formation of the Rio Grande lava plateau aged 80 to 90 million years on the South American plate (Fig. 4.36). At this very time (86 ± 2 million years), the activity of Conrad and Marion plumes manifested with formation of lava plateaus Conrad and Madagascar (see Figs. 4.36 and 4.38).

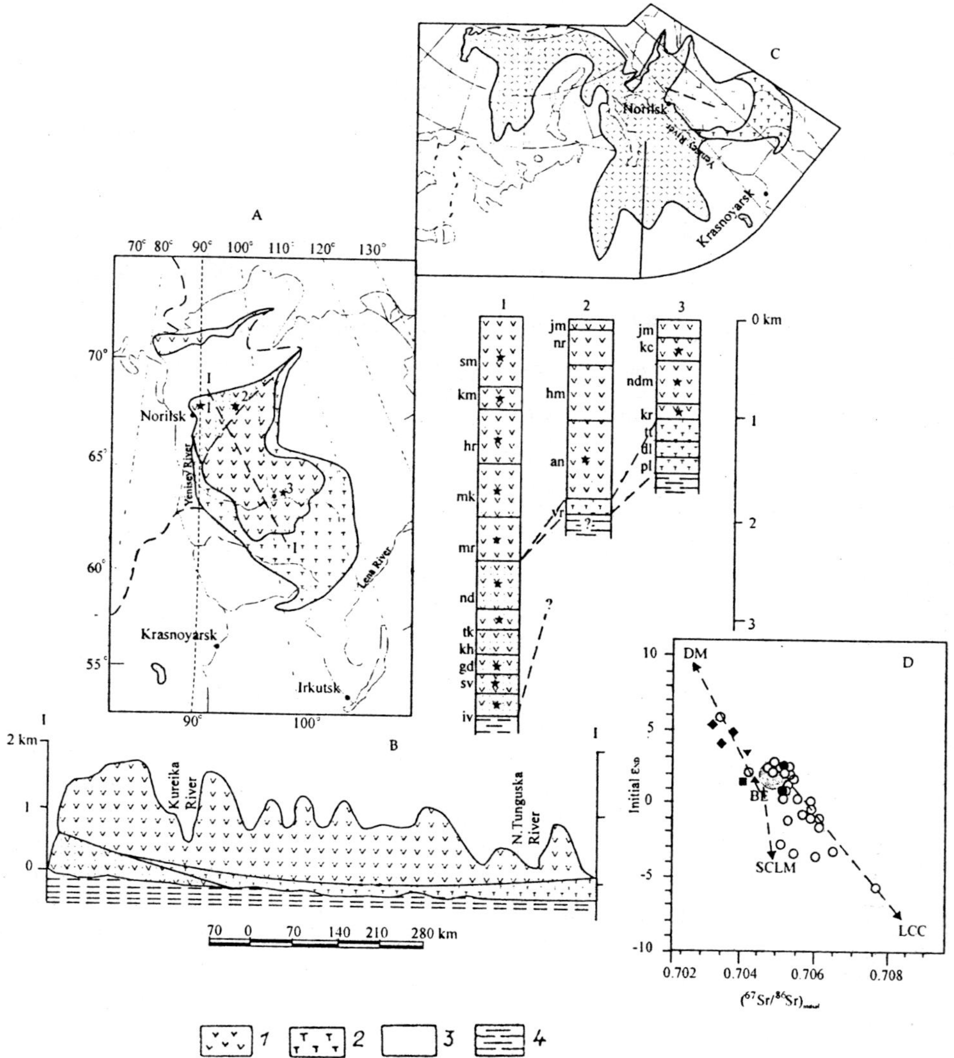

Fig. 4.37. A. Schematic geological map of P-T traps of Siberian platform. Cenozoic sediments are shown in white and symbols are: (1)—basaltic trap, (2)—intrusive traps and tuffs, (3)—subalkaline and picritic basalts, (4)—coloured platform sediments.
B. Lava suites and their approximate thickness, in columns 1 to 3 and cross-section I–I (see A).
C. Distribution of traps and flood basalts in the Siberian platform, West Siberian plain, and Kara Sea and Barents Sea sedimentary basins.
D. Geochemical characteristics of Siberian traps (modified from Al'mukhamedov et al., 1996; C—Basu et al., 1995).
BE—bulk earth; DM—depleted mantle; LCC—low continental crust composition: SCLM—semi-continental lithospheric mantle.

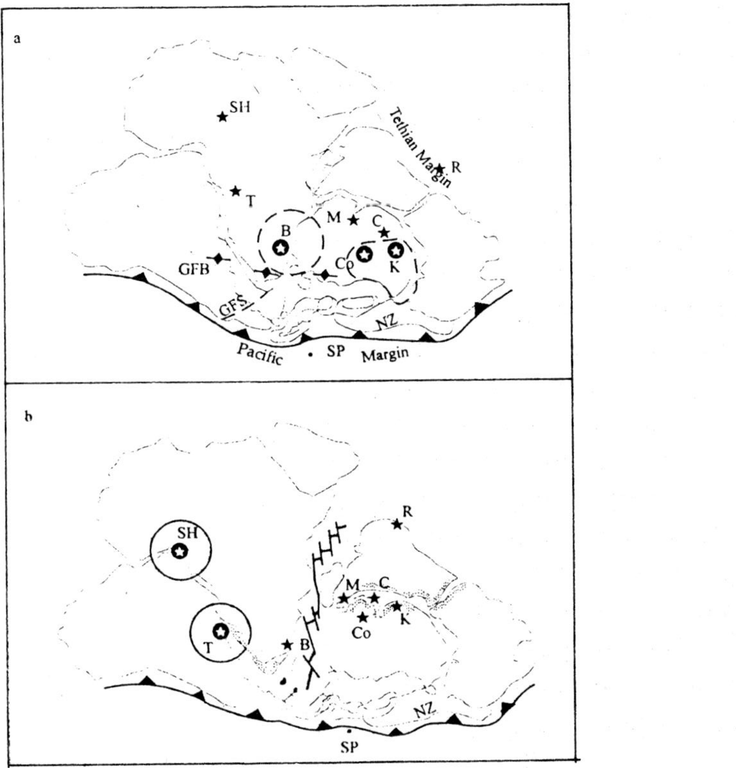

Fig. 4.38. Reconstruction of Gondwanaland around 200 to 180 Ma (a) and 130 Ma ago (b) and summary of the main Gondwanaland break-up of events and correlation with the peaks of blueschist metamorphism (c) (modified from Dobretsov et al., 1987; Storey, 1995). Abbreviations for plumes are as follows: B—Bouvet, Ba—Balleny, C—Crozet, Co—Conrad, GFB—Gondwanian Fold Belt, GFS—Gastre Fault System, K—Kerguelen, M—Marion, NZ—New Zealand block, R—Reunion, SH—St. Helena, SP—South Pole, T—Tristan da Cunha.

178

Finally, in the western part of the Pacific Ocean (Fig. 4.36), the age of formation of the lava plateau on Shatsky rise was estimated in the interval 145 to 138 million years and in Hess rise located farther east and representing a possible course of the same plume, in the interval 110 to 100 million years. Basalt lavas of Ontong Java plateau, the largest after the ST plateau, comprise two age-groups, their composition remaining identical; a) 122 to 120 and b) 90 to 88 million years (Bercovici and Mahoney, 1994). The plume that gave rise to the Emperor-Hawaii volcanic chain became active about 74 million years ago and formed a lava plateau around the Hawaiian islands in the preceding five million years. These largest lava plateaus in the Pacific Ocean are closely correlated with the epochs of the major changes in the movement of Pacific Ocean basin plates (Bercovici and Mahoney, 1994; Storey, 1995). They were not accompanied by additional rifting in the region of lava plateaus but ultimately led to acceleration of the opening up of oceans and global rise of sea level (Larson and Kinkaid, 1996).

Unlike the above, plateau basalt extrusions of Gondwanaland are closely associated with the history of rifting and break-up of Gondwana (see Fig. 4.38) although it is more probable that they do not represent the main cause of these processes but a combination of asymmetric subduction, thermal anomaly under the supercontinent (Anderson, 1994) and action of plumes which promoted rifting and separation of continental plates (Storey, 1995; see also Chapter 5). The largest extrusions of the Siberian and Barents Sea traps initially gave birth to major rises comparable to the corresponding ones in the oceans (Ontong-Java plateau, Shatsky rise and others) but led to different results in each segment: Tunguska syneclise remained elevated (Fig. 4.37a, b), while rift basins in Western Siberia led to the formation of a major oil- and gas-bearing sedimentary basin in the Jurassic-Cretaceous and intense sedimentation synchronous with traps in the Barents and Kara seas.

The high ^{3}He/^{4}He ratio serves as a diagnostic feature of the lower mantle origin for Siberian traps as well as the Hawaiian plume (Basu et al., 1995). Moreover, the composition of primary magmas with respect to Nd and Sr isotopes is proximate to the bulk composition of the Earth (BE) but, in the early stages, the plume melts reacted with the depleted upper mantle (Dm) and, in the late stages, with the lithospheric mantle (SCLM) and lower crust (LCC), as depicted in Fig. 4.37c.

Most definite conclusions can be reached within the framework of the thermal model studied above with respect to mantle plumes originating in the thermal zone of subduction. Volcanoes and volcanic zones associated with subduction zones are displaced little relative to island arcs and continents, i.e., move together with the continents, unlike other plumes where plate movements were seen relative to the poorly mobile mantle jets of deep mantle formation. All the facts point to a fixed position of the source of melting relative to the continent or island arc. Melting sources arise during the down-sinking of the lithospheric plate in a zone of temperature adequate for melting the low-melting crustal part of the plate. This process is studied in greater detail in Chapter 5. Here, it is important to note that the melting chambers under the volcanoes are located at depths of 100 to 250 km (more often 120 to 180 km) where a change of seismic activity is observed (Fig. 4.1; Turcotte and Schubert, 1982; Ueda, 1982).

Let us analyse the energy characteristics of heat sources and the melting time of the island arc mantle from a depth of 120 to 180 km. At $\Delta T_1 = 500°C$ and $\lambda_t = 4$ W/m · deg, the required heat power of the melting centres is equal to 10^6 kW (see Fig. 4.26). In this case, at $\Delta T_0 = 50°C$ (which is 2 to 3 times less than ΔT for other plumes) and $v = 1$ to 10^2 m^2/s; the minimum diameter or width of the zone of intense melting is 1.5 to 4 km. According to eqn (4.18), the maximum depth of melting is 320 km, i.e., at $x = 160$ km, $\bar{x} = 0.5$. At this value, eqn (4.29) gives $t = (2$ to $8) \times 10^4$ years which corresponds to bursts of volcanism extending for 10 to 100 million years in island arcs as well as in the mid-oceanic rifts (Ueda, 1982; Zonenshain et al., 1989; see Sec. 5.8 for more details).

Thus, the geological facts as well as numerical modelling of heat mantle plumes show their reality from the energy point of view and correlation with the upper limit of parameters observed in real plumes. It is especially important that the boundary layer regime and heat conduction regime may be combined in mantle plumes. In the latter regime, instability of the channel and its rotation in the form of a travelling wave occur (see Fig. 4.25). Such a pattern is evident in the growth of many hot spots (Figs. 4.18, 4.33 and 4.34). A combination of rapid plumes (time of ascent 1–5–10 million years) and slow thermal convection in the mantle (with the rotation period of cells ~ 60 million years in the upper and 400 to 450 million years in the lower mantle) explains the basic features of mantle geodynamics and relatively stationary position of the plumes.

4.11. Origin of Lower Mantle Plumes and Their Interaction with the Core

The most complex aspect of estimation of the lower mantle plumes is their origin at the lower mantle-core boundary. This primarily concerns the nature and dynamics of transition layer D_2 (Loper and Lay, 1995).

Figure 4.14 depicts the anticipated rise of temperature in transition layer D_2 of 600°C which should cause a decrease in viscosity and probably a displacement of maximum convection velocity in the lower mantle in this layer. In fact, seismic data show a sharp increase in seismic velocities in layer D_2 of thickness 200 to 300 km (Wysession, 1995; Wysession et al., 1995) and the anticipated increase in density and viscosity in this layer, while the minimum viscosity may be localised between layer D_2 and the maximum viscosity at depth 2000 to 2300 km (Cadec et al., 1995). In this case it would appear that the upper mantle is inverted: layer D_2 = lithosphere, layer 2600 to 2300 km = asthenosphere and 2300 to 2000 km = layer C. Such a layer D_2 is interpreted as the result of the descent of heavy plumes comprising restites, which accumulated in transition layer C as a result of the prolonged action of subduction zones (Fukao et al., 1994; Maruyama, 1994; Zhang and Yuen, 1995; Larson and Kinkaid, 1996).

Figure 4.39 depicts the correlation of hot fields (regions of plume concentration, see Fig. 4.17) and seismotomographic image. It is significant (with confidence level $\geq 95\%$) at depths of 700 to 1200 km and 2500 to 2900 km (Cadec et al., 1995). This interpretation assumes viscosity stratification in the lower mantle with a mirror-reflection of the properties of upper mantle at the bottom of the lower mantle, as pointed out before, and also the descending and ascending plumes with

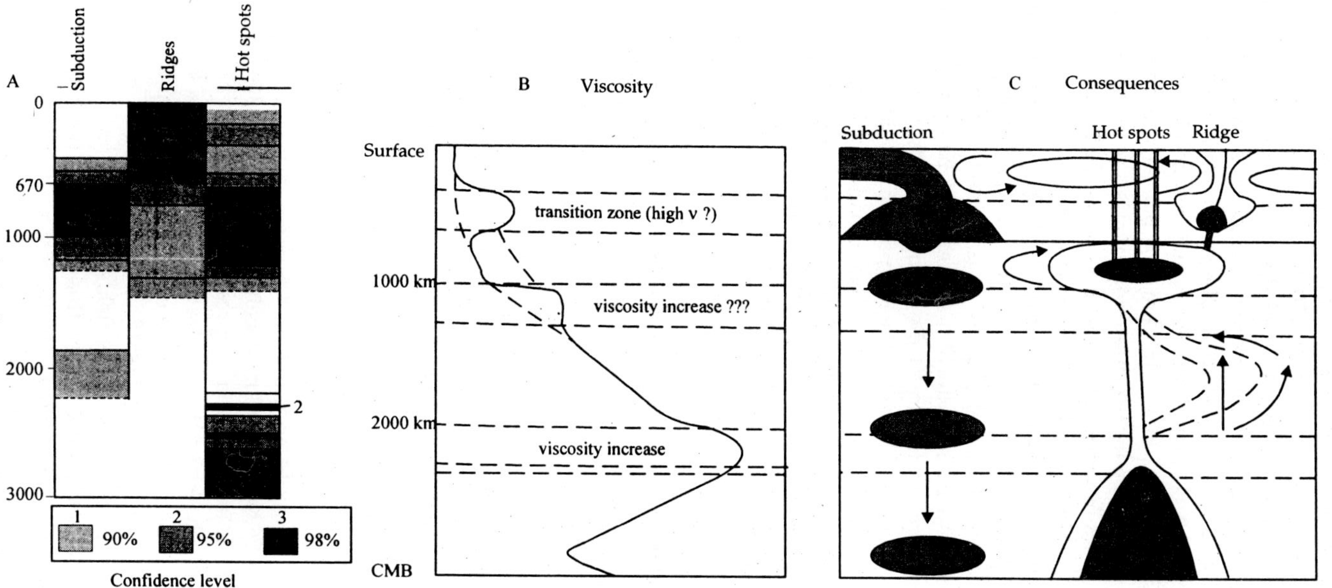

Fig. **4.39.** A—Coherence of tectonic structures on the surface and seismic tomography figures on depth, including the 670 km boundary and two other interfaces at depths of 1200 and 2000 to 2200 km. B—viscosity stratification suggested from A (thick line) and simplified dotted line according to Ricard and Wuming (1991). C—dynamic consequences for both descending cold masses (below subduction, left part) and for megaplumes and associated viscous heating anomalies (dark regions). Hot spots and ridges are surface manifestations of upwellings from the lower mantle; independent convection in the upper and lower mantle is shown schematically (Cadec et al., 1995, modified).

mushroom-like spreading along the boundaries of upper and lower mantles and also of mantle and core.

The absence of correlation between seismotomographic image at depths of 1200 to 2500 km and regions of plume formation may be explained in two ways: 1) constriction of plume in this layer according to experimental data (Griffiths, 1986a, b) and the above-mentioned estimates (100 to 200 km) of plume diameters as a result of which these regions are indistinguishable in the seismotomographic image; and 2) deviation of plume by lower mantle convection, which is possible only for 'slow plumes' and acceleration of convection in depth interval 2300 to 2700 km (Fig. 4.39c).

Independent of this pattern, it may be assumed that layer D_2 performs the role of a heat insulator like the continental lithosphere. In this case the thermal anomalies or thermal 'traps' suggested by Artyushkov (1979, 1993) should occur at the core-mantle boundary. Their minimum sizes as heat sources in the thermal model of the plume (see Figs. 4.28 and 4.31) provide an estimated diameter of about 50 km and height of about 10 km. Only hypothetical considerations can be applied to the nature of these anomalies.

As pointed out in Chapter 1, the outer core has a complex composition in which melt of composition $Fe + Ni + FeO + X$ predominates (Dobretsov, 1980; Boehler, 1992; Allegre et al., 1995). Constituent 'X' may include FeS, SiO_2, a significant amount of dissolved hydrogen, hydrocarbons and K and Na hydrides. In vigorous convection in the outer core, volatiles may accumulate in the boundary layer at the contact with the lower mantle, reducing its density and leading to gravitational instability. At swells in this layer, fluid release of hydride-carbon-hydrogen composition begins and heat is liberated as a result of the reaction of degassing and partial oxidation of the fluid on interaction with the material of layer D_2. Thus melting of layer D_2 may occur on the principle of a 'gas burner'. Material of the lower mantle begins to melt within or above layer D_2 while hydrogen and hydrocarbons, as they ascend upwards, are even more oxidised and converted into a mixture of $H_2O + H_2 + CO_2$. The interaction of this mixture with garnetite layer 'C' of Ringwood's model (Ringwood et al., 1992) gives rise to the kimberlite magma.

Other models associated with the instability of layer D_2 are also possible. In particular, one of the earliest models was that of the separation of a body of 'core matter' $Fe + FeO$ (or Fe_2O) in this layer and its discharge into the core (Monin and Sorokhtin, 1982; Sorokhtin and Ushakov, 1991). There is also the question of the composition of the inner core which was modelled as an intensely compressed plasma by Kuznetsov (1990) on the basis of a 'hot Earth' model. However, based on recent seismic data (Tromp, 1995) it is more likely that the inner core is comprised solid highly anisotropic matter, possibly an aggregate of oriented crystals of metallic Fe.

At great velocities of plume ascent (700 km/million years, i.e., 0.7 m year or even an order slower), convection flows in the lower mantle at a rate of 1 to 2 cm year as well as slow fluctuations of this convection with a period of 400 to 500 million years (see Sec. 3.6) could not have influenced plume stability. On the contrary, mantle plumes in the periods of superplumes may have accelerated more readily and reorganised the flow in the lower mantle. This is reflected in matching the largest plumes and ascending mantle flows (see Fig. 4.17).

Another situation could have prevailed in the early stage of the Earth's history. At this time the heat flux at the Earth's surface may have been 5 to 10 times more than at present, as demonstrated by the origin (in the mantle) of superheated komatiite magmas and formation of high-temperature rocks at a moderate depth (moderate and low pressure granulites) and high temperature ($\geq 100°C$) during siliceous and carbonate sedimentation (Dobretsov, 1980). The consequences of these phenomena were studied above in Section 4.6 (Dobretsov and Kirdyashkin, 1995).

5

Geodynamic Processes in Lithosphere and Asthenosphere

On the basis of the overall global model of movements within the Earth, let us estimate in detail the most important geological processes occurring in the lithosphere and at its surface essentially due to interactions between the lithosphere and asthenosphere. This analysis commences with the interactions of asthenosphere and lithosphere in the oceans leading to sea-floor spreading and formation of oceanic crust. This is followed by an analysis of the genesis of rifting and formation of passive margins of continents. Then comes the analysis of processes occurring in subduction zones leading to the development of island arc—continental crust and specific accretionary—collision complexes. Finally, the zones of continental collision and the formation of orogenic belts, including granitic magmatism and metamorphism, are dealt with at the end of this chapter.

In this analysis we have used our original data and reviews (Dobretsov, 1980, 1981, 1994a, b; Kirdyashkin, 1985, 1989; Kirdyashkin et al., 1987; Dobretsov and Kirdyashkin, 1991, 1992, 1994) as well as numerous publications.

5.1. Ocean-floor Spreading and Origin of Oceanic Crust

It is well known that ocean-floor spreading and origin of new oceanic crust arose at the mid-oceanic ridges (MOR) which form the most widespread structures of the Earth, extending for a length of over 60,000 km (see Figs. 1.1, 1.2). Together with the spreading lines in the back-arc basins and continental rifts, they represent a system of divergent boundaries of lithospheric plates.

Based on the rate of movements and attendant phenomena (relief, magmatism and gravitational effects), these ridges can be divided into the following types (Zonenshain and Kuz'min, 1993a; Small and Sandwell, 1994):

1) fast-spreading (8 to 16 cm/yr); East Pacific rise is a typical example;

2) intermediate (5 to 8 cm/yr); Central Indian and part of Pacific ridges are typical examples;

3) slow-spreading (2 to 5 cm/yr); Mid-Atlantic ridge (MAR) and part of Indian Ocean ridges are typical examples;

4) transitional ridges passing into continental rifts. The spreading rate in them is generally 1 to 2 cm/yr; Gulf of Aden and Red Sea rift are typical examples.

184

The average spreading rate in each type is double the rate in the preceding commencing from the slowest (fourth) type (1.5, 3, 6 and 12 cm/yr). Along with some common features (e.g., similarity in composition of basalts in all the types of mid-oceanic ridges), the above types of ridges reveal many differences.

1) *Fast-spreading ridges*. The main morphological feature of such ridges as the East Pacific rise is the presence of an axial range rising above the abyssal plains (5000 m) up to a depth of 2000 to 3000 m, unlike the slow-spreading type of Mid-Atlantic ridge for which a characteristic feature is the axial rift valley of depth 4000 to 5000 m. Figure 5.1 depicts the typical bathymetry of these ranges. It shows that the axial rise in the East Pacific rise of width about 20 km is manifest on the background of a broad (up to 50 km wide) depression. It can also be seen that the relief of the fast-spreading ridge is more even compared to the slow-spreading type. The profile of the relief away from the ridge is best described by the equation $\Delta H = 0.35 \sqrt{t}$, where ΔH is the increase in depth compared to the axis of the ridge in m and t the age of the ocean floor in million years.

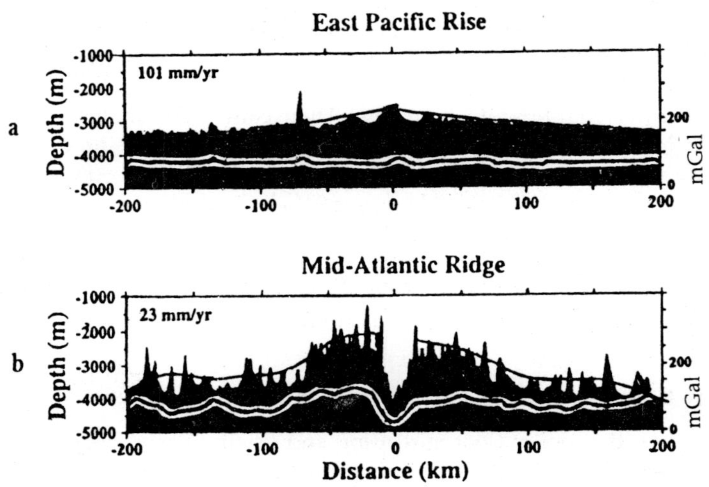

Fig. 5.1. Typical bathymetry and gravitational anomalies (thick line) of fast- (a) and slow- (b) spreading ridges (Small and Sandwell, 1994).

The dependence of relief and gravitational anomalies on the rate of spreading is shown in Figure 5.2. The fast-spreading mid-oceanic ridges of Pacific type are characterised by a fixed relief and presence of small positive gravitational anomalies independent of the rate of spreading in the range 8 to 16 cm/yr. An altogether different pattern is seen in the intermediate and slow-spreading ridges.

The top of the ridge in the East Pacific rise coincides with the chain of volcanic structures, often of the shield volcanoes type. An axial caldera 200 to 300 m wide and 50 to 100 m in depth in generally formed at the top. As can be seen in Figures

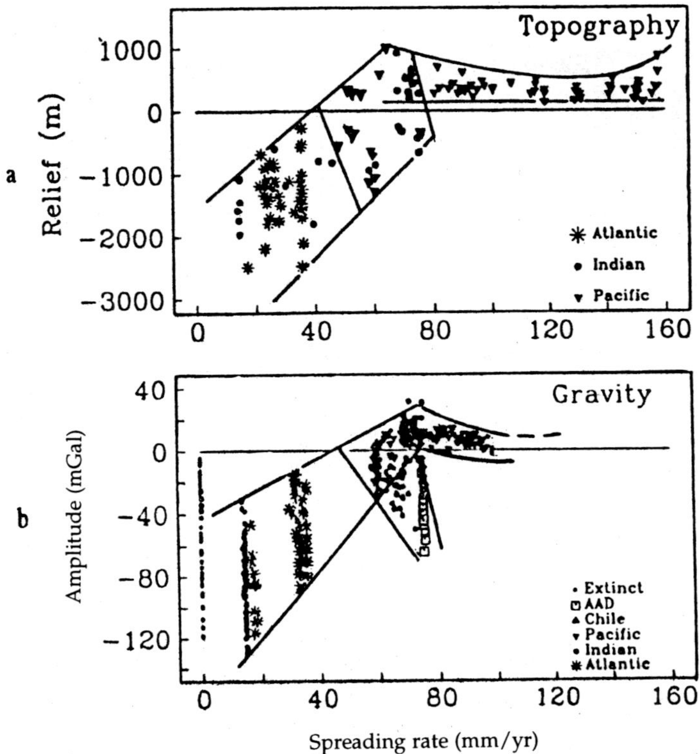

Fig. 5.2. Variation in relief of ridges (a) and gravitational anomalies (b) relative to spreading rate. The cross-section of ridges with intermediate spreading rates (50 to 80 mm/yr) and maximum changes of cited parameters are outlined (Small and Sandwell, 1994).

5.3 and 5.4, however, present-day eruptions occur not only in the axial caldera, but are scattered in the entire axial rise exceeding 10 km in width and revealing a complex and dissymmetric distribution of lavas and dykes (Perfit et al., 1994). Thus, lavas containing MgO > 8% are localised in the north-western section and lavas enriched with basalts of mid-oceanic ridges are in the eastern part (E-MORB) (Fig. 5.3). In a hypothetical section (Fig. 5.4) constructed on the basis of geophysical data (Sinton and Detrick, 1992), the magma chamber including a viscous transition zone has a mushroom-shaped form 5 × 8 km in size and an axial part 5 × 1 km in size saturated with melt. Narrow, dyke-shaped channels for deep melts of the type E-MORB are also assumed. The mixing of these melts with normal basalts of mid-oceanic ridges (N-MORB) as well as differential in the chamber cause wide variations in the composition of lavas and dykes. At the fast-spreading ridges there is little correlation between MORB chemistry and ridge depth. But abyssal peridotite data and available MORB chemistry data suggest that the extent of mantle melting beneath oceanic ridges increases with increase in spreading rate (Niu and Hekinian, 1997).

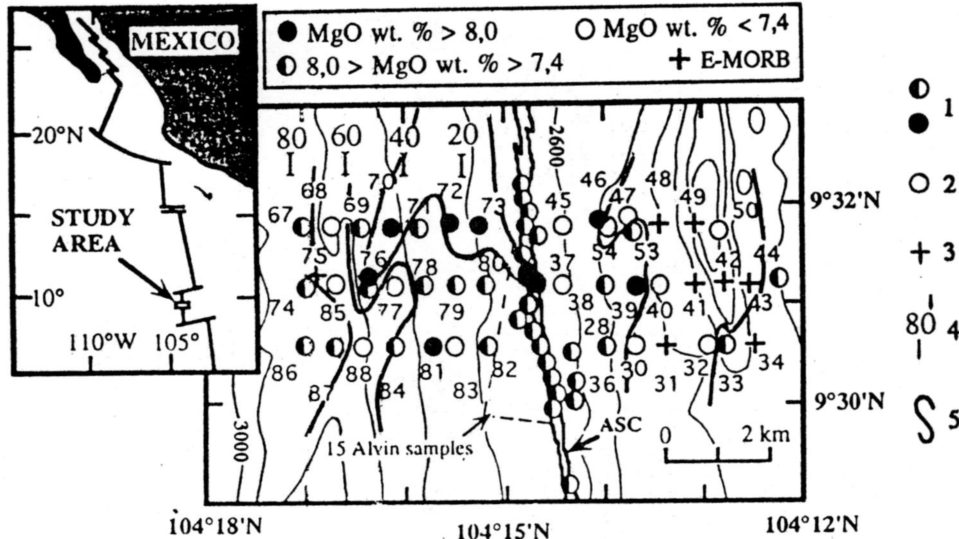

Fig. 5.3. Section of the East Pacific rise round 9° 31′ N lat. studied in detail. Isobaths at 100 m intervals commencing from 2600 m on both sides of the axial caldera (ASC). Chemical composition of each basalt sample with respect to MgO content: 1—MgO > 7.4; 2—normal MORB; 3—enriched basalts; 4—numbers 20 to 80 denote the age of the oceanic crust (in thousands of years) assuming a spreading rate of 11 cm/yr (Perfit et al., 1994); 5—thick lines separate basalt types 1, 2 and 3.

Another characteristic feature of fast-spreading ridges is the wide variation of the form of lavas and extrusions (Fig. 5.4) with a predominance of pillow and cover lavas. Pillow and tube lavas are also extensively distributed in the slow-spreading ridges. Cover (lobate, corded etc.) lavas are characteristic of flat surfaces adjacent to the axial part of the chamber and, together with lava wells and columns, characterise the upper swirling crust of lava lake (Francheteau et al., 1979). Such examples have been described in the territory of Juan de Fuca ridge (Zonenshain et al., 1989a), in the equatorial zone of the East Pacific rise (Francheteau et al., 1979; Perfit et al., 1994) and other sites.

2) *Intermediate ridges* are characterised by a variable spreading rate, sharp transitions from slow- to fast-spreading ridges and their structural instability. Indian Ocean ridges with spreading rates varying from 2 to 8 cm/yr or more represent a typical example (see Fig. 5.2).

Figures 5.5 and 5.6 show typical sections (A, B, C and D) of the Indo-Antarctic ridge in which direct transitions from fast-spreading to slow-spreading sections are observed. Section A shows transition from a slow-spreading ridge in the north-western part in which an axial valley with symmetric, less frequently asymmetric, raised flanks is well depicted, to a fast-spreading ridge with axial rectilinear rises.

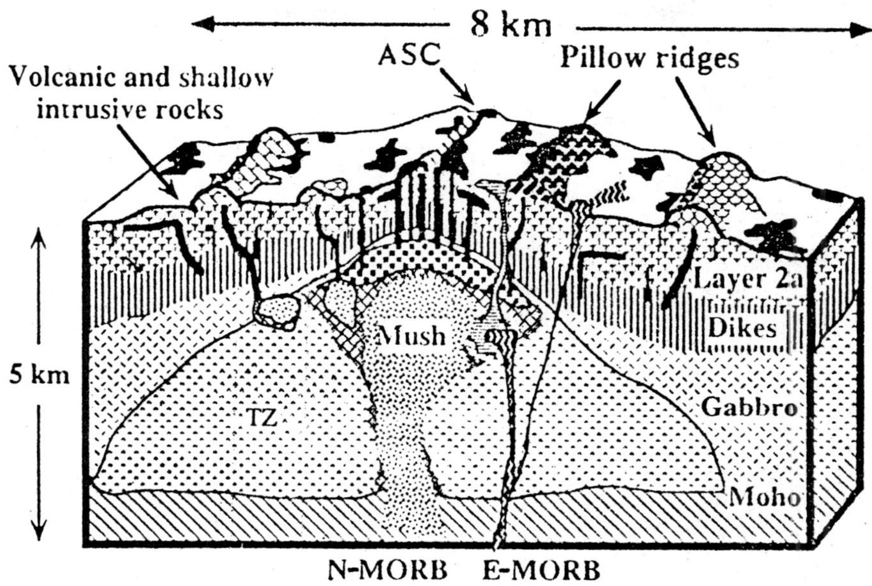

Fig. 5.4. Structure of oceanic crust in section 9°31' (see Fig. 5.3) taking into consideration the model of magma chamber depicted by Sinton and Detrick (1992). The complex and dissymmetric distribution of lavas, dykes and sills outside the axial caldera (ASC) are shown. Part of the lavas arise from the transition zone (TZ) or as result of mixing with deep-level melts of E-MORB type (Perfit et al., 1994).

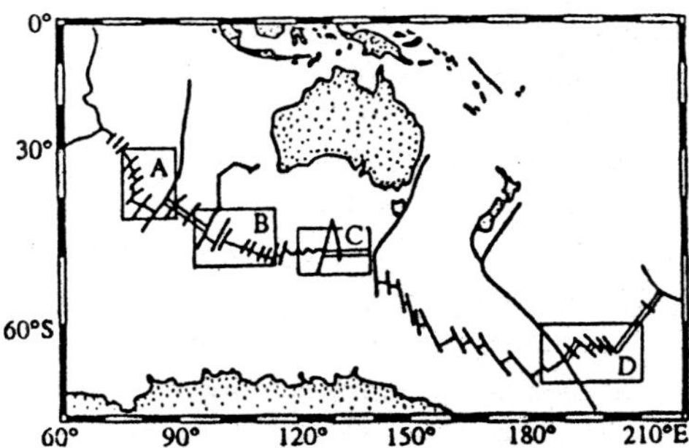

Fig. 5.5. Distribution of sections A to D in the Indo-Antarctic ridge in which fast-spreading sections change into slow-spreading ones.

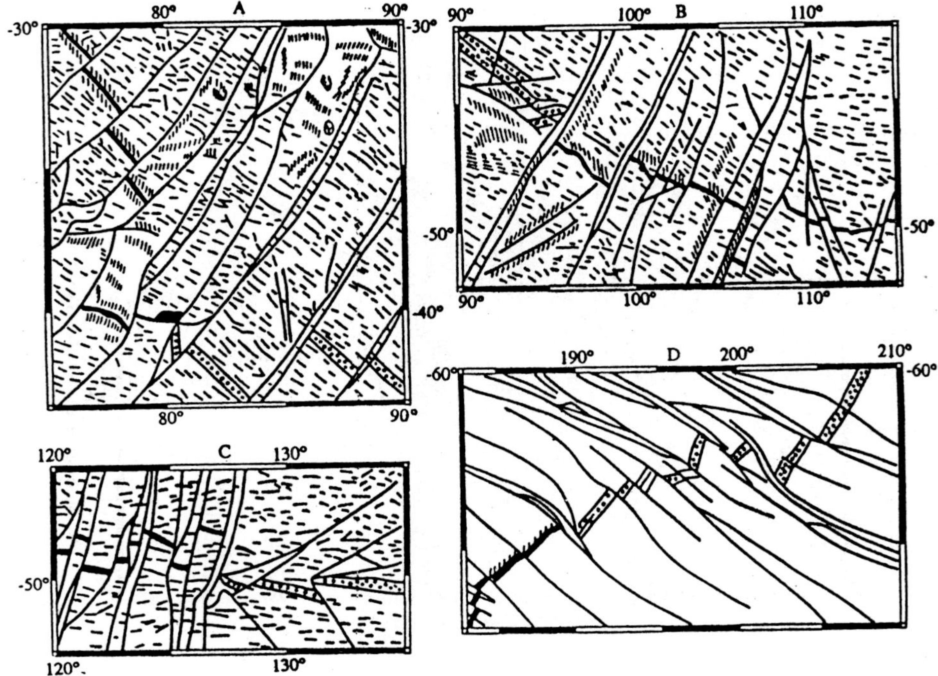

Fig. 5.6. Structure of sections A to D (see Fig. 5.5) based on satellite altimetry. Dotted segments—axial rises of fast-spreading sections and thick lines with symmetrical or one-sided hatching—axial rifts and adjoining rises of slow-spreading sections. Fine hatchings depict the orientation of the microrelief of the ocean floor (Small and Sandwell, 1994, modified).

The transition zone between the two end-member examples reveals many transform faults with branched and oblique hinge faults as also a volcanic chain possibly reflecting the track of plate movement above a hot spot.

In section B, the fast-spreading section again changes into a slow-spreading one and volcanic rises in the zone of transition and a complex system of cross faults, especially in the low-spreading area, are observed. In section C in which the slow-spreading ridge during its eastward passage is transformed once again into a fast-spreading section, cross faults are characteristic of the fast-spreading section while paired faults are characteristic of the slow-spreading ridge. Finally, for section D, in which the slow-spreading ridge in the south-west is transformed once again into a fast-spreading section, paired, less frequently cross faults fare observed in the fast-spreading as well as slow-spreading sections. Volcanic rises associated with hot spots are not seen in sections C and D, however, unlike in sections A and B.

Thus, in the Indo-Antrarctic ridge, from the point of triple junction around 27°S lat. and 70° to 210° E long., three sections of fast-spreading ridge are observed on the background of a predominantly slow-spreading ridge (see Fig. 5.5). All the fast-spreading sections adjoin major transform fault zones, i.e., displacement boundaries of microplates (90° fault in section A; Macquarie fault on the eastern boundary of section C; and fault emerging toward New Zealand in section D).

3) *Slow-spreading ridges* have been studied in most detail on the example of the equatorial zone of the Atlantic ridge. Two types of rift valleys are recognised in them; normal, predominant type and anomalous type developed locally (Zonenshain et al., 1989b; Zonenshain and Kus'min, 1993a).

An example of normal valleys is the Mid-Atlantic ridge (MAR) in the region 37° N lat. which was studied in detail under the FAMOUS project (Riffo and Le Pichon, 1979) and in similar sections in other MAR sites and slow-spreading sections of the Indo-Atlantic ridge (Figs. 5.2 and 5.7). Here, at the centre of the valley is an inner rift 4 to 5 km wide (rarely up to 12 km) and, within it, an even narrower (1 to 2 km wide) neovolcanic zone. The inner rift is framed on both sides by a staircase of 3 or 4 fault scarps of 2 to 5 km wide and 0.2 to 0.5 km high. The total height of the staircase is about 1 km. The neovolcanic zone in the axis of the rift valley varies in height from 100 to 800 m, i.e., at places rises to the height of upper stages of the staircase of scarps. The varying height of the neovolcanic zone, intensity of volcanism and hydrothermal activity depend on the stage of the cycle in which two stages are usually identified: a volcanic (or constructive) stage of duration 200,000 to 300,000 years and a tectonic (or destructive) stage of similar duration. In the tectonic stage volcanic eruptions are absent or extremely rare and the volcanic ridge is destroyed and gigantic debris produced in the form of rock streams along the slopes of the staircase of scarps. In the volcanic phase, lava eruptions occur in short-term impulses interspersed with periods of quiescence of 5000 to 10,000 years. It has been suggested (Gente, 1987) that, in the volcanic phase the neovolcanic ridge rises to the height of the crests, and the tectonic phase it splits and the fault scarps appear much like fragments detached from it.

In the normal valleys, only pillow lavas (predominantly tubular flows) 600 to 800 m in thickness are encountered and the upper part of the complex of parallel dykes is seen only at places.

Anomalous sections of the Mid-Atlantic ridge are characterised by manifestation of high mountains (1.5 to 3 km above the valley floor) along the rift flanks. These anomalous sections are confined generally to the points of intersection of ridges with major transform faults but are also seen far away from faults at points of submergence of the rift valley itself to a depth of 3600 to 3700 m instead of an average depth of 2800 m. Such a section has been studied in the region of TAG around 26° N lat. (Zonenshain et al., 1989a). Here, the axial neovolcanic zone accompanied by fields of hydrothermal activity ('emanating black smoke') has a width of about 5 km (Fig. 5.7). The western flank has the same structure as in the normal valleys (four scarps of total height 700 m). A hilly scarp 1500 m in height rises above the rift valley from the eastern side. In its upper part, the entire section of oceanic crust is opened up: pillow lavas $\geq$ 300 m thick, series of parallel dykes (300 m) and isotropic gabbros. The thickness of the complex (basalts + dykes) is about 600 m, i.e., one-third the normal thickness of this layer. Dykes in this and other similar sections are formed strictly vertically and parallel to the axis of the rift valley, thus pointing to the absence of slopes or block deformations. Pyroxenites and serpentinised peridotites, sometimes forming the apexes of rift mountains, are also encountered in the anomalous sections, in addition to gabbro.

190

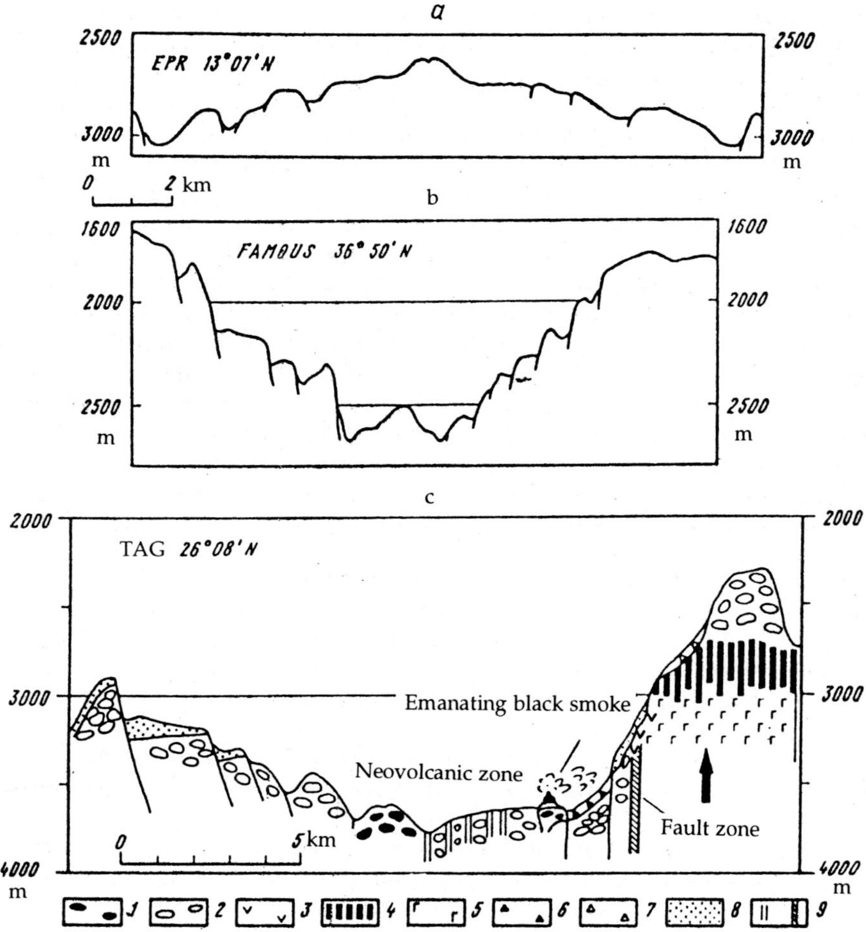

Fig. 5.7. Cross-sections of the axial part of mid-oceanic ridges—fast-spreading East Pacific rise (a) and slow-spreading Mid-Atlantic ridge—in the region of FAMOUS (b) and in the TAG region (c):
Lavas: 1—young; 2—older; 3—altered rocks 'emanating black smoke'; 4—parallel dykes; 5—gabbro; 6—greenstone breccia; 7—block breccia; 8—sediments; 9—various faults (Zonenshain and Kuz'min, 1993a).

The regions of intersection of Atlantic Mid-oceanic ridges with major transform faults exhibit the most complex structure. Such examples are shown in Figures 5.8 (Kane fault) and 5.9 (15°20' fault). The relief of transform fault zones has been explained through different mechanisms. According to Karson and Dick (1983) and Zonenshain and Kuz'min (1993a), the most probable mechanism is that the morphology of transform fault zones with a drop of up to 4000 m in the relief formed at points of their intersection with the rift zone and later moved away passively from the spreading axis. On approaching the transform fault, the floor of the rift valley submerges to a depth up to 5000 m and a nodal basin is formed here along with which the inner angle rises to 4000 m above the submerged floor.

In the zones of intersection, differentiated vertical movements possibly occur 1) submergence of the floor of 'nodal' basins as a result of lithosphere cooling due to contact with the adjoining, already cooled plate (in Fig. 5.8, in the south-eastern corner, contact of the Mid-Atlantic ridge zone in the north with the African plate; in the north-western corner, contact in the south with the North American plate); 2) vertical rise of the 'inside-corner high' due to penetration of water into the base of the crust and serpentinisation of peridotites causing considerable increase in volume. Chains of extinct 'inside-corner highs' and depressions of nodal basins separated from the spreading axis are traced on both sides of the transform fault, resulting in its characteristic relief (Fig. 5.8).

The mechanism of serpentinisation of mantle peridotite probably represents the general cause of formation of anomalous Mid-Atlantic ridge sections and zones of their intersection with transform faults (Bonatti, 1967; Zonenshain et al., 1989a; Cann et al., 1997). This mechanism can be supplemented with others, for example compression as a result of 'oblique' spreading—diverging directions of spreading—and strike of transform faults. Overthrust structures can be anticipated in these cases.

Such an overthrust structure in the scarp of the inside-corner high is fixed from the data of underwater observations in another typical section, i.e., the 15°20′ fault region (Fig. 5.9). Here, extensive sections of serpentinised mantle peridotites have been delineated toward the surface. The rock assemblage in this section reckoned from the results of hundreds of points of dredging implies a predominance of serpentinised peridotites: the proportion of peridotites: gabbro: (basalts + dolerite dykes) is in the ratio 40 : 30 : 30, the proportion of dykes not exceeding 10 to 12% (Dobretsov et al., 1994). It has been observed at many sites that dykes of dolerites and gabbro intruded into peridotites already serpentinised and uplifted and young and fresh basalts extruded directly on their eroded surface. It can be seen from Figure 5.9 that ultrabasites predominate in the transform fault zone and on its southern flank while the complete section of oceanic crust is exposed on the eastern flank of the rift valley in the region of inside-corner highs and on the northern flank of the transform valley. The composition of minerals and melt inclusions in them shows that independent magma reservoirs exist under the rift valley on both sides of the transform fault (Dobretsov et al., 1994).

Another characteristic example of a lithospheric structure similar to that of the Mid-Atlantic ridge has been revealed in King's trough between Spain and the present-day Mid-Atlantic ridge (Fig. 5.10). Unlike the preceding examples, rocks of palaeo-oceanic lithosphere aged 50 to 52 million years (Dobretsov et al., 1991, 1994a) are exposed here. In Palmer ridge located between Freen and Peak valleys and probably representing an old uplift of the inside corner, alternation of blocks with a large amount of deep-level rocks and blocks with normal lithospheric section are observed. The deep-level rocks in this alternation are represented by peridotites, gabbro, gabbro-amphibolites (amphibolitised metagabbro and flaser gabbro) as well as breccia comprising angular fragments of these rocks. The blocks with normal lithospheric section consist of (bottom upwards); gabbro, dyke

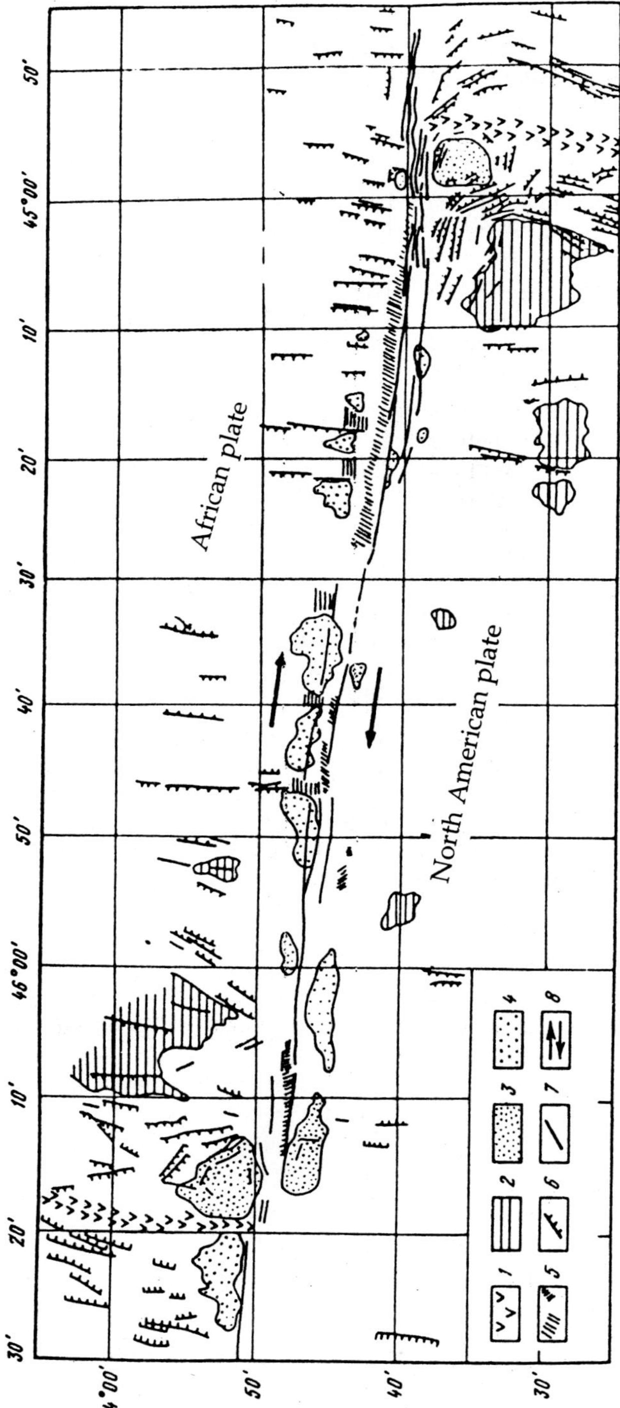

Fig. 5.8. Structural plan of Kane fault in the Atlantic Ocean (Zonenshain and Kuz'min, 1993a).
1—neovolcanic zone, spreading axis, 2—rift mountains (with a bathymetry of less than 2 km); 3—nodal basins; 4—palaeodepressions (3, 4—depth ≥ 5000 m); 5—median range; 6—faults; 7—displacements of main transform faults; 8—directions of plate movements.

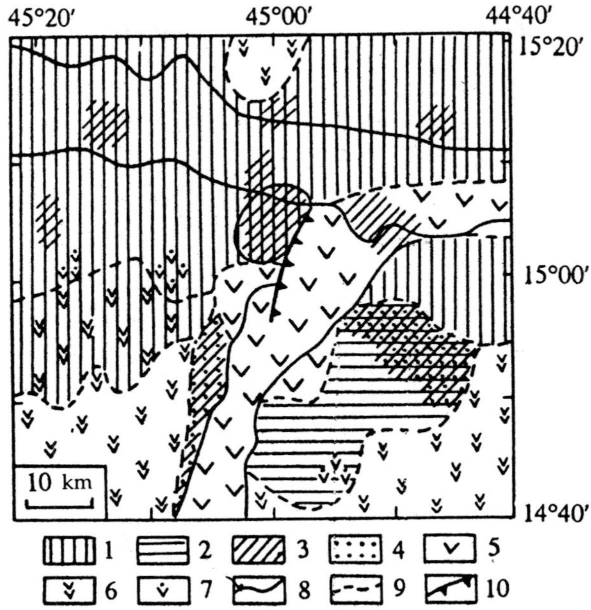

Fig. 5.9. Geological map of the region of intersection between the rift of the Mid-Atlantic ridge and 15°20' transform fault (Dobretsov et al., 1994a). 1—harzburgites and serpentinites with distinct gabbro bodies; 2—lherzolites with varying amounts of gabbro; 3—full section (ultrabasites–gabbro–dykes–dolerites–basalts); 4—full section including metagabbro and blastomylonites after gabbro; 5—young basalts (spreading zones); 6—older basalts; 7—basalts deposited on ultrabasites; 8—boundaries of rift and fault trough; 9—boundaries of rock complexes; 10—deep-level thrust according to the data of the French expedition.

complex, pillow basalts and Eocene sediments. The thickness of the dyke complex is about 1000 m and of basalts 600 m, these values being close to the average thickness values of the oceanic crust.

In the composition of basalts, dyke dolerites and gabbro, apart from normal oceanic basalts (N-MORB), there is a significant admixture of E-MORB type and subalkaline basalts associated with the old hot spot adjacent to the junction of African, North American and European plates and also a large amount of differentiated varieties, which is particularly well reflected in the wide variations of the compositions of clinopyroxenes (Dobretsov et al., 1994a).

In all cases the crust of slow-spreading ridges does not have distinct layered structure as assumed in the averaged model of the oceanic crust and is not maintained even in the fast-spreading ridges. In the slow-spreading ridges, the crust has a 'keyboard-like' character consisting of uplifted blocks including seprentinnites and depressed blocks with 'normal' section of the lithosphere. Another feature is the absence of a permanent magma chamber under the ridge and the pulsed

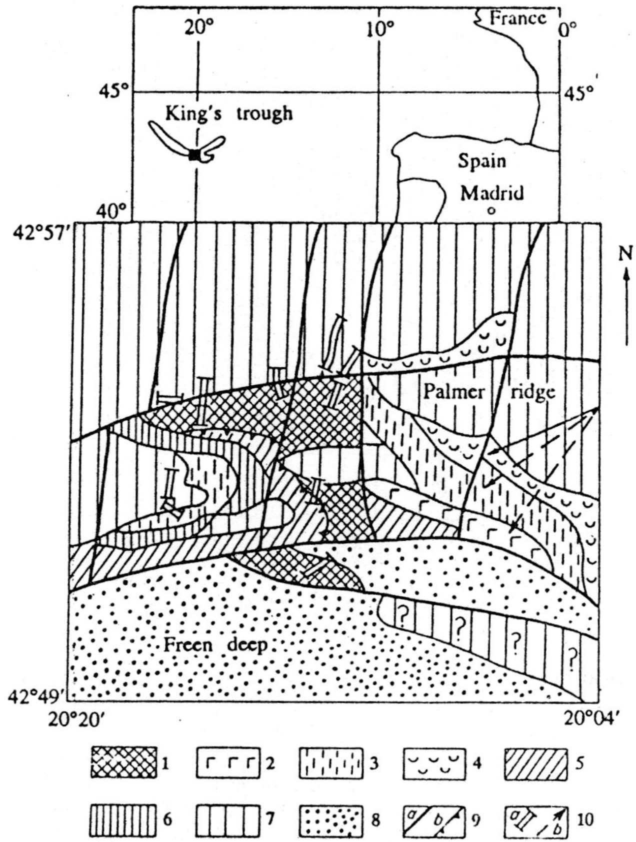

Fig. 5.10. Ancient zone of the joints of Mid-Atlantic ridge and transform fault in King's trough, North Atlantic (Dobretsov et al., 1994a).
Top—location of King's trough; bottom—geological map of Palmer ridge.
1—serpentinised ultrabasites; 2—gabbro; 3—gabbro and dyke dolerites; 4—massive basalts and pillow lavas; 5—metagabbro and amphibolites; 6—breccia; 7 and 8—sediments of: 7—Eocene and 8—present-day; 9—faults (a) and overthrusts (b); 10—draglines (a) and scanning lines (b).

nature of magmatism in the intervals between which are observed only tectonic activity and hydrothermal processes caused by the seepage of ocean water into the cooling fissured rocks of the crust. These features are associated with the inadequate inflow of magma into the spreading zone where the oceanic lithosphere formed finds itself on a 'hungry solder'.

However, formation of oceanic lithosphere, averaged through millions of years, in the slow-spreading ridge is quite stable and is essentially no different (with respect to the principal kinematics, composition of predominant basalts and non-deformation of sediments on the oceanic palaeolithosphere) from the fast-spreading ridges and the lithosphere being formed here. This stability was disturbed evidently only in the early stage of development of the Atlantic Ocean when many parallel rifts originated and fluctuations of spreading zones are observed. In the northern Atlantic, such fluctuations have been observed in the region of Baffin Bay, in the Caribbean basin and in the northern part of the Arctic Ocean. Contrarily, in such fast-spreading zones as the East Pacific rise, fluctuations of spreading zones have been observed even in the already developed oceanic lithosphere.

In order to evaluate the specific features of the initial stage of slow-spreading ridges, the structure and development of transitional rifts of the Red Sea type need to be studied.

4) *Transitional rifts.* It can be seen from Figures 1.1 and 1.2 that the Mid-Indian ridge north of the triple junction, shown also in Figure 5.5, transits directly into Sheba ridge representing the spreading centre in the Gulf of Aden and later, through Tadjoura, Afar and Danakil rifts, joins the Red Sea and Ethiopian rifts (Fig. 5.11). Afar represents the zone of triple junction of Aden, Red Sea and Ethiopian rifts. The Aden and Red Sea rifts, however, exhibit all the characteristics of slow-spreading oceanic rifts: the rate of spreading in them is 2.0 and 1.6 cm/yr respectively, and typical mid-oceanic ridge basalts are formed in the axial zone. The Ethiopian rift is typically intracontinental; at a spreading rate of 0.4 cm/yr, there was an extension of 40 km and thinning of continental crust in it by 10 km (Kaz'min, 1987). Acidic volcanites and alkaline basalts predominate in the Ethiopian rift and basalts of transitional series extrude only in the axial zone 4 to 5 km wide adjoining Afar and in Afar itself. These basalts differ from those of mid-oceanic ridge by their higher content of potassium and rare earth elements.

The western extremity of Aden rift adjoining Afar, i.e., Azal rift, began opening up only 10 million years ago. In the neovolcanic zone, basalts similar to E-MORB extrude here. These basalts differ only in insignificantly higher content of Ti, K, Ba, Sr, Rb and Zr. At this same time (8 million years ago), Danakil microplate resting in-between the Arabian and African plates commenced detachment. Until then, the Red Sea rift evidently joined straight with the Aden rift through the Tadjoura rift (Fig. 5.11). The Red Sea rift per se underwent prolonged evolution in the course of 40 million years (Coleman, 1984). In the first stage, after the arched uplift stage, this rift was similar to the Ethiopian rift. In the second stage, after 25 million years when the system of East African rifts began forming, the centre of intense spreading was situated on the north-eastern flank of the Red Sea rift where the Tihama-Azir complex was formed. Coleman (1984) interpreted this complex as a special type of ophiolites. Here, parallel dykes and stratified gabbro bodies intruded into the stretching continental crust. Sialic blocks (screens) of sialic crust are preserved between parallel dykes, which melted and transformed into acidic dykes. Thus, combinations of primary basalts close to MORB, secondary acidic melts and subalkaline basalts (hybrid or deep-level melts) are noticed in dykes and relict lava covers. Finally, 8 to 10 million years ago, simultaneous with the formation of Azal rift, continental crust in the Red Sea rift split and oceanic crust began to form; the spreading axial shifted south-westwards to the present position of the axial valley of the Red Sea rift where typical MORB basalts are formed (Al'mukhamedov et al., 1985). Another feature of evolution of these rifts is the formation of a system of transform faults. These are not seen in the Ethiopian rift and only began to be formed in the Red Sea rift while a dense network of transform faults (25 such faults on 11°) with minor displacements (20 to 30 km) of the axial valley was formed in the rifts of Azal, Tadjoura and Sheba. In the better developed system of slow-spreading ridges (western part of Sheba ridge, Mid-Indian and Mid-Atlantic ridges), some transform faults have became extinct and major displacements manifested (see Figs. 5.6 and 5.11) although the average distance between the transform faults was 0.5° (50 to 60 km). Finally, in the fast-spreading ridges, the average distance between the transform faults is 3 to 5°, i.e., 350 to 550 km (see Figs. 5.3 and 5.6). For the Mid-Atlantic ridge in the zone between the southern and northern tropics, the average distance between the transform faults is about 250 km, i.e., intermediate between slow- and fast-spreading ridges, but the distance here too is uneven, with modes of about 60 to 100 and 300 to 350 km.

Rift valleys of mid-oceanic ridges, occupying about 0.01% of the area of oceans, hold about one-third of the active hydrothermal springs with which the sulphide and oxide ore deposition is associated. On the whole, the system of spreading ridges comprises about 30% of the area of oceans and houses two-thirds of the hydrothermal fields and ore bodies. The rest is confined to transform faults beyond the limits of mid-oceanic ridges or to marginal seas (Zaikov and Zaikova, 1994).

Most of the productive hydrothermal fields are confined to the zones spreading at the rate of 4 to 8 cm/yr. Here are seen hydrothermal springs of different intensities and all types of ore bodies, including those 'emanating black smoke' with a temperature of up to 350 to 400°C depositing massive sulphide ores, and those 'emanating white smoke' depositing sulphates, carbonates and opal (Lisitsyn et al., 1993). At a low spreading rate, hydrothermal activity is less intense, is usually characterised by low temperatures (below 150°C) and is manifest as vein-impregnated sulphide mineralisation in basalts and gabbro, oxide-silicate ore crusts and ore glasses (Sharapov, 1994). Manifestations of massive sulphides have also been found, however, in some sections with low spreading rates in the Mid-Atlantic ridges and in Gord range (Zaikov and Zaikova, 1994).

It has been suggested that the deep-level portion of hydrothermal systems where the seeping water reacts with the hot rocks around the magma chamber has a temperature of up to 400°C and pressure 200 to 500 atm (Lisitsyn et al., 1993). The temperature of ore deposition depends on the degree of dilution of the original fluid with marine water (Fig. 5.12). Sulphides of zinc and copper are deposited, usually in the form of cones rising to a height of up to 70 to 100 m, from nearly undiluted solutions (temperature 200 to 350°C). On dilution and reduction of temperature (below 150 to 200°C), they are replaced by sulphates of barium and calcium (barite and anhydrite) associated with secondary sulphides of gold and silver. On further dilution (temperature about 100°C), nontronite, oxyhydrates of iron and manganese (often with carbonate), opal and oxyhydrates predominate.

A very similar zoning has also been established in some ore hills deposited from a long-existing hydrothermal spring, for example in Juan de Fuca ridge in the Pacific Ocean (Fig. 5.13). The biggest ore deposit on the margin of the hill rises to a height of about 100 m with diameter exceeding 150 m (Goodfellow and Franclin, 1993). At the centre, the ore lodes consist of a network of hexagonal pyrrhotite which is associated with sphalerite, isocubanite and chalcopyrite with

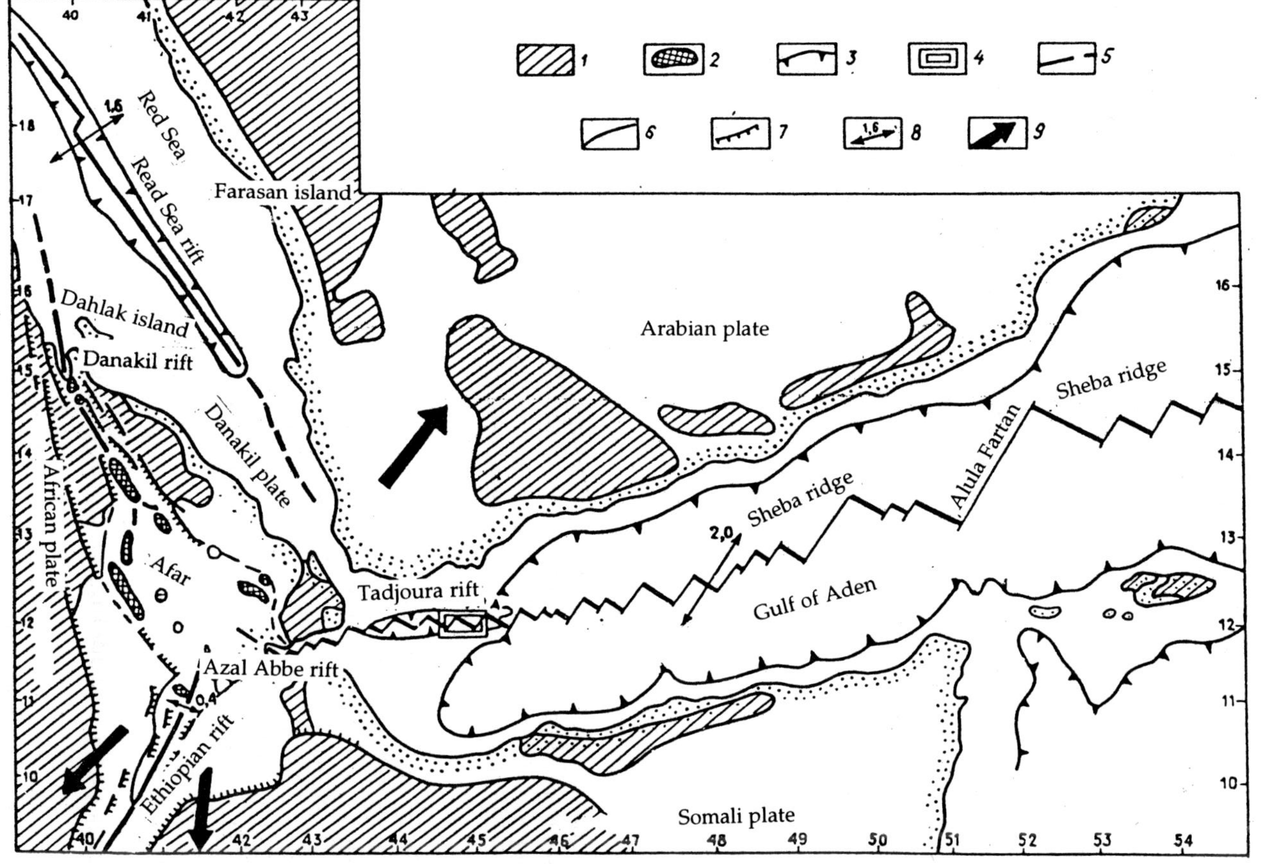

Fig. 5.11. Triple junction of Red Sea, Aden and Ethiopian rifts (Zonenshain and Kuz'min, 1993a).
1—Pre-Miocene basement; 2—axial volcanic ridges of Afar; 3—1000 m isobath; 4—Tadjoura range; 5—divergent boundaries of plates; 6—faults; 7—fault scarps; 8—spreading direction and rate; 9—direction of plate movement.

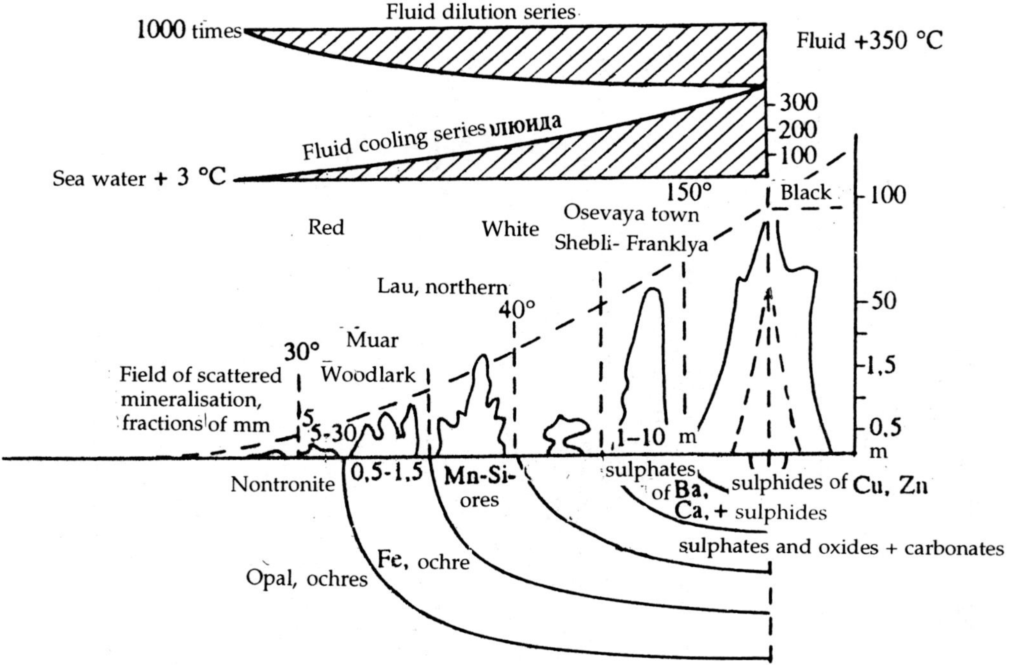

Fig. 5.12. Dependence of size, shape and composition of hydrothermal structures on temperature and cooling and dilution sequence of fluids (Lisitsyn et al., 1993). Spatial zoning of different types of structures with increase in distance (and cooling) from the hydrothermal centre, also observed in some bodies, are shown below (see Fig. 5.13).

a zinc content up to 11 and copper up to 1.2 wt, %. On the periphery of the lode, these minerals are replaced by pyrite, marcasite, late sphalerite, covellite and iron oxide. Zoning of hydrothermal non-ore minerals is also observed as depicted in Figure 5.13. On the periphery, hydrous silicates are replaced by anhydrite, illite, barite, pyrite (zone 3) and calcite, illite and pyrite (zone 4).

Ores and hydrothermal alterations in MOR resemble the hydrothermal systems in islands arcs and back-arc basins although some differences do exist (see Sec. 5.5).

5.2. Thermophysical Modelling of Spreading Zones in Mid-oceanic Ridges

The following characteristics of mid-oceanic ridges are most important for thermophysical modelling:

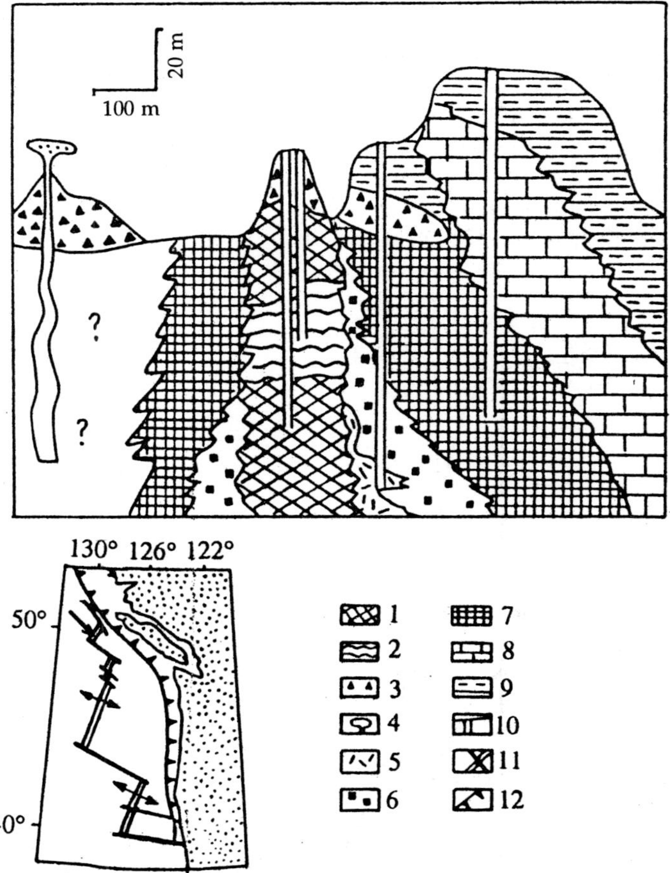

Fig. 5.13. Profile through Bent Hill ore hill in the hydrothermal fields of Central Rift Valley, Juan de Fuca ridge (Goodfellow and Franclin, 1993; cited by Zaikov and Zaikova, 1994). 1—massive sulphides; 2—massive sulphides with an admixture of quartz and chlorite; 3—sulphide breccia; 4—active hydrothermal source; 5 to 8—hydrothermally altered rocks: 5—quartz + chlorite + muscovite + rutile + chalcopyrite + pyrrhotite, 6—albite + chlorite + muscovite + pyrite, 7—anhydrite + barite + pyrite, 8—calcite + pyrite; 9—hemipelagic deposits; 10—boreholes; 11—rift valleys and transform faults; 12—zone of subductions.

1) Constant average reduction of ridges height with increasing distance from its axis which is expressed by the following empirical equation for fast-spreading ridges:

$$\Delta H = 0.35 \sqrt{t},$$

where ΔH is the increase in depth in metres and t the age of the ocean floor in million years (proportional to the thickness of the lithosphere). The size of the ridge normal to its axis is 3000 to 5000 km.

2) Transform faults occur normal to the ridge axis and repeat at intervals of 50 to 60, 100 to 150 or 300 to 400 km. A typical zone of faults occurs as a linear depression 1 to 2 km deep and 10 to 30 km wide.

3) Measurements of heat flux reveal their increase by 2 to 5 times on the ridge axials from the average values at the bottom of the oceanic plateau (Parsons and Sclater, 1972; Sorokhtin and Ushakov, 1991; Stein and Stein, 1992) (Fig. 5.14). Moreover, reductions in heat flux are associated with zones of transform faults (Sclater and Francheteau, 1970; Stein and Stein, 1992).

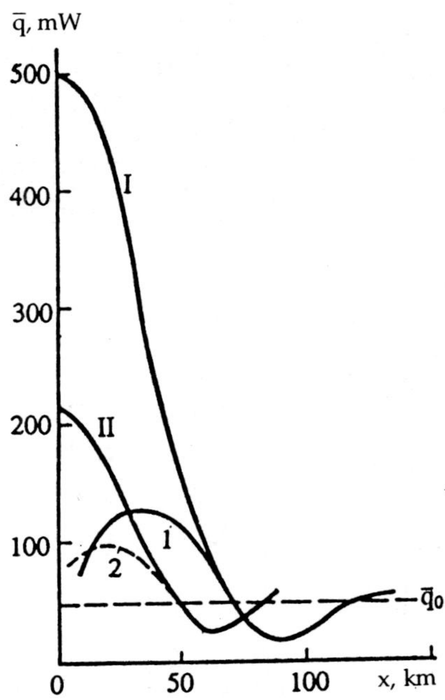

Fig. 5.14. Average experimental profiles of heat flux across the strike of the rift valley of the Mid-Atlantic ridge:
1—from Researchers' faults to Azores faults; 2—from the region of Azores islands to Gibbs fault. I and II—profiles of total heat fluxes were obtained as the sum of convective components of heat flux calculated respectively from curves 1 and 2 (Lyubimova et al., 1988).

4) Seismic observations show that the thickness of the lithosphere is least on the ridge axis (5 to 15 km) and increases to 70 to 80 km with increase in distance from the ridge axis (Karson, 1988; Sorokhtin and Ushakov, 1991; Zonenshain and Kuz'min, 1993a; Perfit et al., 1994). The lithosphere is underlain by asthenosphere of low viscosity at depths of up to 400 to 420 km. From experimental investigations of the melting temperature of basalt rocks, it may be suggested that the boundary

of phase transition (partial melting) corresponds to a temperature of 1100 to 1200°C (Ringwood, 1974; Zharkov, 1983).

Thus the asthenosphere under mid-oceanic ridges can be represented in the form of a flat horizontal layer with average thickness $2\,l = 420 - 80 = 340$ km (about 400 km in the ridge axis) and extent $2\,x_0 \geq 4000$ km, i.e., $x_0/l >> 1$. The monotonic decrease in ridge height is associated with reduction in average temperature in the layer, which causes: a) crystallisation at the bottom of the lithosphere and increase in its thickness and weight; and b) change in density of the asthenosphere. The change in pressure in the asthenosphere beneath the mid-oceanic ridge can be determined using the following equation:

$$(1/\rho)\,(dP/dy) = \beta\,gT\,(x, y). \tag{5.1}$$

Here, P is the pressure reckoned from the lower boundary of the asthenosphere at $y = -l$, where $P_{-l} = $ const; ρ is the density; β the coefficient of volume expansion; g the gravitational acceleration; and T the temperature.

Equation (5.1) is integrated for the thickness of layer $2\,l$:

$$P = \int_{-l}^{+l} \beta g\,\rho T\,(x, y)\,dy. \tag{5.2}$$

Equation (5.2) can be rewritten as:

$$\Delta P = P - P\,(x_0) = \beta g\,\rho(\overline{T}(x) - \overline{T}(x_0))\,2l, \tag{5.3}$$

where ΔP is the difference in pressure in sections x and x_0; $P(x_0)$ the pressure in section $x = x_0$ and $y = -l$; $\overline{T}(x)$ the mean integral temperature across the layer in section $x = $ const; and $\overline{T}(x_0)$ the mean temperature along the thickness of the layer at $x = x_0$.

Pressure deficit ΔP should be compensated by the height of the lithospheric column:

$$\Delta P = g\,(\rho_l - \rho_{H_2O})\,\Delta y = g\,\beta\rho\,(\overline{T}(x) - \overline{T}(x_0))\,2l, \tag{5.4}$$

where ρ_l is the average density of the lithosphere in the mid-oceanic ridge and ρ_{H_2O} the density of water.

In this case, we find:

$$\Delta y = [\beta\rho\,(\overline{T}(x) - \overline{T}(x_0))]\,2l/(\rho_l - \rho_{H_2O}), \tag{5.5}$$

where Δy is the height corresponding to the pressure deficit in the section $x = $ const and characterising the uplift of the ridge relative to section $x = x_0$ (Fig. 5.15).

Equation (5.5) and Fig. 5.15 are simplified as they do not take into account the change in density of lithosphere, $\rho_l = f(T, x, y)$, during crystallisation of its bottom and $\Delta\rho$ due to change in relief of the top of the asthenosphere. Nevertheless, they indicate a definite influence of the monotonic temperature variation in the asthenosphere on the relief of the mid-oceanic ridge. Since height is maximum at $x = 0$, the value of average temperature is maximum at $x = 0$. Measurements of heat flux confirm this dependence.

The gradient of average temperature along x is less than zero ($d\overline{T}/dx < 0$). The foregoing discussion clearly shows that even for an extremely small value of horizontal temperature gradient in the gravitational field, horizontal flows arise in the horizontal liquid layer. These flows proceed with no stability threshold with vigour proportional to the value $d\overline{T}/dx$ in the layer (Kirdyashkin et al., 1983). From the monotonically decreasing average height of the ridge, it follows in accordance with eqn (5.5) that the horizontal flow arising in the asthenosphere is one-celled (on one side of the ridge axis, Fig. 5.15). In a multicell flow along x, the average temperature, relief and heat flux would change periodically.

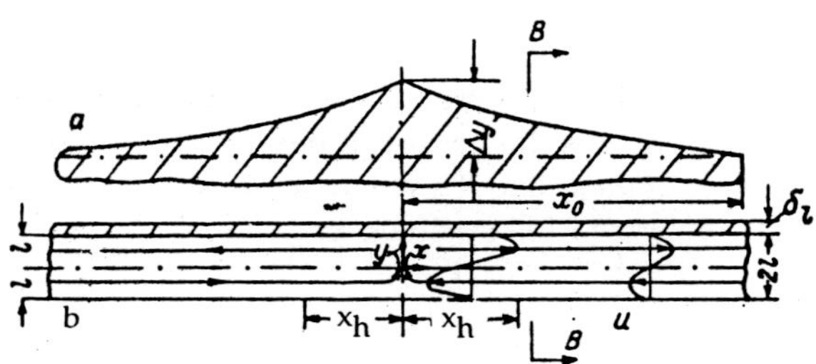

Fig. 5.15. a—Sketch showing relief of mid-oceanic ridge and b—model of asthenospheric flow beneath mid-oceanic ridge (magnified along y).

Conditions of unstable stratification may arise at the cooled upper surface of the asthenosphere-lithosphere interface since the temperature in the asthenosphere increases with increasing distance from this boundary. Close to this surface, the origin of convection rolls is possible with the axis of rolls along the direction of the mean horizontal flow. The presence of longitudinal rolls with axis normal to the axis of the mid-oceanic ridge should influence the relief of the ridge and explain the origin of transform-fault basins, at least of the so-called 'zero transform' faults along which there are no displacements. If such are present, they are insignificant while variations in the relief are of the same order (1 to 2 km). The point is that, in the region of the descending flow of a roll, the least temperature is observed and hence the surface of the ocean floor will descend. The maximum temperature occurs in the region of the ascending flow in the roll and thus a relative rise of the floor will be observed (Fig. 5.16).

These premises were confirmed by experimental and theoretical investigations (Kirdyashkin, 1989) which also helped in quantitatively evaluating the aforesaid effects. Before taking up their brief description, it should be emphasised that the above premises, experimental modelling and their consequences pertain to a simplified situation in which there is no movement of lithospheric plates or it has not commenced. Nevertheless, it is important to understand that the flow in a thin horizontal layer of the asthenosphere and the corresponding relief arise even in the absence of plate movements as a consequence of local heat flux in the axial zone of the mid-oceanic ridge and as a consequence of a horizontal temperature gradient. The subsequent plate divergence (as a consequence of flows as well as

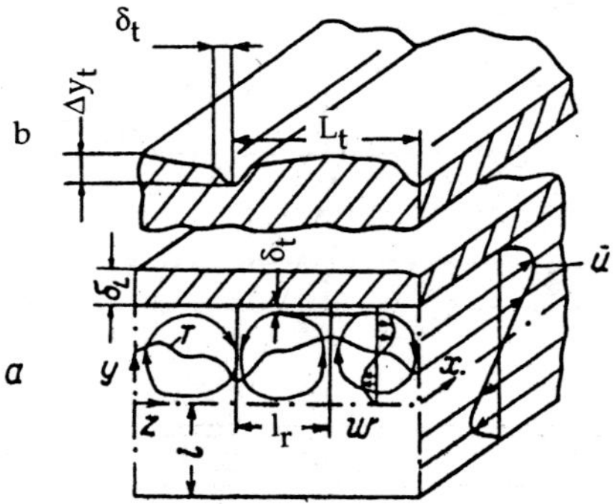

Fig. 5.16. Model of free-convection flows in the asthenosphere.
a—flow structure (section along BB of Fig. 5.15) and b—effect of temperature field on relief of the floor.

processes in the subduction zone) introduces significant corrections to the estimates obtained (Kirdyashkin, 1989).

As pointed out earlier, the height of an oceanic ridge diminishes due to a reduction in the average temperature of the asthenosphere with increase in distance from its axis. The question arises about the conditions requisite for a plane one-cell flow in a flat layer cooled from above. Experimental studies were carried out to understand this subject.

One-cell flow in a layer $x_0/2l >> 1$ can be generated, for example, by heating the layer at one end and cooling it at the other; in a general case, one-cell flow in the layer $x_0/2l >> 1$ is possible under conditions of a longitudinal temperature gradient that remains constant along the axis as well as changes monotonically (without change of orientation of temperature gradient vector $\partial T/\partial x$).

In the horizontal layer $x_0/2l >> 1$ under conditions of thermogravitational convection and isothermal cooling from above, one-cell flow (along x) arises in the case of lateral heat input and thermally insulated lower surface (Fig. 5.17a). In this case, the flow in the upper half of the layer is oriented towards the temperature gradient $\partial T/\partial x$ (away from the heater) and parallel to the vector of longitudinal temperature gradient in the lower half of the layer (toward the heated end). At distance x_1 from the heated end, the temperature in the layer becomes equal to the temperature of the cooled surface.

Experimental investigations of the structure of thermogravitational flows were carried out in a flat layer $x_0/2l >> 1$ while heating from below and cooling from above. In a cuvette of organic glass (with horizontal dimensions 65×200 mm), a flat heater ($2\,x_h$) was placed symmetrically relative to section $x = 0$ and its size ($2\,x_h$) varied relative to the layer thickness ($2l$). The height of the layer was recorded by calibrated plates which also restricted the length of the layer ($2\,x_0$).

The upper transparent heat exchanger (cooler) rested on the calibrated plates. The lower bounding surface and lateral ends were heat insulated.

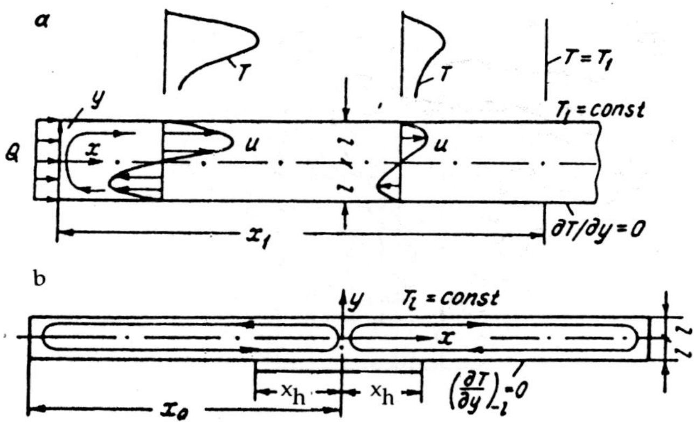

Fig. 5.17. Fluid flow in a layer.
a—lateral heat input and cooling on top; b—cooling on top and heat input from below.

The nature of liquid flow in the layer when the heater size $x_h \leq 4\,l$ and $x_0 < x_1$ is shown in Figure 5.17b; in these cases, two symmetrical cells are formed with a plane of symmetry and ascending flow in section $x = 0$. At $x_0 < x_1$, turning of the flow by 180° is observed adjacent to the end; at $x_0 > x_1$, in the region $x > x_1$, there is no flow and the liquid temperature is equal to the temperature of the cooler surface (Fig. 5.17b).

The flow pattern in a layer of ethanol was visualised using aluminium particles dispersed in the liquid. Adjacent to the cooled surface, stable rolls arise in the region around $x = 0$. The average horizontal size of two rolls is $2\,l$ (see Fig. 5.16). Thus, rolls arise in the upper part of the layer at the cooled surface and are situated in the liquid flow moving away from the heater.

For a relative size of heater $x_h \geq 4l$, in a flat layer cooled from above and adiabatic lower surface, at $x > x_h$ $((\partial T/\partial x)_{-l} = 0))$, the flow structure is complicated in the region $0 < x < (x_h - 4l)$ (Fig. 5.18).

In the region $+(x_h - 4l) > x > -(x_h - 4l)$, polygonal cell flows are developed with random disposition of cells (see Fig. 5.18). In the region $(x_h - 4l) < x < x_1$ (to the right and left of the heater), a one-cell flow is formed and peters out at distance

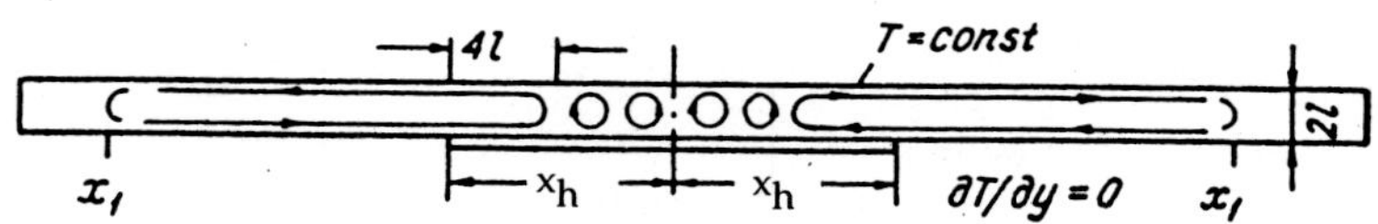

Fig. 5.18. Flow in a flat layer for $x_h > 4l$.

x_1. At $x > (x_h - 4l)$, the flow is one-cell on average with longitudinal convection rolls superposed on it at the upper cooled surface. At $x_h > 4l$, these longitudinal flows are unstable because of perturbations arising in the region of the polygonal cell structure $(0 < x < (x_h - 4l))$. The stable character of longitudinal rolls is observed when $x_h < 4l$ and a polygonal cell structure cannot exist around $x = 0$ (see Fig. 5.17b). Thus, the size of the heater can also be judged from the nature of development of longitudinal rolls: $x_h < 4l$ when these flows are stable and polygonal cell structures are absent in the vicinity of $x = 0$. Experimental investigations with visualisation of flow using aluminium particles 510 μm in size showed that the horizontal roll size is equal to one-half of the thickness ($l_c = l$). The distance between the descending flows representing the size of two convection rolls is equal to the thickness of the layer (see Fig. 5.16).

At $x_h < 4l$, the heat supplied to the flow on the lower heating surface in the region of $x = 0$ is carried by vertical flow to the upper cooled surface in the narrow zone $0 \leq x \leq l/2$ (Figs. 5.19 and 5.20) and it can be thought of as heat transmitted to the end of the layer $x = 0$.

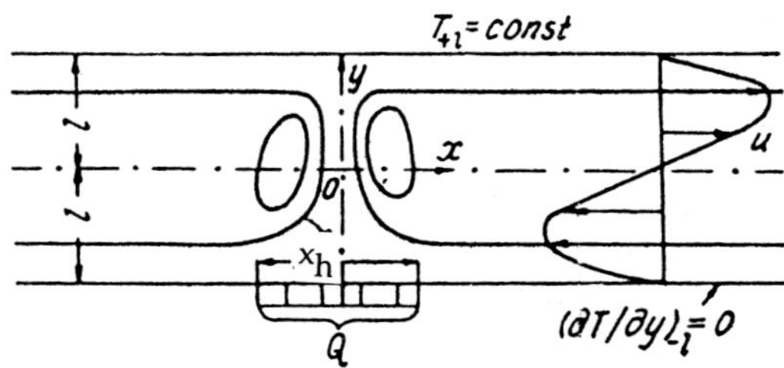

Fig. 5.19. Fluid flow in a horizontal layer on local heat input at the bottom (x_h) and cooling from above.

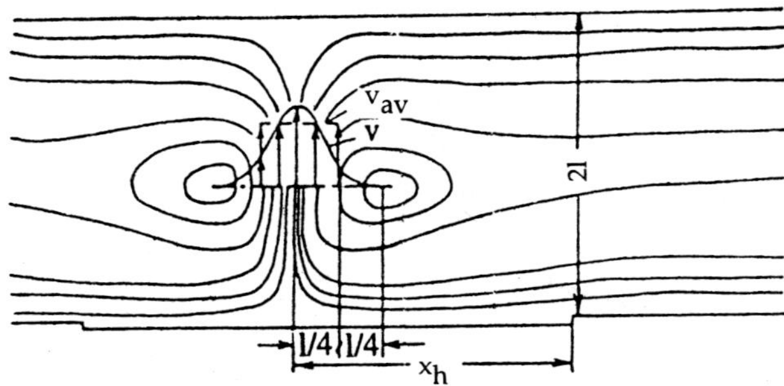

Fig. 5.20. Experimental flow lines in a glycerine layer ($2l = 19$ mm; $x_h = 1.4\,l$; $Q = 190$ W/m). Also shown is the profile of vertical velocity v and average velocity v_{av}, the latter adopted in the calculations.

The size of heater placed symmetrically relative to $x = 0$ on the lower surface influences the nature of distribution of heat transfer coefficients in the vicinity of a plane of symmetry $x = 0$.

For a relative size $x_h < 4l$ the specific heat flux has the maximum value around $x = 0$ and decreases as the value of x increases.

The heat flux on the upper cooled surface was measured in a layer of octadecane at $2l = 20$ mm, its melting point being 28.2°C. The upper cooled surface (duralumin) was isothermal, $T_l = 19.0$°C. The heater ($2x_h = 10$ mm; $l = 0.5$) was placed on the lower heat-insulated surface.

The temperature fields and profiles of longitudinal velocity component were measured; for known thickness of crystallised octadecane on the cooled surface (δ_l), local values of specific heat flux were determined: $q = (T_1 - T_{melt})/\delta_l$. The thermal conduction coefficient of solid octadecane at $T = 19$°C was equal to 0.342 W/m · K and at $T = 28.2$°C, $\lambda_{melt} = 0.335$ W/m · K (Kirdyashkin, 1989).

As the heat transfer regime was established, the upper cooled plate together with the octadecane crystallised at its surface was raised and local and average thicknesses of crystallised octadecane were measured for different values of x.

The average thickness of the octadecane layer for $x = $ const and calculated values of average specific heat flux for the corresponding values of x are shown in Fig. 5.21. It can be seen that the ascending flux is localised in the region $0 \le x \le l/5$ where the maximum value of heat flux has been observed. Changes of temperature profiles in the section $y = $ const in the vicinity of $x = 0$ also show that the change of temperature in section $y = 0.4$ occurs in the region $0 \le x \le l/6$ and heat fluxes decrease sharply in the region $0 \le x \le 0.4\,l$.

A maximum heat flux much greater than the average value $q_{max}/\sqrt{q} = 3$ (Fig. 5.21) was recorded at heater size $x_h/l = 0.5$. The relative size of heat source x_h/l

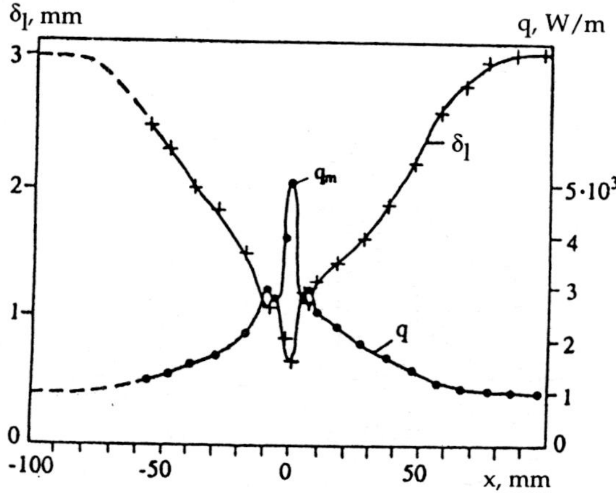

Fig. 5.21. Average thickness values of crystallised octadecane (δ_l) along the width of the cooled plate and values of heat flux (q) for different values of x at $2l = 20$ mm, $x_h = 0.5\,l$, $T_1 = 19$°C and $x_0 = 100$ mm.

can be judged from the nature of variation of heat flux in the region $x = 0$ and relative increase of $q_{max}/\sqrt{q}$.

Measurements of heat flux on the ocean floor of spreading oceanic ridges showed that heat flux at the axis of the ridge are 2 to 5 times more than the values away from it. The nature of variation of heat flux is such that the heat flow are maximum on the axis of the ridge in a zone of ~ 50 km and decrease rapidly with increase in distance from it to $x > 100$ km when the values are comparable to the average heat flux. This indicates the localisation of heat input adjacent to the axis of oceanic ridges.

The above experimental results show that $q_{max}/\sqrt{q} = 3$ for relative heater size $x_h = 0.5l$ on the lower surface and that the region of sharp reduction in heat flux is observed at $0 < x < 0.3\,l$ (see Fig. 5.21). Thus, the zone of heating on the lower surface even for the Atlantic mid-oceanic ridges should be comparable to the values of l.

Based on the discussion presented above and geophysical data, the model of thermogravitational flows in the asthenosphere in the region of oceanic ridges can be represented.

1) Let us study the steady free-convection heat transfer in a chamber of size x_0 depending on the wideness of the ridge (see Fig. 5.15).

2) The total heat flux transmitted from the asthenosphere to the lithosphere is equal to $Q = \bar{q}x_0$, where $\bar{q}$ is the average value of specific heat flux from the ocean floor to water and x_0 the horizontal site of the asthenosphere under the ocean.

3) Based on the monotonic decrease in average temperature in the layer of the asthenosphere (i.e., monotone decrease in average height of the ridge with increasing distance from its axis) and the nature of variation of heat flux along the ridge axis (q_{max} on the ridge axis, see Fig. 5.21), the mean flow is one-celled.

4) From the nature of variation of heat flux at the ridge axis (for the Atlantic mid-oceanic ridges, $q_{max}/\sqrt{q} \approx 3$ and zone of sharp reduction of heat flux at $x > 100$ km) it follows that for a one-cell flow, the heat enters the vicinity of the ridge axis and its amount is equal to the amount of heat transferred through the lithosphere and transmitted to the ocean, $Q = \bar{q}\,x_0$ (see Fig. 5.21).

5) On input of heat to the asthenosphere in the vicinity of the ridge axis $x = 0$ and cooling on top at the lithosphere-asthenosphere interface, conditions of unstable stratification arise adjacent to the cooling surface. According to the experimental data, in this case at Ra $= \beta g \Delta T l_r^3/a\nu > 1700$ near the cooled surface, rolls are formed with axes parallel to the flow direction (see Figs. 5.18 to 5.20). In the descending flows of rolls a reduction in temperature occurs, which is correspondingly reflected on the relief of the ocean floor around the ridge axis as reduction of floor level (see Fig. 5.16). The zone of reduction of ocean-floor level can be identified with the 'transform faults'. The distance between two adjacent faults (along the ridge axis) L_{tr} corresponds to the distance between two longitudinal rolls $2l_r$. It follows from the aforesaid experiments that $2l_r$ corresponds to the layer thickness, i.e., $2l_r = 2l$. For the Mid-Atlantic ridge, in the region along the width between the subtropics, $L_{tr} = 250$ km on average; thus, the asthenospheric zone with free-convection flow is equal to $2l = L_{tr} = 2l_r = 250$ km (see Fig. 5.16). For more typical modes of 100 and 350 km, the asthenospheric thickness variates from 100 to 350 km.

208

The temperature along the asthenospheric layer is variable and, in the zone of high temperature, viscosity is less than in the boundary zone of the asthenospheric layer (as also, probably, the velocity of transverse seismic waves which decreases in the zone of higher temperatures).

When analysing the structure of thermogravitational flows, effective viscosity is adopted as an average for the entire layer since the rheological properties of the asthenosphere are not known.

6) For the Mid-Atlantic ridge, $q_{max}/\sqrt{q} \approx 3$ and the zone of sharp reduction of heat flux is ≈ 100 km, i.e., the relative dimension for which heat fluxes decrease rapidly is comparable to the half-thickness of layer l. The stable nature of transform faults is seen along the ridge, i.e., stable nature of longitudinal rolls.

According to experimental investigations, the above characteristic patterns of mid-oceanic ridges show that the heated zone along the lower surface of the convecting part of the asthenosphere is comparable to thickness l (at the least, $x_h < 4l$). In this case, as can be seen from the experiments, the zone of ascending flux around $x = 0$ is $0.5\ l$ and the flux in the asthenosphere in the zone $x_0 > x > l$ may be regarded as flux in a flat layer $x_0/l >> 1$ heated from one end ($x = 0$) and cooled from above; the lower limiting surface for $x_0 \geq x > x_h$ is adiabatic where $(\partial T/\partial y)_{-1} = 0$.

The lower surface is regarded as adiabatic because the heat flux from the mantle through this surface is much less than the heat flux at the cooling lithsophere-asthenosphere boundary and much less than in the region of the ridge axis. The lower surface, according to petrological and geophysical data, represents the boundary of phase transition and the temperature at depth 420 km, i.e., the isotherm is ~1560°C.

The problem henceforth lies in a study of thermogravitational flows in a flat horizontal layer with bounding steady (or non-steady) surfaces, the upper one of which is cooled (isothermal) and the lower one adiabatic when the heat input is in the neighbourhood of one of the ends (see Fig. 5.19).

The solution to the problem for a flat layer $x_1/2\ l >> 1$ has been detailed in a monograph by Kirdyashkin (1989). The origin of co-ordinates is on the ridge axis $x = 0$ and at the centre of the layer; the co-ordinates of the bounding horizontal surfaces are $y/l = +1$. Hence the following relations were derived to determine the velocity and temperature profiles in a layer, taking into account rolls on the upper cooled surface:

$$u = (2T_{max}\ \beta g l^3 /x_1\ v)\ (1 - x/x_1)\ u(y/l), \tag{5.6}$$

$$T = T_{max}\ (1 - x/x_1)^2\ \theta(y/l). \tag{5.7}$$

with

$$T_{max} = 3.19\ K_{ef}^{-1/3}\ \mathrm{Ra}_0^{-1/3}\ (Q/\lambda), \tag{5.8}$$

$$x_1 = 0.244\ l K_{ef}^{-2/3}\ \mathrm{Ra}_0^{1/3}, \tag{5.9}$$

and $\mathrm{Ra}_0 = \beta g l^3\ Q/a\ v\lambda.$

The effective value of the thermal conduction coefficient (λ_{ef}) was determined using the following equation:

$$\lambda_{ef} = K_{ef}\lambda = \lambda\{1 + 0.038\,\text{Ra}^{1/4}\,[1 - (\text{Ra}_{cr}/\text{Ra})^{3/4}]\}, \tag{5.10}$$

where λ is the thermal conduction coefficient and $\text{Ra}_{cr} = 1700$. Equation (5.10) is applicable for $\text{Ra}_{cr} < \text{Ra} < 5 \times 10^5$.

The value of K_{ef} was determined by successive approximations: in the first approximation, from eqn (5.8) for $K_{ef} = 1$, $\text{Ra} = \beta g T_{max}l^3/a\nu$ was determined, and from eqn (5.10) the value of K_{ef}. In the second approximation, $\text{Ra} = \beta g T_{max}l^3/a\,\nu K_{ef}^{4/3}$ and using its value, K_{ef} was determined from eqn (5.10). Three approximations proved adequate for determining the value of K_{ef}.

The values of functions $u(y/l)$ and $\theta(y/l)$ are shown in Figure 5.22. The heat flux in the stabilised zone of flow (far away from the ridge axis) was determined using this equation:

$$q = (3.86\,\lambda_{ef}\,T_{max}/l)\,(1 - x/x_1)^2. \tag{5.11}$$

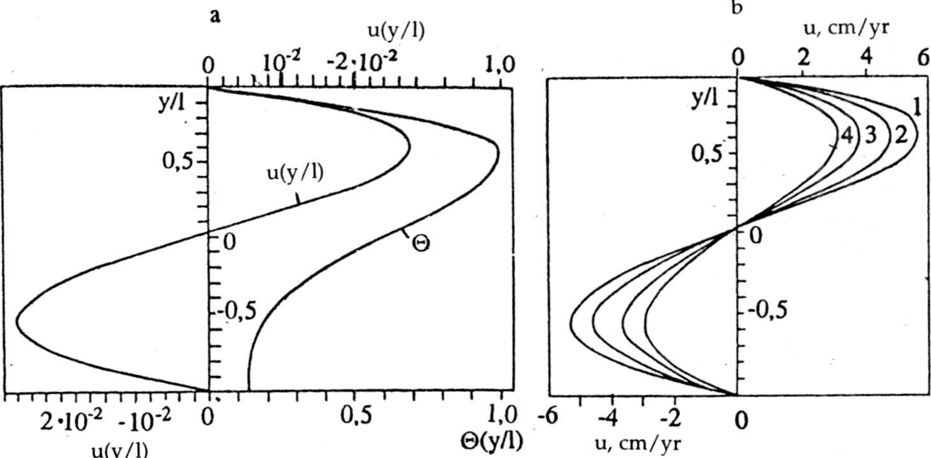

Fig. 5.22. a—Profiles of velocity (u) and effective temperature (Θ) in the horizontal layer on local heat input and cooling from above; b—velocity profile in the asthenosphere at $Q = 187.7$ kW/m; $x_0 = 2732$ km; $\nu = 2.2 \times 10^{15}$ m²/s; $x_1/l = 46.1$; and $2l = 250$ km. At x (km); 1—200, 2—1000, 3—2000 and 4—2700.

A comparison of theoretical premises with the experimental results for fluid layers differing in viscosity by three orders showed satisfactory agreement. It can be seen from the above eqns (5.6) and (5.7) that u and T are independent of Prandtl number $\text{Pr} = \nu/a$.

210

Experimental and theoretical investigations were also carried out at the upper cooled surface (around the ridge axis, $x = 0$).

The heat flux in the neighbourhood of the ridge axis was determined using the following equations:

$$q = (T_{max}\lambda K_{ef}^{2/3}/1.855)(3V_{av}/4al^2)^{1/3}[8.333 \times 10^{-2} + 0.5(x - 0.25)]^{-1/3}, \qquad (5.12)$$

$$V_{av} = \beta g l^2 T_m B_1/3x_1 \, v \text{ (for profile of } V_{av}, \text{ see Fig. 5.20).} \qquad (5.13)$$

At $x \leq 0.25 \, l$, the heat flux was constant.

The thickness of the steady lithosphere was determined from the following equation:

$$\delta_l = \lambda_k \, xT_1/q, \qquad (5.14)$$

where $T_1 = 1200°C$.

The following physical properties were accepted for the asthenosphere: $\beta = 3 \times 10^{-5} \, K^{-1}$; $\lambda = 2.51 \, W/m \cdot °C$; $a = 10^{-6} \, m^2/s$; $Q = \bar{q}x_0$, where $\bar{q} = 59.5 \, mW/m^2 = 1.42 \times 10^2 \, cal/m^2 \cdot s$.

Compressibility and disssipative processes could be ignored while analysing free convection flows in the conditions under study. The adiabatic temperature gradient was taken into account during plotting of the temperature field.

As pointed out before, the heat flux delivered to the ridge axis could be determined from the equation $Q = \bar{q}x_0$, where x_0 is the distance from the axis of the ridge to the continental slope. Assuming average temperature at x_0 to be higher than the temperature of the extant asthenosphere, equal to 1200°C, calculations were done using the above equations for $x = x_1$.

Calculations were made for different values of v and those values for which the calculated height of ridge uplift corresponded to the real values Δy for the region under consideration were accepted. As a result of calculations, the range of kinematic viscosity (v) was found to be $(2 \text{ to } 2.5) \times 10^{15} \, m^2/s$. According to the calculations of Zharkov (1983), $v = 6 \times 10^{14} \, m^2/s$.

The velocity profiles in the asthenosphere for different distances from the ridge axis are shown in Figure 5.22b. Velocity decreases linearly with increasing distance from the ridge axis, its maximum value being 5.7 cm/yr. In the region of continental slope ($x = 2730$ km), $u = 3.12$ cm/yr. The coefficient of friction $\tau = (v/\rho)(\partial u/\partial y)$ is equal to $6.2 \times 10^5 (1 - x/x_1)$, N/m^2.

The average value of the coefficient of friction at $x_0 = 2730$ km is: $\tau = 4.57 \times 10^5 \, N/m^2$ and the friction for the layer as a whole (x_0) per running metre along the ridge axis corresponds to $P = \tau x_0 = 1.25 \times 10^{12} \, N/m$.

The Ra number determined from T_m is equal to 10^5 and hence stable rolls prevail adjacent to the upper cooled surface. This denotes that origin of transform faults should be stable commencing from the ridge axis.

Heat flux and thickness of the lithosphere beneath the ocean for different values of x_0 and Q are shown in Figure 5.23. Values of heat flux correspond to the measured values of heat flux q at the ocean floor at different distances from the ridge axis. Heat flux shown in Figure 5.23 for $x > x_0$ pertains to the asthenosphere under the continent. It can be regarded as heat flow from the lower mantle of the Earth. In the continental lithosphere, the temperature field exists under the total influence of heat input from the asthenosphere to the continental lithosphere and in the continental crust under the influence of heat flux from the internal heat sources.

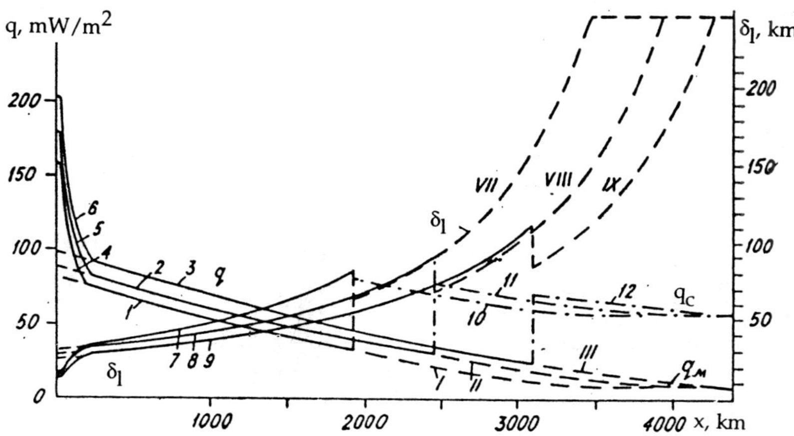

Fig. 5.23. Heat flux and thickness of the lithosphere of ocean and continent at $l = 125$ km. Heat flux to the oceanic lithosphere was calculated using eqn (5.11):
1—$v = 2.5 \times 10^{15}$ m²/s, $x_0 = 1932$ km, $Q = 136.74$ kW/m; 2—$v = 2.2 \times 10^{15}$ m²/s, $x_0 = 2448$ km, $Q = 166.63$ kW/m; 3—$v = 2.05 \times 10^{15}$ m²/s, $x_0 = 3009$ km, $Q = 196.2$ kW/m; heat flux curves 4, 5 and 6 were calculated using eqn (5.12); I, II and III—lower mantle heat flux q_M; 10, 11, and 12—heat flux at the continent. Thickness of the lithosphere beneath the ocean for x_0 (in km): 7—1932, 8—2448 and 9—3009; thickness of continental lithosphere for x_0 (in km): VII—1932, VIII—2448 and IX—3009.

The results of calculating the thickness of the oceanic and continental lithosphere are shown in Figure 5.23 for different values of Q and x_0. In this Figure, in the region of continental lithosphere, q_M and δ_l have been depicted as broken lines while the heat flow at the continental surface ($q_c = q + q_M$) is shown as dot-dash lines. It can be seen from Figure 5.23 that the thickness of the continental lithosphere attains the maximum value at $q_M = 7.75$ mW/m², becoming constant thereafter (calculations were carried out for average values of distribution of internal sources in the continental crust $Q_{in} = 2.65 \times 10^{-6} \exp(-y/1.5 \times 10^4)$ W/m³). With increasing distance from the continental slope, $x > x_0$ (in the neighbourhood of the continental and shelf margins), an increase in heat flux is observed because of internal heat sources and hence a reduction in thickness of the continental lithosphere in this region compared to the oceanic lithosphere. In this region a very high temperature level in the continental crust and more active geochemical processes should

212

be expected. It is well known that oil and gas pools are found in the continental shelf and slope.

For example, depths at which the interface lithosphere-asthenosphere occurs were calculated for the South Atlantic (Fig. 5.24). Five points (1 to 5) are shown in the Figure for which temperature profiles were calculated with due consideration of the adiabatic temperature gradient $(\partial T/\partial y)_a = 0.6°C/km$, shown in Figure 5.25. The calculated maximum temperatures in the mantle from the zone of mid-oceanic ridge (point and profile 1) to the continental crust (point and profile 4) range from 2000 to 1400°C.

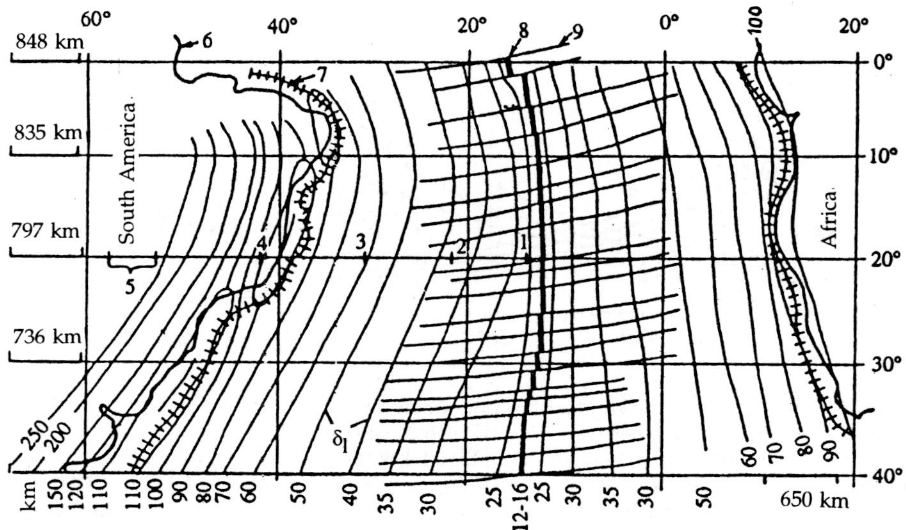

Fig. 5.24. Thickness of lithosphere of the South Atlantic and the South American continent: 1 to 5—regions for which velocity profiles (see Fig. 5.22) and temperatures (see Fig. 5.25) are shown: 6—coast; 7—continental slope; 8—axis of mid-oceanic ridge; 9—transform faults.

The high calculated temperatures (Fig. 5.25) compared to the estimated values reported in petrological studies (Dobretsov, 1980; Sobolev, 1996; Sobolev and Chaussidon, 1996) may be associated with the low value of $\lambda = 2.51$ W/m · °C adopted in the calculations. According to the estimates of Lyubimova (1968 and others) shown in Chapter 1, the thermal conductivity coefficient λ varies from 2.3 W/m · °C on the surface, decreases slightly at a depth of about 30 to 40 km and later rises to $\lambda = 20$ W/m · °C at a depth of 400 km, remaining constant at greater depths.

According to eqn (5.8), $T_m \sim \lambda^{-1/3}$. If $\lambda = 6.5$ W/m · °C is accepted for the asthenosphere, the temperature level decreases by 35 to 40%, i.e., remains at 1600 to 1000°C, this being close to the petrological estimates relative to H_2O content (Sobolev and Chaussidon, 1996).

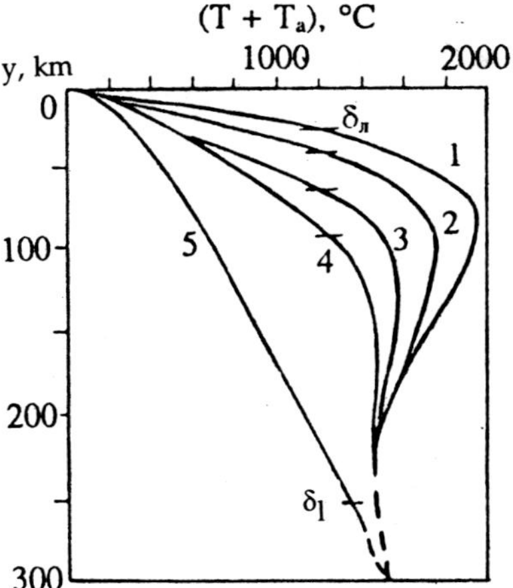

Fig. 5.25. Temperature distribution in the asthenosphere and oceanic (1 to 3) and continental (4 and 5) lithosphere (see Fig. 5.24) for x (km): 1—200, 2—1000, 3—2000, 4—3100 and 5— > 4140.

The values of the thickness of lithosphere (δ_l) presented in Figure 5.24 show that the continental lithosphere attains maximum thickness (200 to 250 km) at a relatively small distance from the continental slope.

As pointed out above, at the upper cooled surface (at the lithosphere-asthenosphere boundary), convection rolls arise due to conditions of unstable stratification. This flux extends to the half-thickness of the layer with one-directional flow and for a length up to x_1 at which $\text{Ra} > \text{Ra}_{cr}$.

The presence of longitudinal rolls with axes normal to that of the mid-oceanic ridge affects the floor relief. The maximum temperature in the zone of an ascending flow in rolls is comparable to T_{max} for section $x = \text{const}$. In the zone of an ascending flow in which the temperature distribution is uniform, a relative rise of the floor will be observed. In the zone of descending flow in roll, the least temperature comparable to that of the cooled surface will be at the lithosphere-asthenosphere boundary and hence a transform fault is formed. The width of the transform fault will be comparable to $2\delta_t$, i.e., the thickness of 'displacement' of the thermal boundary layer which is estimated (Fig. 5.26) by the following eqn (Kirdyashkin, 1989):

$$2\delta_t = [(l^3/8.64)\,(\text{Ra} - 1700)^{1/2}]^{1/3}. \tag{5.15}$$

For case $x_0 = 1500$ km, $v = 2.8 \times 10^{15}$ m^2/s and $\text{Ra} = 6.64 \times 10^4$ from eqn (5.15), it follows that $2\delta_t = 9.6$ km. This estimate of $l_{tr} = 2\delta_t$ has been derived without taking into account the temperature distribution along the width of the ascending flow in the roll and should be regarded as overestimated. The depth of transform fault is estimated from the condition of isostatic state of the floor surface:

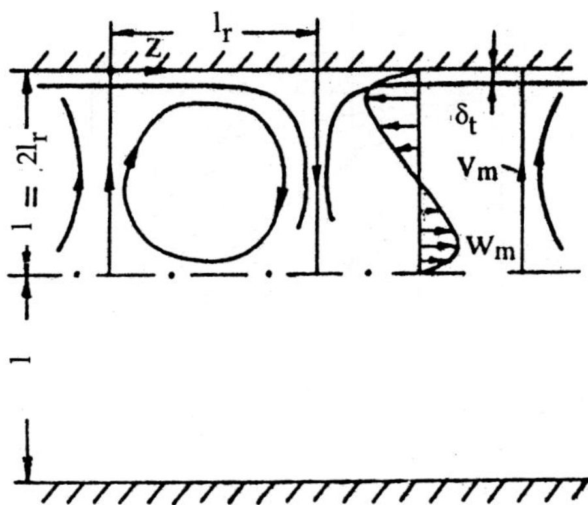

Fig. 5.26. Convection rolls at the upper cooled surface.

$$\Delta y_t = \beta \rho l \, (T_{z=0} - T_{z=1})/(\rho_t - \rho_{H_2O}). \tag{5.16}$$

The difference in average temperatures between ascending and descending flows in the roll $(T_{z=0} - T_{z=1})$ is estimated as $0.5\Delta T_r = 0.29 \, T_{max}$.

Equation (5.16) is rewritten as:

$$\Delta y_t = \beta l \rho 0.29 T_{max}/(\rho_t - \rho_{H_2O}). \tag{5.17}$$

In the region of the ridge axis at $x_0 = 1500$ km, $\Delta \rho = 1.85 \times 10^3$ kg/m^3 and $\Delta y_t = 1158$ m, the estimates obtained were high since the temperature drop between the ascending and descending flows was assumed as constant along their width and close to the maximum. Nevertheless, the derived values of l_{tr} and Δy_t correspond to the observed values.

Assuming that isostatic conditions of equilibrium are satisfied everywhere, the depth of transform fault will vary with increasing distance from the ridge axis as:

$$\Delta y_t = \Delta y_{t; x=0} \, (1 - x/x_1)^2. \tag{5.18}$$

As pointed out above, transform faults with displacement and 'zero faults', 'quiescent faults' and those filled partially with sediments are known. The first of these arise during displacement of the ridge axis on both sides of faults and is not a random event. In the zone of descending rolls, shear stresses develop simultaneously and cause displacements in the lithosphere. If the ridge axes are not displaced for two adjoining rolls, the zones of descending flow in the longitudinal rolls appear only as descent of the floor and represent the second type of fault.

, Thus, on the basis of analysis of the geophysical structure of the oceán floor and asthenosphere as well as experimental laboratory modelling, the flow in the asthenosphere can be represented as flow in a horizontal layer with heat rejection at the upper surface and heat input at the lower surface in the vicinity of the axis of the mid-oceanic ridges, even when the lower boundary surface is adiabatic ($\partial T/\partial y = 0$).

The solutions obtained helped in analysing the structure of thermogravitational flows in the asthenosphere and quantitative determination of the average viscosity of the asthenosphere (for known floor profile); the zone of existence of asthenosphere; temperature field in the asthenosphere and heat transfer coefficients at the lithosphere-asthenosphere boundary (see Fig. 5.25); thickness of the lithosphere under the ocean (see Figs. 5.23 and 5.24); velocity field in the asthenosphere (see Fig. 5.22) and coefficients of friction at the lithosphere-asthenosphere boundary and hence the forces acting on the continental lithosphere from the side of the oceanic lithosphere; as well as depth and width of transform faults. The thickness of the continental lithosphere and temperature distribution along its thickness were also found for average values of internal heat sources in the continental crust (see Figs. 5.23 and 5.25).

The solutions were obtained under the assumption that the lithosphere is stationary so forces acting on the continental lithosphere were not determined. To understand the possibility of continental displacements under the action of these forces, a knowledge of solid phase transition processes (eclogitisation) during subduction of the lithosphere is necessary. This knowledge enables estimating the buoyancy forces, friction forces and, as a first approximation, the rate of displacement of continents. These aspects are discussed in part in Section 5.7.

5.3. Passive Margins of Continents and Intracontinental Rifts

On the example of the Red Sea rift, we find (see Fig. 5.11) that development of oceanic rifts commenced with extension of the continental crust and formation of an intracontinental rift. These phenomena gave rise to secondary rifts (later becoming extinct) and transitional thinned lithosphere on which the passive ocean margins developed. However, continental rifts generally did not reach the stage of opening up of the oceans; they have their own characteristic mechanisms of development, studied here and in the next section.

The transition of intracontinental rifts into oceanic rifts and formation of passive margins were studied in most detail on the example of Atlantic rifts. Figure 5.27a shows rifts originating from the opening up of the South Atlantic (Burke and Dewey, 1973; Zonenshain and Kuz'min, 1993a). Apart from the main rift parallel to the future Atlantic Ocean, there were also Early Mesozoic rifts diverging at an angle. They were active in the initial period of the opening up to the Atlantic but later became extinct later (Tanatu, Amazonka, Benue, Parana etc.). Of these, the most characteristic is the Benue rift at the apex of the Gulf of Guinea. At the moment of opening up, when Africa and South America were together, three rifts converged at one point like the present-day joint of the East African, Red Sea and Aden rifts (compare Figs. 5.11 and 5.27).

Burke and Dewey (1973) reached the conclusion that the initial break-up of the lithosphere resulting in the opening up of the ocean occurred at one or many points in which a three-branched rift system arose at an angle of 120°. Cleavage

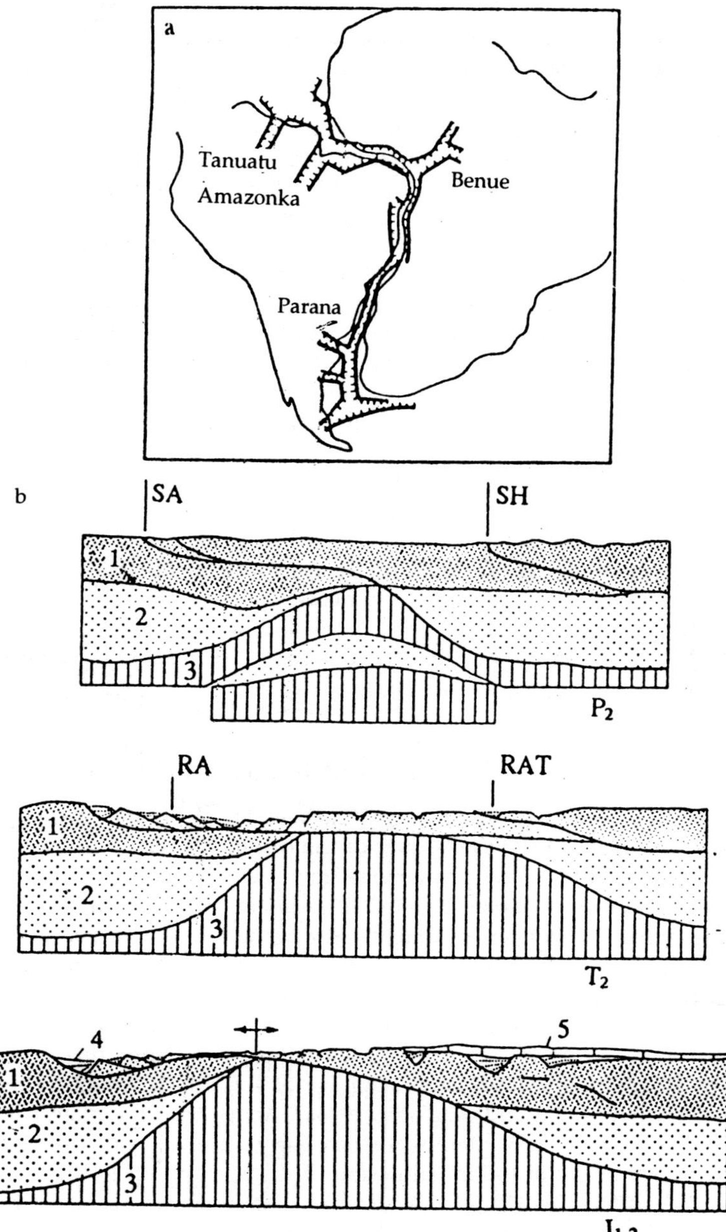

Fig. 5.27. Development of southern and central Atlantic.
a—three-branched rifts that prevailed during the opening up of the Atlantic. Extinct rifts terminate blindly inside the continent (Zonenshain and Kuz'min, 1993a); b—asymmetric rifting in the central Atlantic (Picque and Laville, 1993). Primary sutures SA and SH and Atlantic (RA) and Atlas (RAT) type rifts, the latter two arising in the Triassic, are shown. In the Jurassic, the opening up of the Atlantic Ocean commenced: 1—crust; 2—lithospheric mantle; 3—asthenosphere; 4—rift sediments; 5—post-rift carbonate sediments.

and detachment of the continents occurred later only along two of the three branches; the third became extinct and diverged sideways together with the continent. The Benue rift, formed during the opening up of the Atlantic, was such an extinct branch. In the geological future, the East African branch of the Afar system may become such a branch when a wide oceanic space arises in the course of spreading at the site of the Red Sea and Aden rifts. Such examples of extinct three-branched rifts existed even during the opening up of other present-day and palaeo-oceans: Wachita system on the North American platform during the opening up of the North Atlantic in Triassic; Donetsk rift during the opening up of the Palaeo-Tethys in the Devonian; and Viluy aulacogen during the opening up of the Palaeo-Asiatic Ocean in the Riphean (McKenzie and Morgan, 1969; Burke and Dewey, 1973; Zonenshain and Kuz'min, 1993a).

Apart from three-branched rifts, there are, and have been, rifts and basins parallel to the main 'oceanic' rift. Examples are the rift of the Atlantic coast of North America—Newark graben, Caroline basin etc. (Klitgord and Hutchinson, 1988, and others), passive margin of northern Africa, the Bay of Biscay (Le Pichon and Barbier, 1987) and other sites.

Figure 5.27b depicts the development of the central part of the Atlantic Ocean in the Permian-Early Jurassic represented as a model of asymmetric rifting (Picque and Laville, 1993). In the Late Permian, thickening (doubling) and uplift of the asthenosphere occurred in the central part of the Permian Pangaea (see Fig. 1.13) and caused some oblique detachment faults in the lithosphere. In the Triassic, there was further uplift of the asthenosphere and movements along gentle slopes led to the formation of a rift system of the Atlantic type in the west and the Atlas type in the east. In the Early Jurassic, Atlas type rifts became extinct and were covered by shallow subplatform sediments while Atlantic type rifts continued to develop until and easternmost of them situated above the axis of the asthenospheric uplift became the spreading axis of the Atlantic. The other rifts became extinct but are seen in a fossil state in the Atlantic passive margin of North America together with thick formations of the passive margin (see Figs. 5.29 and 5.30).

Figure 5.28 shows the general model of asymmetric rifting leading to the formation of different types of passive margins (Wernicke, 1985; Le Pichon and Barbier, 1987). In this model, the 'effectively hard' (brittle) crust experienced simple shear over the entire thickness of a single fissure (zone) dipping at an angle of 10 to 20°, along which the acute margins of the crust diverged, as a result of which the crust between them became thin. Smaller listric-type faults developed in the overhanging plate. Gravitational sliding of inclined crust blocks occurred along such faults, as a result of which the crust became thin over an extensive area. Apart from this model, perhaps applicable to the opening up of several oceans, there are alternative models of symmetrical extension and rifting which are more characteristic of intracontinental rifts. These are discussed below.

During the formation of passive margins, thick sedimentary prisms formed and often house major oil and gas deposits. A general mode of formation was 'avalanche sedimentation' i.e., the rapid redeposition of shelf deposits downwards along the continental slope by longitudinal and transverse flows with the formation of different types of turbidites (Lisitsyn, 1978, 1991).

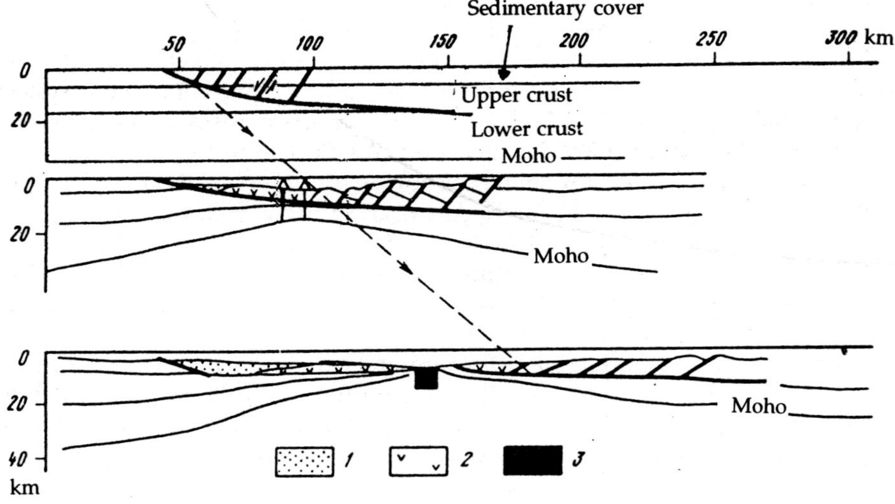

Fig. 5.28. Model of the formation of intracontinental rift and passive margin according to Wernicke's model (Wernicke, 1985).
1—synrift sediments; 2—synrift volcanites; 3—oceanic crust.

In the tectonic plane, many cases of such massive sedimentation can be distinguished: 1) development of parallel of secondary basins with migration of sediments in the direction of the ocean under development (Atlantic coast of North America); 2) development of deep sedimentary basins at the junction with major transform faults (Nova Scotia basin south of Newfoundland (Erickson, 1993)); and 3) delta formation of major rivers such as the Bengal, Amazon and Mississippi cones (Meyer et al., 1992).

Examples of the first and second types are shown in Figures 5.29 and 5.30. The thickness of deposits of passive margins goes up to 10 to 12 km (Fig. 5.30). Gradual progressive accretion of sedimentary prisms towards the ocean (Blake Plateau and Central Baltimore Canyon) as well as erosion contacts and redeposition in the deepwater part of the continental slope (Georges Bank and Caroline trough) are observed. Models of such redeposition have been studied in the monographs of Lisitsyn (1978), Klitgord and Hutchinson (1988), Meyer et al. (1992) and Erickson (1993). These troughs with a sharp structural asymmetry on the cold lithosphere of the passive margins are unstable in many cases and may lead to the development of subduction zones (on transition to compression) or a new rifting and/or shifting of the spreading zone (see below).

Turning to intracontinental rifts, it is important to note that there is wide variety in this respect and the term 'rift' is often used as a synonym for the 'extension zone' which, in many cases, is inadequate or not real for explaining the observed system of sedimentation and volcanism. The following can be distinguished as a first approximation:

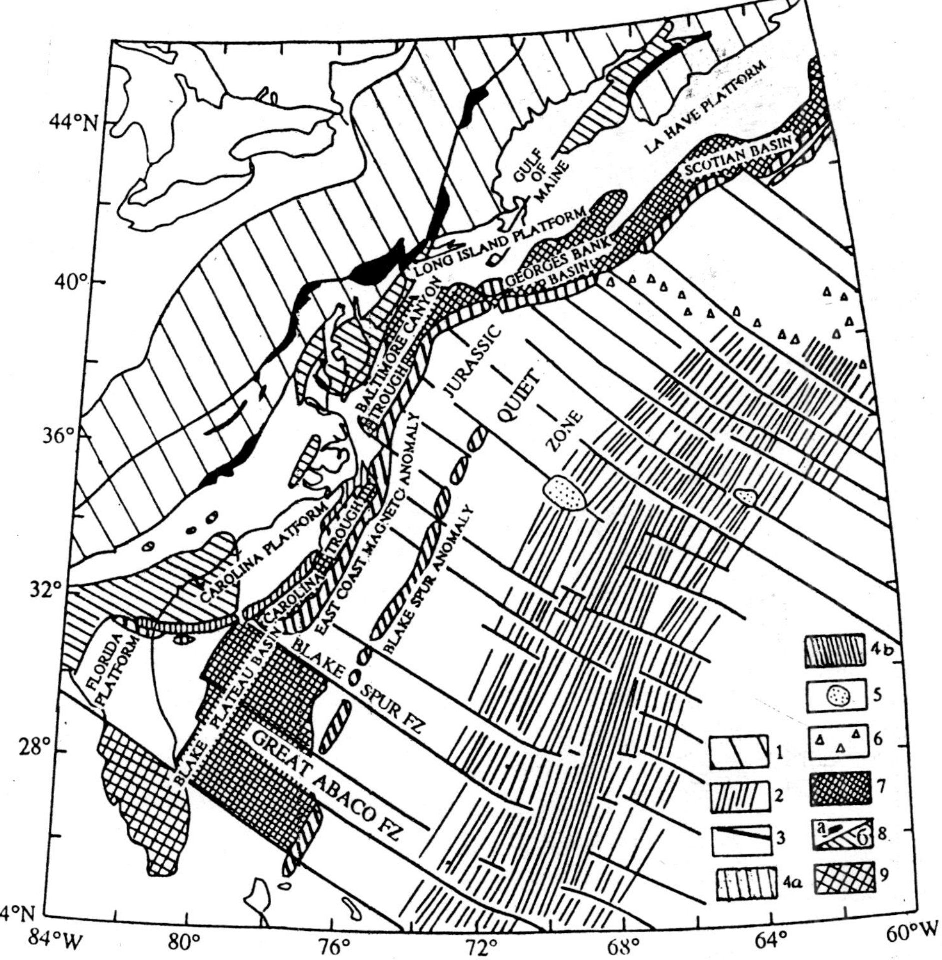

Fig. 5.29. West Atlantic continental margin and adjacent part of the Appalachian folded region (Klitgord and Hutchinson, 1988).
1—Appalachian folded region; 2—Mesozoic oceanic floor with linear magnetic anomalies; 3—transform zones; 4—magnetic anomalies indicating a) uplifts and b) troughs; 5—basement salients; 6—New England chain of seamounts (track of hot spot); 7—Mesozoic basins; 8—Triassic basins: a) exposed and b) their assumed continuation; 9—Triassic and Jurassic volcanites.

a) circumoceanic rifting not leading to the formation of ocean but forming initially a diverging system of rifts and later a major sedimentary basin; an example is the West Siberian sedimentary basin with a system of Triassic rifts in its base (Fig. 5.31) and the North Sea (see below);

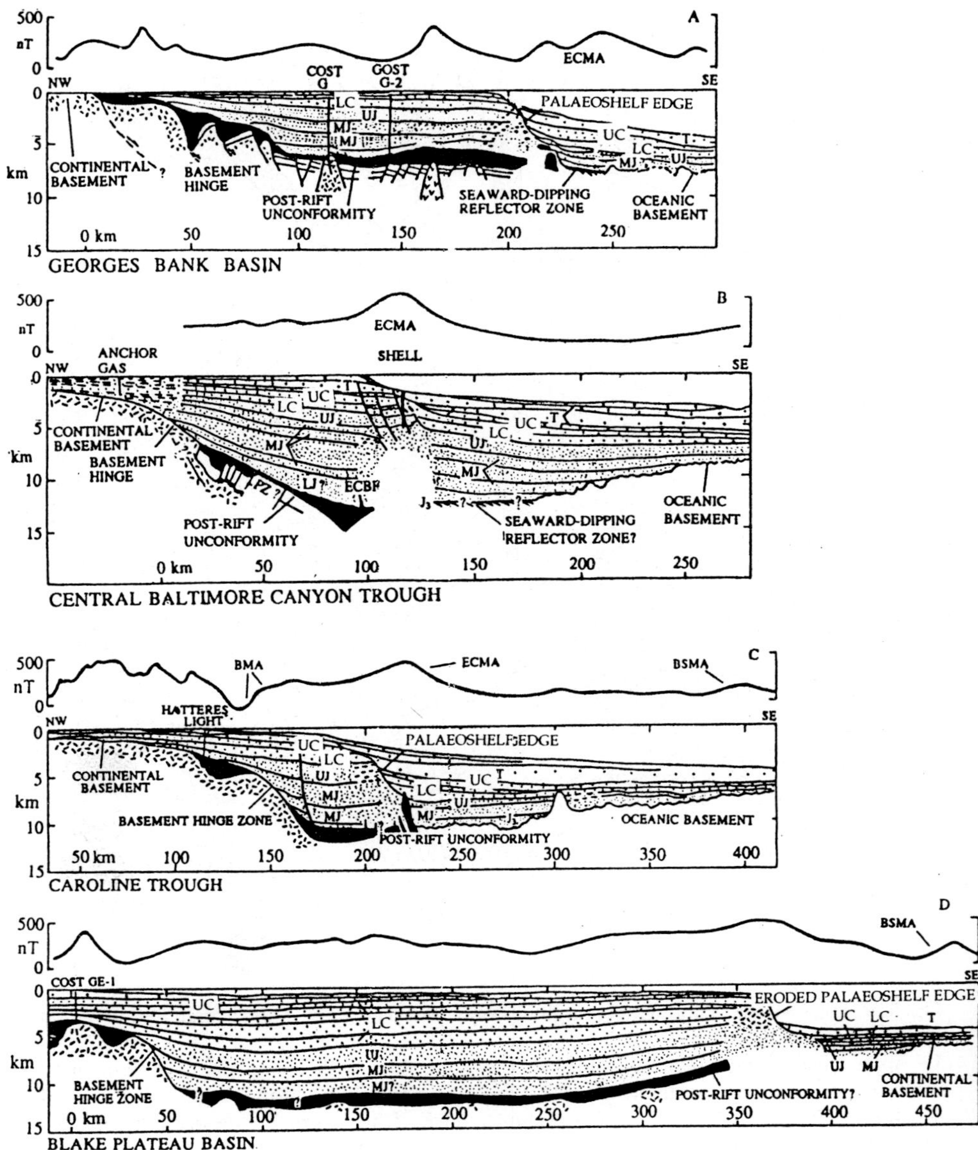

Fig. 5.30. Sections of the Atlantic coast of the USA (Klitgord and Hutchinson, 1988, modified). A—section of Georges Bank basin; B—Central Baltimore Canyon; C—Caroline trough; D—Blake plateau basin (for localisation, see Fig. 5.29). Black portions at the bottom show the Triassic sediments of the rift stage. Sediments: J3, LJ, MJ, UJ—Jurassic; LC and UC—Cretaceous; and T—Tertiary.

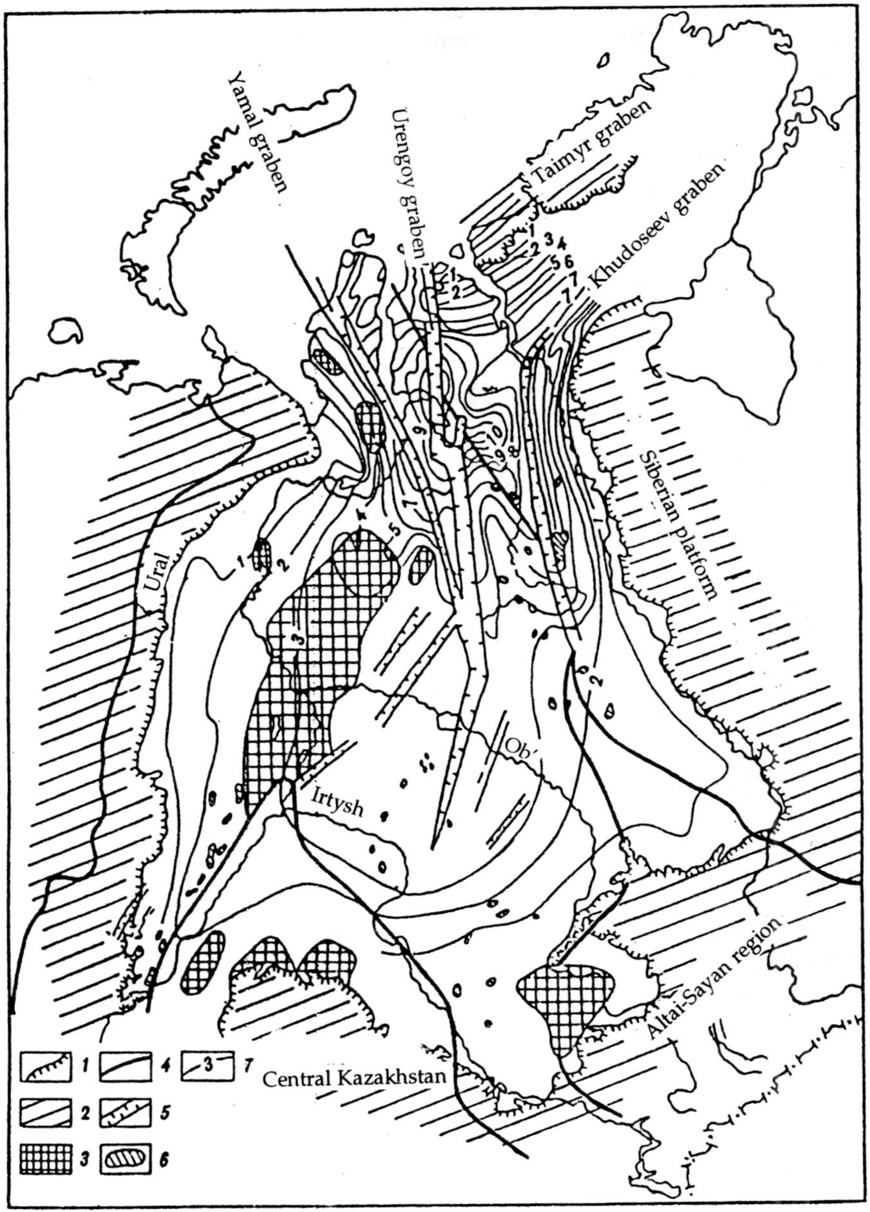

Fig. 5.31. Structure of the West Siberian sedimentary basin (Zonenshain and Kuz'min. 1993a): 1—boundary of the basin; 2—outcrops of pre-Mesozoic basement; 3—old massifs on the basement of the basin; 4—main Palaeozoic sutures; 5—Triassic rifts; 6—Triassic basalts at the base of the cover; 7—isopach lines (km).

b) dispersed spreading accompanied by active volcanism associated with mantle plume (Tunguska syneclise) or buried continuation of the mid-oceanic ridge, i.e., mantle plumes of a different type (Basin and Range province—Rio Grande rift, North America);

c) simple linear extension structures (Oslo graben type);

d) displacement type rifts where an oblique extension and pull-apart type structures are combined (Baikal rifts, Fig. 5.32);

e) back-are rifts at the rear of subduction structures of the Andean type (Altiplano rift, South America) or west Pacific type where rift structures accreted into marginal seas (Sea of Japan); and

f) sedimentary basins including linear ones adjoining the preceding but formed under compression conditions—foredeeps (Himalayan, Carpathian and Po river basin), ramp structures (Tarim, Tien Shan basins) and others.

Rifts and basins of the last two types (e and f) are studied later in the appropriate sections. In the first four groups are seen mutual transitions, for example simple rifts of the Oslo graben type may be present in the composition of more complex structures of types (a) and (b). Structures associated with mantle plumes (b) may transform into displacement zones of type (d).

Let us examine the evolution of complex structures on the example of the West Siberian basin (Fig. 5.31) and Baikal rift (Fig. 5.32).

Geophysical methods revealed a system of Triassic rifts at the basement of the West Siberian basin. This system is similar in age to the Triassic coal-bearing grabens of Transurals (Chelyabinsk graben) (Surkov et al., 1982). Rifts in the basement of the northern and central part of the West Siberian basin have a similar graben-like form but are accompanied by massive extrusions and sills of traps similar to those in Tunguska syneclise. It is possible that both the basins began forming simultaneously by a similar mechanism but the submergence of the entire wide area of the West Siberian plate, with a thickness of sedimentary cover up to 10 km, commenced from the Jurassic. In the Tunguska syneclise there was no such submergence but a more intense one occurred in the Triassic with the accumulation of the coal-bearing basin. One possible explanation for these differences is that the northern part of the West Siberian basin in the Triassic represented a transitional rift, or Ob' palaeo-ocean, as suggested by the residual linear anomalies of a high-amplitude magnetic field (Aplonov, 1987).

Similar examples of rifts at the base of large sedimentary basins were detected in the Pechora-Barents Sea basin (the same Permian-Triassic rifts), in Bonaparte basin in north-western Australia and Pripyat basin of the European platform (Devonian-Carboniferous rifts) (Zonenshain and Kuz'min, 1993a) and in the North Sea (Cenozoic rift).

Baikal rift (Fig. 5.32) is often compared with east African rifts but differs markedly from them in the compositions of volcanites and structure. The main difference is that the volcanic chain extending from Udokan through Vitim plateau to eastern Sayan does not coincide with the rift basins, except Tunka, where the zone of rift basins is intersected by a volcanic chain (Zonenshain and Kuz'min, 1993a). In Tunka basin, the volcanites are situated even in the lower part of its section as well as at its surface over an interval of more than 20 million years. Tunka basin apparently represents a pull-apart type basin, like the three basins in

Baikal itself: southern with a thickness of sediments up to 10 km, central (≥ 6 km) and northern (3 km). These represent half-grabens with steep north-western flank and gentle south-eastern flank with inclined listric faults. The kinematics of movements show a predominance here of shear displacements accompanied by oblique spreading with an amplitude of up to 20 km over 15 to 20 million years (Hutchinson et al., 1994).

Baikal rift joins the Mongolian and is situated on the periphery of a major Mongolian superplume (Windley and Allen, 1993). Mantle upwelling of the asthenosphere has been established not only under the Baikal rift but from over a much wider area. The centre of this plume with maximum heat flux is located in the Hantay hills near the Mongolian-Russian border. Mantle xenoliths with maximum temperatures up to 1500°C have been collected here (I.V. Ashchepkov, 1994; pers. comm.). Local asthenospheric upwellings developed ·as mantle diapirs on a background of the overall plume have been confirmed on the basis of xenoliths in other volcanic regions, for example in Vitim plateau (see Fig. 4.34).

If the activity of the plume leads later to manifestation of new volcanic zones or to displacement of the Baikal volcanic zone under the rift basins, an altogether different scenario can be anticipated: acceleration of extension and thinning of the lithosphere, together with transition into a structure of the type Basin and Range province in North America.

5.4. Models of Rift Formation

A general geodynamic model of the formation of rifts structures in the continental crust was developed for the first time by McKenzie (1978) and Jarvis and McKenzie (1980), many aspects of which had been considered before. This model assumes the extension and thinning of the lithosphere, upwelling of the hot asthenonsphere, restoration of isostatic equilibrium and cooling. It has been suggested in various alternative models that the lithosphere expands and becomes thin instantaneously (McKenzie, 1978); with exponential acceleration until the ultimate value of extension is attained, after which the process ceases quickly (Jarvis and McKenzie, 1980). The ultimate extension is realised at constant rate (Sheplev and Reverdatto, 1994). As a consequence, linear depressions arise in the Earth's crust and are constantly filled with sediments.

Subsequently, it was shown that these simplified models should be related to the varying importance of tectonic stress caused by changes in the relative movements of lithospheric plates (Cloetingh et al., 1985; Cloetingh, 1992). Variations of tectonic stress play an important role not only in the formation of rift basins, but also in their subsequent evolution, especially in the fluctuations of sea levels exceeding the 'normal' eustatic fluctuations (Embry, 1990; Aubry, 1991; Cloetingh, 1992).

Let us first consider the simplified model (Sheplev and Reverdatto, 1994; Fridenger et al., 1991).

The equation of heat transfer from the asthenosphere into the lithosphere expanding at a constant rate is described as follows (Jarvis and McKenzie, 1980; Sheplev and Reverdatto, 1994):

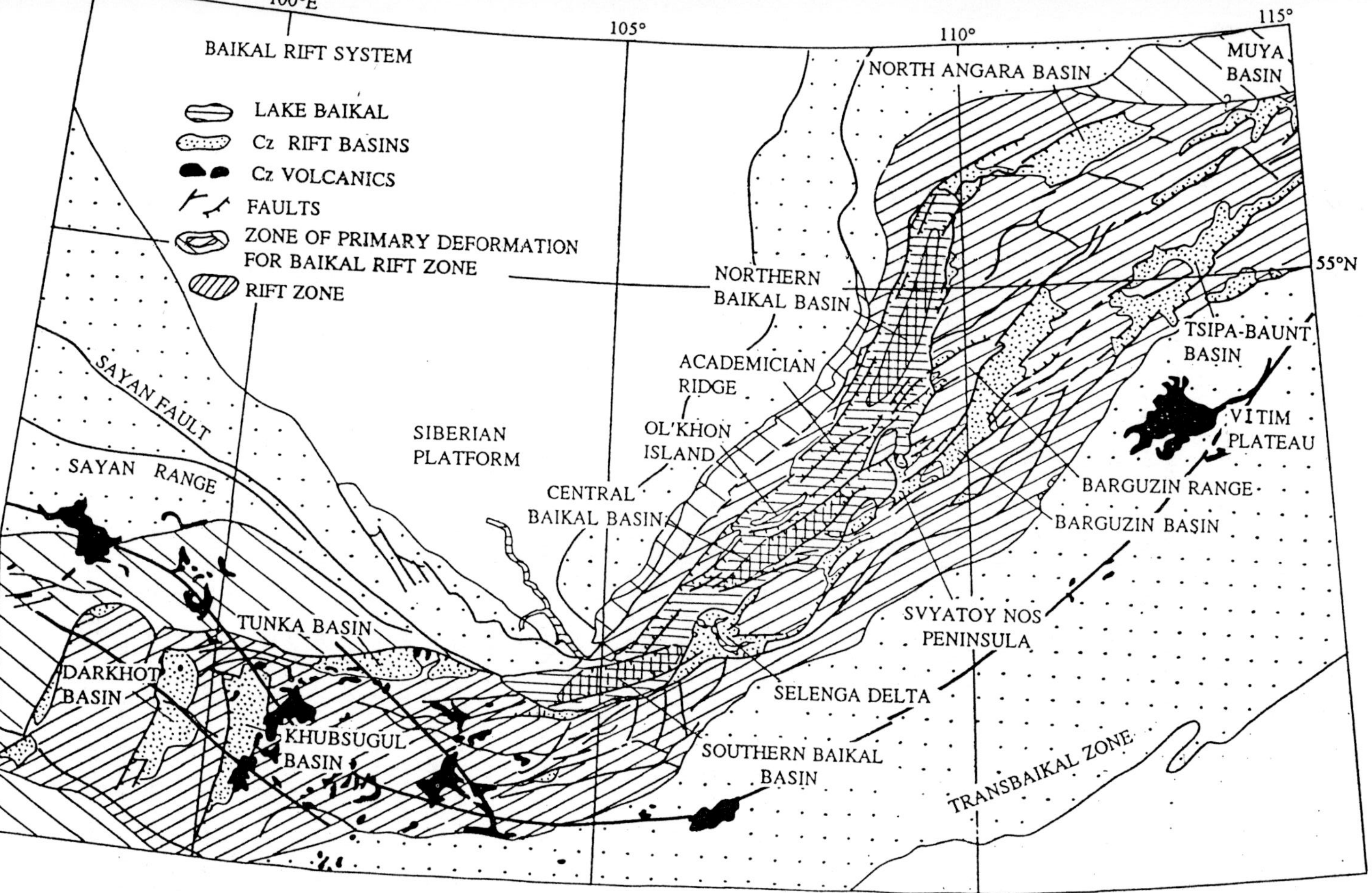

Fig. 5.32. Baikal rift system (based on the data of Logachev, 1993; Zonenshain and Kuz'min, 1993a).

$$\partial T/\partial t = a\, \partial^2 T/\partial z^2 - v_1(t)\,(1 - z/l)\,\partial T/\partial z, \tag{5.19}$$

for the initial and boundary conditions:

$$t = 0;\ T = T_1\,(1 - z/l);$$

$$t > 0,\ z = 0;\ T = T_1;$$

$$z = 1;\ T = 0,$$

where z is the vertical co-ordinate, $T(z)$ temperature of the lithosphere; T_1 temperature of the upper mantle; a thermal diffusion coefficient; t time; l initial thickness of the lithosphere; and v_1 rate of reduction of lithospheric thickness.

Let us introduce dimensionless values:

$$x = (1 - z/l);\ \theta = T/T_1;\ \tau = at/l^2;\ G = v_2 l^2/ba;$$

$$v(\tau) = G/(1 + G\tau);\ \overline{\beta} = 1 + Ga\,\Delta t/l^2 \approx b/b_0, \tag{5.20}$$

where b is the length of the lithospheric plate; b_0 initial length; $\overline{\beta}$ extension coefficient of the lithosphere; and v_2 rate of extension of the lithosphere.

In the case, eqn (5.19) may be represented as:

$$\partial\theta/\partial\tau = \partial^2\theta/\partial x^2 + v(\tau)\,\partial\theta/\partial x, \tag{5.21}$$

at $\tau = 0,\ \theta = x$; at $\tau > 0,\ x = 0,\ \theta = 0;\ x = 1,\ \theta = 1$.

The solution to eqn (5.19) for the case where $v_1 = \mathrm{const}$ may be obtained through confluent hypergeometric functions, and for the case $v_2 = \mathrm{const}$, numerically by the trial-and-error method.

The depth of submergence $S(\tau)$ was calculated taking into account the weight of the lithosphere which can be expressed as

$$S(\tau) \approx [P(\tau) - P(0)]/[\rho_0\,(1 - \overline{\beta}T_1) - \rho_w], \tag{5.22}$$

where $P(\tau)$ is the weight of the lithosphere at moment τ; $P(0)$ initial weight of the lithosphere; ρ_0 density of asthenosphere; ρ_w density of water; and $\overline{\beta}$ extension coefficient of the lithosphere.

Maximum submergence of the lithosphere is shown in Figure 5.33 as the function of extension coefficient of the lithosphere $\overline{\beta}$ against growing thickness of the crust. The actual maximum submergence goes up to 10 to 12 km for $\overline{\beta} = 3$ to 1.5 and crustal thickness 30 to 40 km. Figure 5.34 shows the calculation of this submergence over time (dynamics) for two cases and different extension rates of the lithosphere. It can be seen from this Figure that calculations for accelerated extension and expansion at constant rate (Fig. 5.34, curves 2 and 3) practically coincide while the results of sharp extension differ significantly although the ultimate submergence (7 to 8 or 10 km) is the same.

A special computer program for modelling, BASTA (Fridenger et al., 1991), enabled comparison of the development of basins and accumulation of sediments in them on the basis of stratigraphic-lithological information, the modelled submergence dynamics of the basin floor based on geophysical data and the model given above. The following average parameters were adopted for the lithosphere in the calculations: $\rho = 2.8\ \mathrm{g/cm^3}$; thermal conductivity $\lambda = 7.5\ \mathrm{mcal \cdot cm^{-1} \cdot s^{-1} \cdot {}^\circ C^{-1}}$;

226

heat flux through the lithosphere $q = 1.5 \times 10^{-6}$ cal $\cdot$ cm$^{-2} \cdot$ s^{-1}; surface temperature 10°C; for the upper mantle $\rho_M = 3.33$ g $\cdot$ cm^{-3} and $T_1 = 1350$°C. The thickness of the lithosphere varied from 70 to 200 km.

The results of calculations for four typical rift structures for which detailed information was available are shown in Figure 5.35.

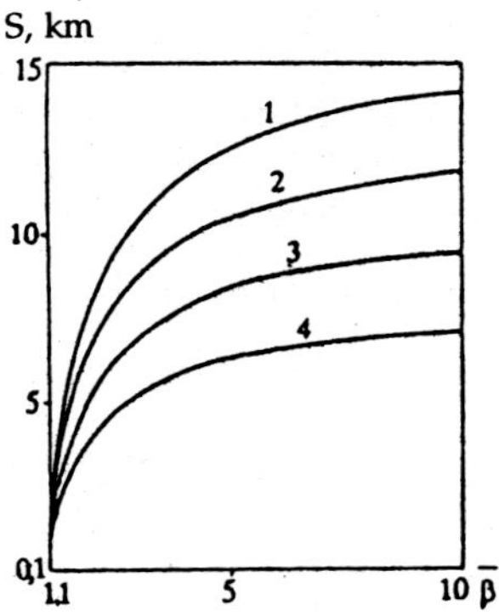

Fig. 5.33. Calculated curves of the thickness of sediments (S) against the index of extension $\bar{\beta}$ (Reverdatto et al., 1995). Numbers 1, 2, 3 and 4 correspond to the thickness of the Earth's crust 30, 40, 50 and 60 km respectively. Thickness of lithosphere: 125 km.

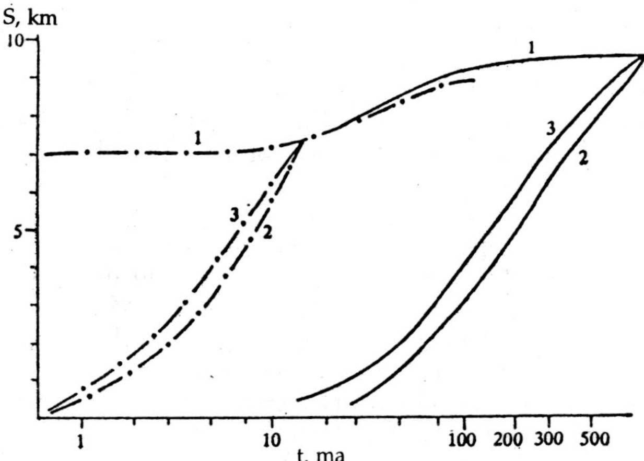

Fig. 5.34. Depth of subsidence (S) for different variants of extension of the lithosphere (numbers on curves): 1—sharp (McKenzie, 1978); 2—accelerated (Jarvis and McKenzie, 1980); 3—at constant rate (Sheplev and Reverdatto, 1994). It was adopted that $l = 125$ km; $a = 0.00804$ cm$^2 \cdot$ s^{-1}; coefficient of thermal expansion 3.28×10^{-5} s^{-1}; and t time in million years. Dot-dash lines correspond to cases at $\bar{\beta} = 3$ and $G' = 50$; continuous lines, for $\bar{\beta} = 3$ and $G' = 2$.

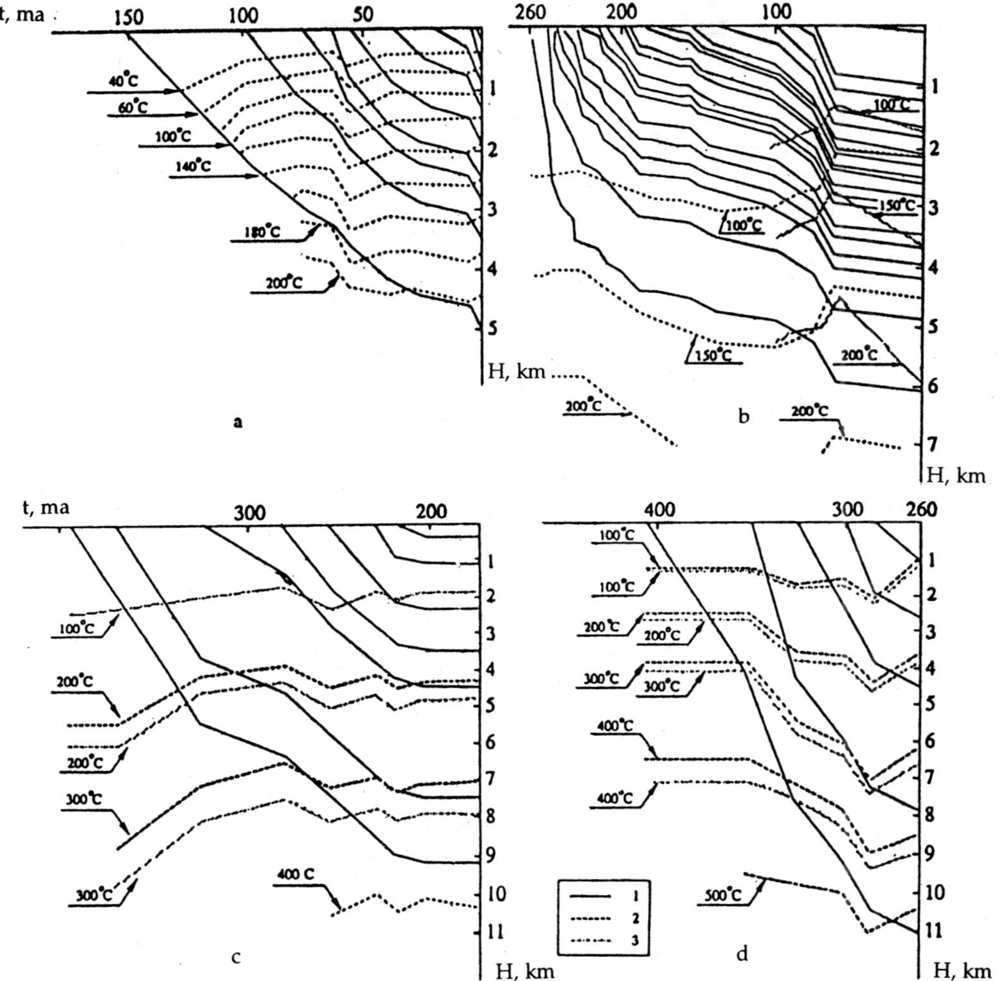

Fig. 5.35. Reconstructed sections of rift basins and distribution of temperature in them (Reverdatto et al., 1995).
a—North Sea graben, region of Konoko borehole 15/30 = 1 (Sclater and Christie, 1980); b—Danish basin (Vejbaek, 1989); c—Reichenhogen basin; d—Dneipr-Donets aulacogen (for explanations, see text). 1—Submergence profiles; 2 and 3—isotherms.

In the first example (Fig. 5.35a), submergence dynamics (up to 5 km) is shown for the Central graben in the North Sea, the length of the graben being about 500 km and its width 45 to 75 km (Sclater and Christie, 1980). On the basis of seismic data, the thickness of crust in the graben region is 15 to 25 km, increasing up to 35 km away from the central depression. These parameters were adopted for modelling: thickness of the lithosphere 125 km, initial thickness of crust 35 km, $\bar{\beta}$ = 1.4, commencement of rifting 150 Ma ago and commencement of cooling of the asthenosphere 100 Ma ago. The difference between model calculations and observed submergence dynamics (on the basis of stratigraphic information) was less than 0.1%. Palaeoisotherms were calculated for 'standard' values of $q = 1.6 \times 10^{-6}$ cal $\cdot$ cm$^{-2} \cdot$ s^{-1} and average heat conduction of lithosphere and maximum temperature in the most submerged part of the graben equal to 210°C.

Another example (Fig. 5.35b) relates to the axial part of the Danish basin which is in the form of a trough exceeding 600 km in length and width 25 to 50 km, filled with sediments to a thickness of about 6 km. On the basis of seismic data, the thickness of the crust in the trough is 26 km rising to 40 km or more north of it (Vejbaek, 1989). The following values were adopted for computation of the models: thickness of lithosphere 94 km, initial thickness of crust 37 km, and $\bar{\beta}$ = 1.44. In this case there was agreement between the observations (on the basis of stratigraphic data) and calculated dynamics. The difference in the position of curves was 0.3%. The calculated position of the palaeoisotherms (in the two variants) showed the same range of temperatures (180 to 220°C) on the floor of the trough.

In two other examples, Reichenhogen deposits south of Rugen island (Fig. 5.35c) and Dneipr-Donets aulacogen (Fig. 5.35d), the selection of parameters (thickness of lithosphere 110 km, initial thickness of crust 37 and 45 km and $\bar{\beta}$ = 1.6 and 1.8) yielded a lower but satisfactory accuracy (19% for Fig. 5.35c and 11% for Fig. 5.35d). It is significant that the calculated temperature amounted to 400 to 500°C in the most submerged (deepest) part of the rifts. This is due to the fact that in the same period (150 to 200 million years), the depth of the rift compared to the preceding examples doubled (9.2 and 11 km), which is also indicated by a high value of $\bar{\beta}$ and high ascent of the top of the asthenosphere.

Tectonic stress, according to recent studies (Cloetingh et al., 1985; Cloetingh, 1992), not only influences the rate of submergence of the basin caused by thermal factors (cooling of the asthenosphere), but also lead to additional vertical movements which disturb its configuration also. For example, compressive stresses cause a relative uplift of the basin flanks, additional submergence of the central part, i.e., migration of the shoreline toward the sea, and accelerated rate of sedimentation (Fig. 5.36a).

Increase in tensile stresses induces extension of the basin, lowering of flanks and hence trangressive migration of the shoreline and local rises at the centre of the basin (Fig. 5.36a). The change in stratigraphy of the rift basin caused by 1 kb compression after the principal rift stage is shown in Fig. 5.36b.

A study of the Mesozoic rift basins in the North Atlantic and North Sea (Cloetingh et al., 1989, 1990; Cloetingh, 1992) showed that sedimentary filling of the basins is most satisfactorily modelled by taking into account fluctuations of the stress field on the background of prolonged thermal evolution. Further, stress variations correlate with the change in stress field in the course of spreading and interactions of plates in the Atlantic and Mediterranean, especially with the phase of intense collision between the African and Eurasion plates (Cloetingh et al., 1990).

An even greater effect of stress is detected during the formation of ramp basins under compressive conditions and also complex basins of the transitional type, for example, in the for-montane basins of the Pyrenees, in Beltic Cordillera where the

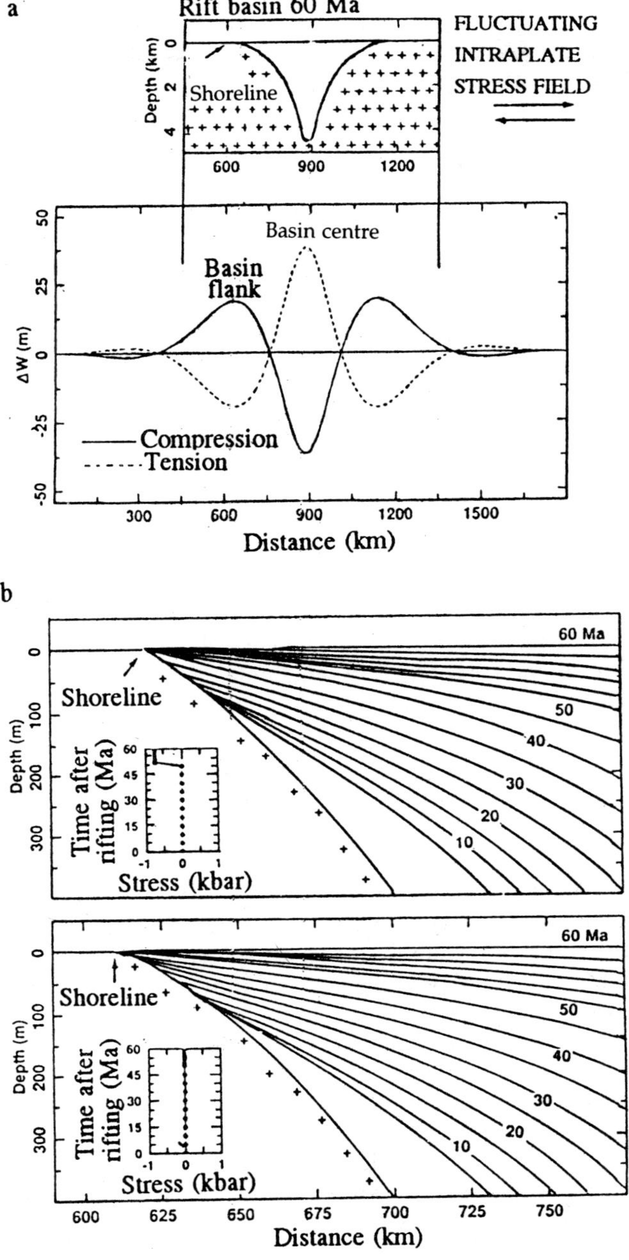

Fig. 5.36. Stress-induced submergence (uplift) in rift basins (Cloetingh et al., 1985). a—differential submergence or uplift (in m) induced by compression of 1 kb (continuous line) or tension of 1 kb (broken line); b—influence of the effect of 1 kb compression on the stratigraphy of the rift basin when accepting the classic model of loading by sediments.

early rifting stage is replaced by compressive conditions and foreland formation (Zoetemeijer et al., 1990; Cloetingh, 1992). We shall examine these cases while describing collision conditions.

Finally, the actual process of extension in the lithosphere occurs differently in the upper brittle zone and plastic lower crust and lithospheric mantle. One of the earliest to describe this feature was Ivanov (1978), who even recognised a special metamorphism of the extension zones, and later interpreted the Conrad boundary in the Earth's crust as an interface in the form of overthrust (detachment over-thrust) and the boundary of brittle zone and plastic deformation zone (Ivanov and Ivanov, 1993).

The actual pattern may be even more complex. Figures 5.37 and 5.38 provide a comparison of the actual structure of the extension zone of the Basin and Range in Arizona state (USA) obtained by seismic and geological data (Krüger and Johnson, 1994) and the results of experimental modelling of extension in a multilayer model: brittle on top and plastic below (Brun et al., 1994). As the Figure 5.28, a characteristic feature in the combination of the main detachment fault and listric faults in the upper brittle crust, of which one may be the principal, accommodating one (LAF—Listric Accommodation Fault, Fig. 5.37).

Unlike the simplified models studied above, however, a dome-shaped uplift of the lower plastic into the upper one arises and later 'freezes down' and manifests itself as a granite-gneiss dome or metamorphic core of the Cordillera type. Even a simple 'softening' (reduction in viscosity) of granite-gneiss of the lower or intermediate crust is adequate to form an metamorphic dome. If, however, there is concentration of granitic light melt in the dome, the additional effect of buoyancy and reduction in viscosity would provide maximum effect (Lister and Baldwin, 1993).

5.5. Island-Arc Type Subduction Zones

Subduction zones in which the oceanic lithosphere is submerged and a new lithosphere forms, including crust of andesite composition, represent the most important structural elements of the Earth and the most important active zones in which 90 to 95% of seismic energy is released and the majority of most active volcanoes concentrated. Many mineral deposits are also formed here. However, geodynamic processes in these zones are very complex and have not yet been deciphered. Therefore, subduction zones represent key elements, important for the successful development of a geodynamic theory and geology.

The present-day subduction zones are concentrated in two global belts—around the Pacific Ocean and the Alpine belt (or Indonesian-Mediterranean-Caribbean). The situation in these belts is complex, however, since only part of the subduction zones of the former continuous Mesozoic subduction belt is preserved in the latter. The Pacific framework is divided into three nearly identical segments, of which the southern (from New Zealand to Scotia zone, see Fig. 1.2) does not contain the present-day active subduction zones while other two segments differ in characteristic features.

The north-western part of the Pacific Ocean from Alaska to New Zealand is surrounded by island-arc type subduction zones which are young, i.e., of Cenozoic

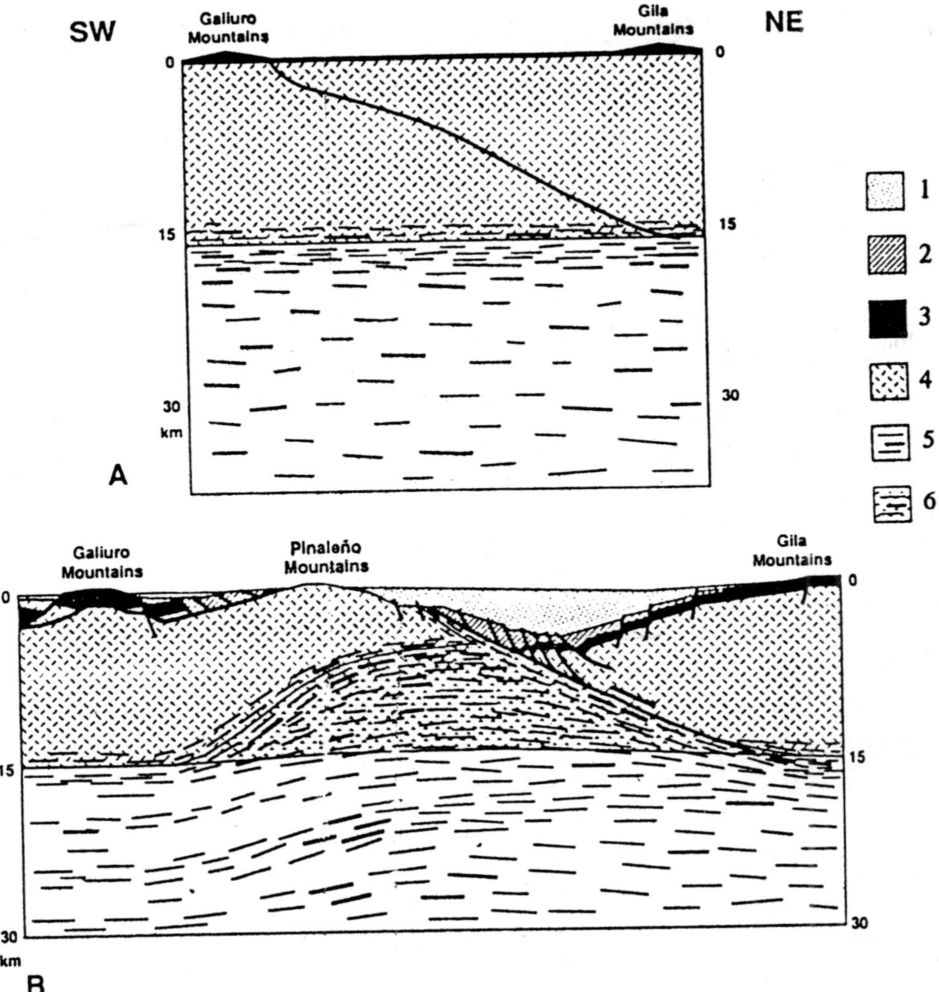

Fig. 5.37. Section from Galiuro mountains to Gila mountains, western states of the USA (Krüger and Johnson, 1994).
A—before movements along the detachment fault; B—present. 1—Late Cenozoic sedimentary basin; 2—Middle Tertiary sediments synchronous with extension; 3—Middle Tertiary volcanites; 4—zone of brittle deformation in the pre-Eocene crust; 5—plastically deformed crust; 6—crust deformed plastically during extension and elevation of present-day boundary of brittle deformation.

age (or underwent significant change of structure at the end of the Cretaceous to the beginning of the Cenozoic). Similar subduction zones are present in the Alpine belt but here they represent relics of an older Mesozoic belt. Most of the Mesozoic

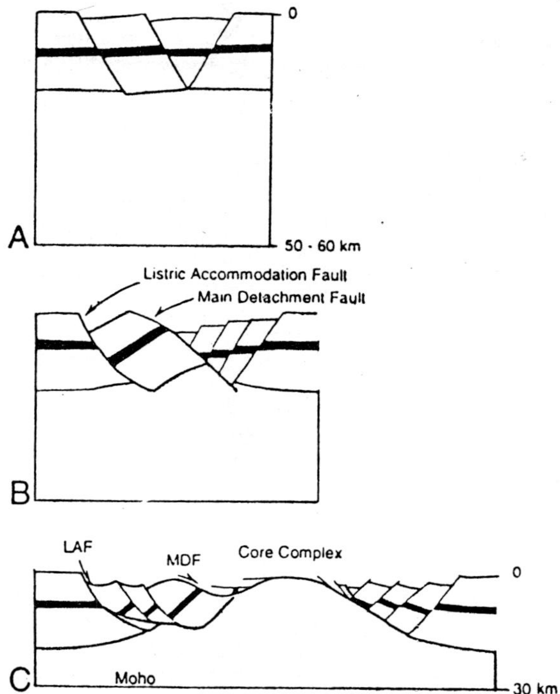

Fig. 5.38. Evolutionary model of system with detachment fault based on the results of experimental modelling (Brun et al., 1994).

subduction zones in the Alpine belt were either significantly restructured or disappeared at the end of the Cretaceous to the first half of the Cenozoic, right up to the collision of India with the Asian continent 40 Ma ago.

The eastern part of the Pacific framework is surrounded by the Cordilleras of South and North America, most of which together with the adjacent troughs represent Andean-type subduction zones (also called the Andean type active continental margins). The northern fragment of this zone, from California to Alaska (see Figs. 1.2 and 1.3) has a more complex structure, combining fragments of the Andean-type zone (Cascade mountains), submerged spreading zones (Basin and Range zone) and major fault zones (San Andreas and others). However, a continuous zone of the Andean type prevailed even at the end of the Cretaceous and thus 'Californian-type conditions' can be regarded as a particular case of evolution of Andean-type zones.

Mutual transitions are seen in all these types of active zones. We have already referred to one such case of the Cordilleras of North America (from California to Alaska). In the southern part of South America (in Chile), conditions of island arc and marginal sea prevailed up to the middle of the Cretaceous, after which the marginal sea was sealed. In island-arc systems of the north-western part of the Pacific Ocean, there are fragments (Kamchatka and New Zealand) which actually correspond to the Andean-type, and in the geological past (Cretaceous-Early Palaeogene) an active Chukchi-Okhotsk belt of Andean type extending up to Japan existed here. Outside the present-day active zones within the framework of palaeo-oceans (Palaeo-Asiatic, Palaeo-Tethys, Japhetus and Mongolion—Okhotsk), conditions of island arc, Andean and more complex types combined and interchanged in an intricate manner during the Late Precambrian, paleozoic and Early Mesozoic.

Nevertheless, as a first approximation, zones of subduction of island arc and Andean types can be distinguished with confidence. The main characteristics of present-day subduction zones of the island-arc type are examined below.

Kuril, Japan and Izu-Bonin represent typical island arcs and are the best studied. Variations in seismicity and deep-level structure in these arcs are depicted in Figure 4.1. These arcs have several other significant differences. In the rear part of the Kuril and Japanese arcs can be seen young troughs with oceanic-type crust—southern Kuril and central Sea of Japan—that originated in the Miocene 20 to 25 Ma ago. Before this, an Andean-type active margin prevailed at the site of Japan while the Kuril arc restructured after collision with the active margin of Japan, the result of which was also the opening up of the two troughs referred to above.

In the region of joining of the Japan and Kuril arcs on Hokkaido island, a sharp change is observed in the nature of seismic zones (see Fig. 4.1). Under the Kuril arc, the seismofocal zone marks only the lower boundary of the lithospheric plate while under the Japan arc, the upper and lower or only the upper boundary (Hasegawa, 1994). In other words, the nature of sliding of the subducting lithospheric plate with the overlying island arc and the underlying asthenosphere differs significantly.

Judging from seismic data and the distribution of volcanoes, the slope of the island-arc type subduction zones is quite steep (45 to 65°), sometimes almost vertical (in Mariana arc). This feature correlates with the relatively weak and variable 'coupling' in the steeper zones of subduction occurring only along the lower or upper boundary (Kuril and Japan arcs) or disappearing almost completely as in the Mariana arc. Here, between 17° and 25° N (see Fig. 4.1) can be seen a subvertical localised distribution of deep-focus earthquakes which can be explained by the sumbergence of eclogitised blocks under conditions of weak linkage of the subducting plate. The breakaway of such eclogitised blocks is also possible in the steeply inclined section c and d on 29° and 28° N and in Tonga arc, judging from isolated clusters of earthquake centres at depths of 400 to 700 km (see Fig. 4.1).

A change of slope and seismicity is the first distinctive feature of subduction zones of the island-arc type.

Izu-Bonin and Mariana arcs at the rear have an oceanic structure of the Philippine Sea which underwent prolonged evolution commencing from the Cretaceous (*Geology of the Philippine Sea Floor...*, 1980). Figures 4.1 and 5.39 show the extinct Kyushu-Palau island arc which was active in the Cretaceous-Palaeogene west of Izu-Bonin and Mariana on the floor of the Philippine Sea. Between these two lies the younger (Miocene) Bonin basin (B) and to the west of Kyushu-Palau, the older (Mesozoic) residual western Philippine basin (WP). The Mariana-Bonin arc repre-

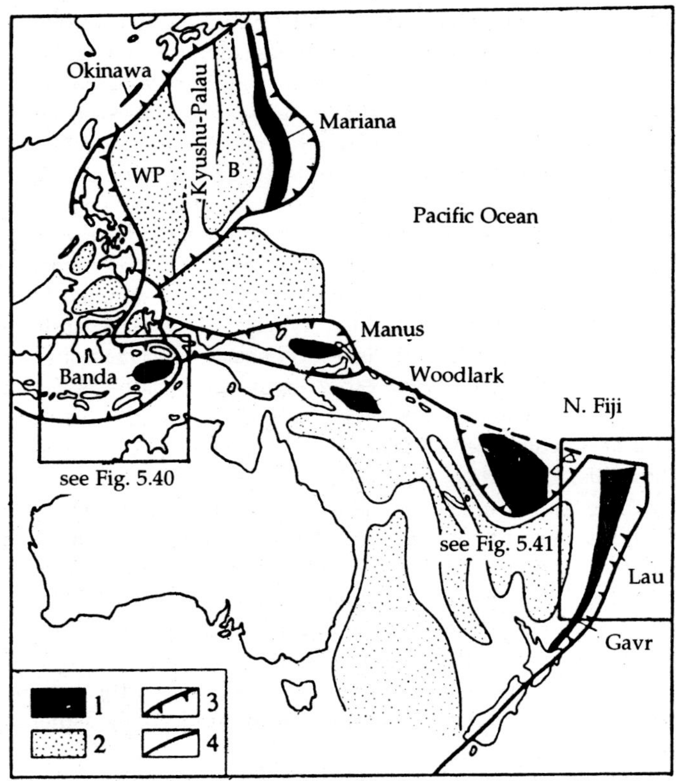

Fig. 5.39. Back-arc basins in the south-western Pacific Ocean (Zonenshain and Kuz'min, 1993a). Basins: 1—active; 2—extinct (B—Bonin, WP—Western Philippine, with extinct Kyushu-Palau arc between them); 3—subduction zones; 4—major transform faults.

sents a double arc: Mariana interarc basin which is developing actively at present occurs between the most recent (eastern) and more western arc (Fig. 5.39).

The basins between the Tonga-Kermadec arc and Australia are similar in structure (Fig. 5.39). The easternmost basins, Lau-Gavr and North Fiji, adjoining the active arcs, are presently active while the westernmost ones adjoining Australia (Tasman Sea) are the oldest. Thus, most of the arcs of the north-western margin of the Pacific Ocean (Kuril, Japan, Bonin, Mariana and Tonga) were per se successively displaced towards the ocean and increasingly younger back-arc basins arose successively in the same direction, with local back-arc zones of spreading (Zonenshain and Kuz'min, 1993a).

The active basins Woodlark, Manus and Banda (Fig. 5.39) originating in the joint zone of the Melanesian-Indonesian chain and the arc of the western framework of the Pacific Ocean 'squeezed' by northward movement of the Australian continent are more complex in structure. Some of these basins (Woodlark, Manus and Gavr) contiune in the form of continental rifts with alkaline bimodal valcanism in Papua New Guinea and North Island of New Zealand. A similar bimodal alkaline volcanism aged 19 to 9 million years is seen in northern Hokkaido in the pinched-out part of South Kuril basin (Nakayama, 1994).

The structure of the collision zone itself (Fig. 5.40) shows that the collision of the Indonesian arc and the Australian plate in Alor-Ramong sector (A-R, Timor Island) and the south-western part of the Banda arc (SW B-A) commenced 5 Ma ago and led to the northward displacement of the trough and arc and pushed the margin of the Australian continent under Timor Island for a distance of over 150 km. The origin of counterthrusts of Flores and Wetar and the opening up of the young Banda basin at the rear of the Indonesian arc shows that transition of the subduction zone has already commenced in a southerly direction with manifestation of an Andean-type active margin.

A study of the composition of volcanism and in particular of the ratio of Nd and Pb isotopes helped ascertain (Van Bergen et al., 1993) that sediments of average thickness 2700 m were subducted over 5 Ma ago in the A-R sector. One-third of these sediments (850 m thick) was remelted and drawn into the igneous material. Sediments of average thickness 2300 to 1750 m were drawn into the adjoining zones (Flores, A-P and SW B-A; see Fig. 5.40). A quarter of these sediments (about 530 to 480 m thick) contributed to the magmatic source. In the common island arcs, the contribution of sediments to the composition of basalts and andesites under melting is insignificant.

Thus, the second important element of island arc structures of the western Pacific Ocean type is the presence of back-arc basins along with present-day active spreading zones.

The active basins measure up to 1000 km in cross-section. Active volcanism of variegated composition occurs in them and produces mid-oceanic ridge-type basalts (in the northern Fiji basin, youngest basalts of Lau basin and others), hybrid basalts of back-arc basins (BB) and andesite-dacite series (in Bonin basin, Lau, Woodlark, etc.). Basalts of back-arc basins are characterised by high (compared to those of mid-oceanic ridges) contents of K, Rb and Be, isotope ratios $^{85}Sr/^{86}Sr$ and similar or reduced contents of Ti, Nb and Hf and $^{143}Nd/^{143}Nd$ ratios (Hawkins, 1989; Zonenshain and Kuz'min, 1993a).

Two basic factors probably influence the composition of magmatic rocks of back-arc basins: back-arc spreading ensuring the supply of magma from the depleted asthenosphere (mid-oceanic ridge basalts) and subduction ensuring the supply of water-saturated magma separated from the downsinking oceanic plate. The effect of the subduction component is reflected in the extensive development of andesites-dacites and manifestation of highly vesicular varieties. Back-arc basin basalts are characterised by high vesicularity and enrichment with some lithosphile elements and hence they can be regarded as hybrid (mid-oceanic ridge-type melts from the asthenosphere + fluid additives from the subduction zone). However, associations with bimodal alkaline series at the site of pinching of back-arc basins (northern Hokkaido, Papua New Guinea and North Island of New Zealand) reveal the possible effect of deep-level (lower mantle) convection flows. A similar interaction is restored during movement of the Yellowstone hot spot in North America (Chapter 4).

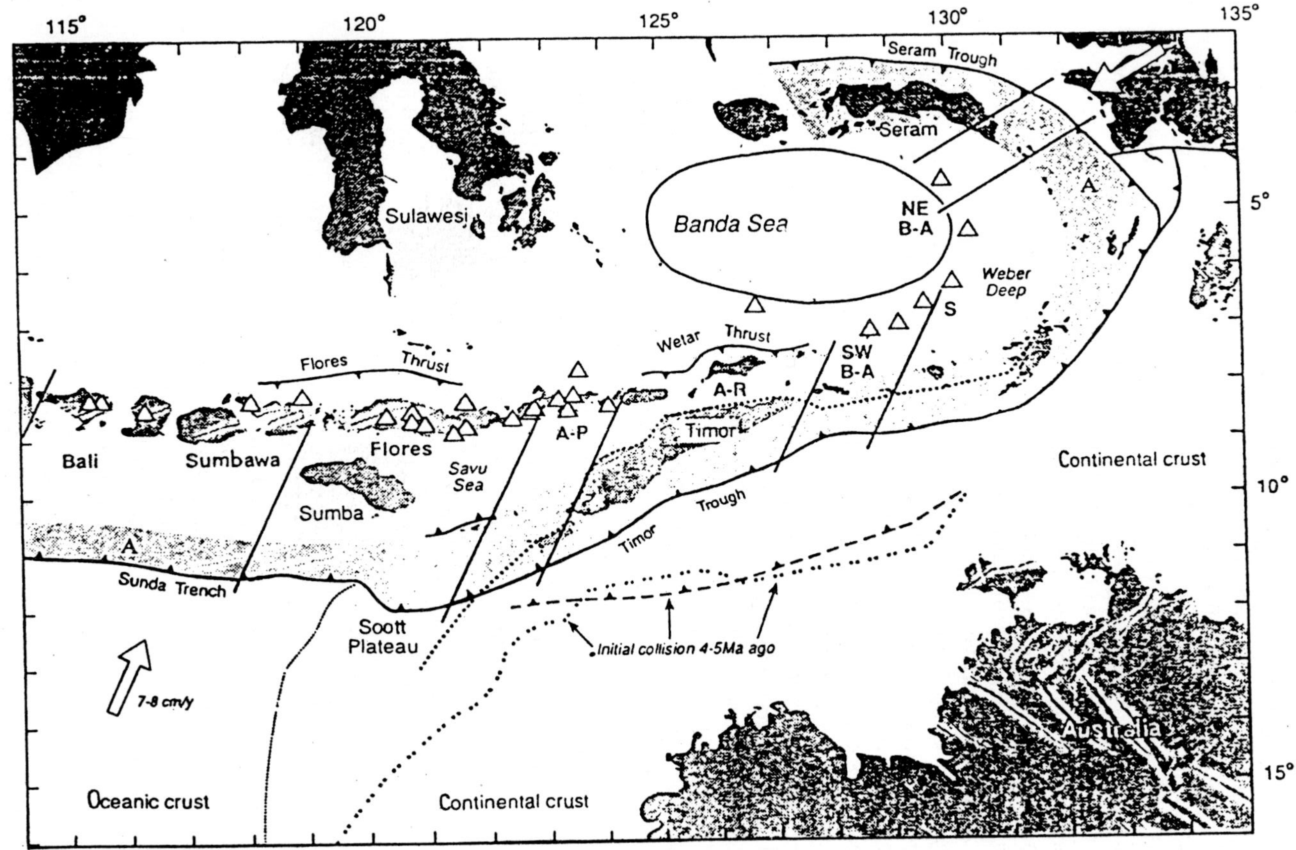

Fig. 5.40. Zone of collision of the arc and the Australian continent in western Indonesia (Van Bergen et al., 1993). A-P, A-R and SW B-A: sectors of the subduction system zone in the collision front. Broken dotted lines: boundary of the continental crust at present and 5 million years ago; continuous dotted line: initial position of Timor trough; triangles: present-day volcanoes; smudged zone (A): accretionary complex.

It is also important to note the relation of volcanites with the sedimentary series in the back-arc basins. In the incipient back-arc basins (Okinawa, early stage of the Sea of Japan and others), splitting occurs within sedimentary (turbidite) filling and sill-basalt complexes associated with turbidites often originate here. Fossil examples of such strata have been described as specific ophiolites of Hungary (Balla and Dobretsov, 1984) and Shimokawa ophiolites in Hidaka belt (Komatsu et al., 1992). After extinction of spreading axes, back-arc basins were rapidly filled with sedimentary strata as in the southern Okhotsk basin and Sea of Japan. In the section of the southern Okhotsk basin close to the Kuril arc, granular turbidites containing the erosion products of volcanic rocks of the arc play a major role. In the central part of the basin, thin distal turbidites accumulate together with organic-siliceous diatomic silts. In the region close to Sakhalin, a considerable amount of coarsely fragmented turbidites is confined to submarine canyons. Gas hydrate deposits with which gas flares are associated, are also formed here, and oil deposits occur in the central part. In the basins of the tropical zone, reef limestones (in the shallow-water zone) and in the extinct volcanoes and carbonate muds (in the deep parts) play a major role (Zonenshain and Kuz'min, 1993a).

Hydrothermal activity and sulphide mineralisation similar to that in mid-oceanic ridges are seen in the back-arc basins close to the spreading zones but as a result of the effect of sediments in subduction zones, Pb, Ag and Au (i.e., polymetallic mineralisation) occur here although copper-zinc ores (e.g., in the northern Fiji basin) are also present.

In Bonin basin (see Fig. 5.39) adjoining the dacite dome, stratified ores of the Kuroko type are deposited. Here itself, in the caldera of one of the andesite volcanoes, the Au and Ag contents in pyrite deposits at a depth of 400 m are 140 and 150 g/ton respectively (Urabe et al., 1987).

In Lau basin (Fig. 5.41), hydrothermal fields were detected in the northern and central parts and, in the spreading Waluf ridge, a long belt of hydrothermal fields which have their own names: White Church, Van Lily and Hine-Hina (Fouquet et al., 1991; Bortnikov et al., 1993; Zaikov and Zaikova, 1994). Most of these, as in the other back-arc basins, are of low temperature origin and iron-manganese and barite-opal and sulphate occurrences are associated with them. There are also high-temperature springs (up to 400°C) with which are associated copper rich sulphides deposited from acidic (pH = 2.5) solutions (Fouquet et al., 1991).

In Manus basin (see Fig. 5.39), in a hydrothermal field called 'Green Forest', three stages of sulphide mineralisation have been delineated (Shadlun et al., 1992). The early high-temperature copper-zinc association (chalcopyrite-pyrite-wurtzite-sphalerite) has been replaced by polymetallic (galena-sphalerite-pyrite-marcasite-barite-opal) with characteristic reniform and colloform structures. A sulphide-opal-barite association including silver minerals (acanthite, polybasite, proustite and pyrargyrite) represents the closing stage.

The following main elements have been recognised in the structure of the volcanic arc (Fig. 5.42):

1) trench in which oceanic sediments and turbidites (transported by currents along the trench) occur together; specific volcanites are possible;

2) accretionary prism (or wedge) comprising a series of slices and highly deformed rocks;

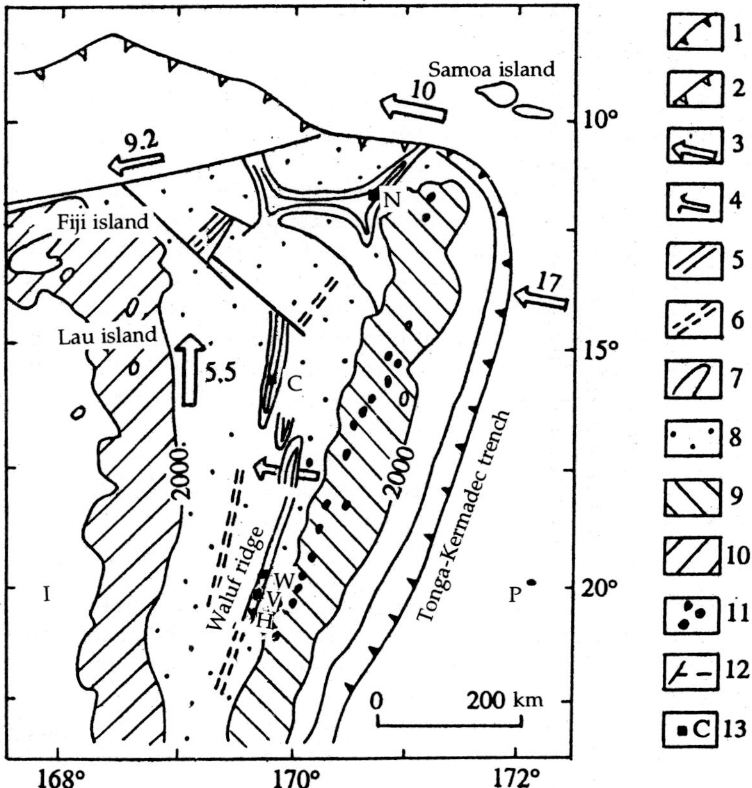

Fig. 5.41. Map of Lau tectonic basin (Fouquet et al., 1991; Bortnikov et al., 1993): *Subduction zones*: 1—active and 2—extinct; *directions of plate movements*: 3—absolute and 4—relative; *spreading axes*: 5—active and 6—extinct; 7—neovolcanic ridges; 8—young oceanic crust of Lau basin; *island arcs*: 9—active and 10—extinct; 11—submarine island arc volcanoes; 12—faults; 13—hydrothermal zones (C—central, H—Hine-Hina, N—northern Lau, V—Van Lily, W—White Church); Pacific (P) and Indo-Australian (I) plates.

3) forearc basin (sometimes two basins) formed on ancient volcanites or ancient accretionary complex; their sediments comprise essentially turbidites originating due to the erosion of volcanites; and

4) active volcanic arc.

Geophysical anomalies—negative gravitational (up to − 250 mGal) low heat flux— have been established for the trenches. The ocean-side slope of the trench is gentler and entirely belongs to the oceanic plate; the island-side slope is steeper and is related to the accretionary prism. Some trenches are filled to different extents with turbidites deposited horizontally in the axial valley and on the marine trench. The trench in front of the Cascade mountains on the coast of North America is

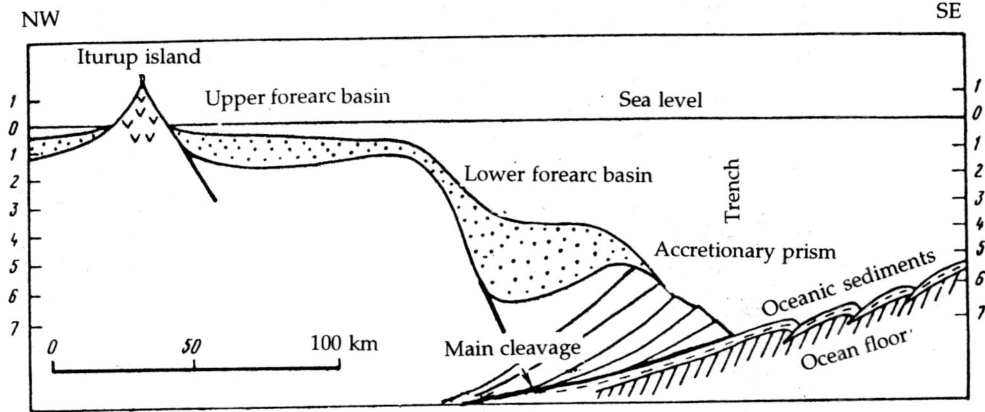

Fig. 5.42. Section across Kuril island arc showing the principal elements of the arc (Zonen-shain and Kuz'min, 1993a).

wholly filled with sediments and can be traced only be geophysical methods. The trench sediments often reveal olistostromes due to sliding and overthrusting from the island slope.

The most important element in the island arc structure is the accretionary wedge (prism). As is shown below, the presence of this wedge helps in under-standing the conditions of the stable existence of subduction zones and/or the conditions of instability and back transport of material at depth. The lower bound-ary of the wedge represents the main cleavage or the main tectonic zone. The sizes and structure of the accretionary wedge vary widely, its size in the Kuril arc (Fig. 5.42) being 100 × 40 km (outcrop on the slope of the arc does not exceed 20 to 25 km). The accretionary wedge around Barbados island has a width of 250 km (with a gentle dip of the subduction zone) and a large number of mud diapirs and mud volcanoes, which mix up the wedge material, have been established in its lower part. Intensely deformed lenses have been observed on the hanging wall side (upper zone) and within the wedge with a large number of seismic reflection horizons (Brown and Westbrook, 1988). The overthrust separating the lower and upper zones is a 15-m thick zone with a high level of fluid activity (Shipley et al., 1994).

The deep-level structure of a typical accretionary wedge and the entire upper part of the subduction zone have been studied in great detail on the example of Japan (Fig. 5.43). Here can be seen a gradual increase in the inclination angle of the main slope from 5 to 10° close to the trench to 45° at depths of about 90 km. The upper boundary of the wedge has been determined on the basis of geological data about the occurrence of undeformed sediments of the forearc trough (see also Fig. 5.42) and on the basis of seismic data about the presence of earthquake foci grouped mainly within the wedge and partly in the underlying oceanic plate. At a depth of about 50 km at the base of the island arc lithosphere, the change in the

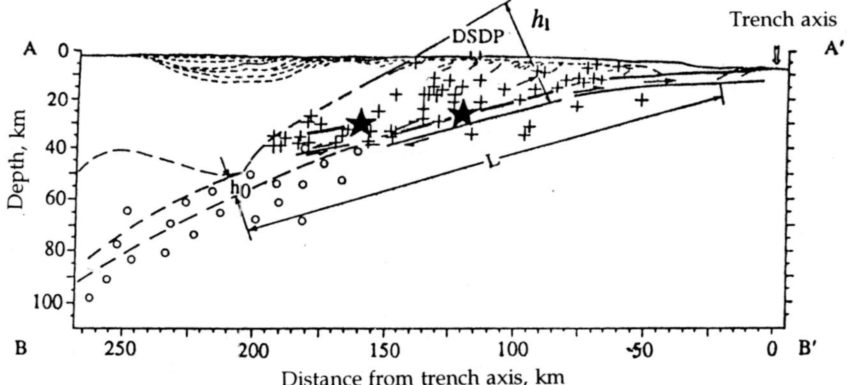

Fig. 5.43. Section across the Japan trench and subduction zone (Magee and Zoback, 1993, modified). See Figure 4.1 for the position of the section. For dimensions of the accretionary wedge h_0, h_1 and L, see the text. Stars and crosses show the major and normal earthquakes, and circles the aftershocks. The geological section is shown on the top. DSDP: Deep-sea drilling programme.

nature and grouping of earthquakes illustrates the transition from wedge h_1 to the narrow slit-like zone h_0 (see Section 5.7).

Geophysical data show that a compression mechanism together with displacement parallel to the subducted slab acts at the bottom of the wedge in the zone of the main cleavage (Fig. 5.43). This feature was established long ago (Isacks et al., 1974) and, together with a high Q factor (strength) in the subducted slab differing sharply from the asthenosphere, points to relatively low viscosity (rigidity and strength) of the lithospheric plate. The conditions of extension in the upper part of the wedge and detection of active seepage of pore waters in this section of the wedge are new phenomena responsible for the formation of the above-mentioned mud diapirs and overthrusts as well as transition to other mechanisms of compression—detachment in the deep-level part of seismofocal plane. These transitions depend on the relative rate of plate movements (Magee and Zoback, 1993; Shipley et al., 1994).

Most of the parameters discussed above vary significantly in subduction zones of the Andean type.

5.6. Andean-type Subduction Zones

Andean-type subduction zones differ from those of island arcs in the following principal characteristics:

1) gentle slope of the subduction zone accompanied by 'tectonic erosion' of the continental plate and separation of magmatic front from the trench (Fig. 5.44);

2) absence of back-arc basins, instead of which back-arc continental rifts are seen in many cases in the background of major arches;

3) degraded nature of accretionary wedge; and

4) special character of magmatism (high role of acidic effusions and presence of granite batholiths) suggesting the involvement of sediments and/or slices of continental crust in melting.

These features can be explained by the rapid counterwise movement of oceanic and continental crust. Before examining this hypothesis, let us first briefly discuss the aforesaid characteristics.

1) A comparison of seismofocal zones and different types of magmatic arcs is shown in Figure 5.44. The magmatic front under the active margins (NC and P) is situated 100 to 200 km farther than in the typical island arcs as a result of the gentle slope of the subducting plate. Island arcs, as already pointed out, are divided into two sub-types: the usual arcs, with a plate inclination of 45 to 60° (field with points NZ, IB, T and KK) and the anomalous arcs with a steep slope of 60 to 90° (NH, CA, ALK, AL and M) and not infrequently with an intense curvature, which manifest themselves in a gentle inclination of the plate adjacent to the trench (ALK, AL and NZ). For Andean active margins, such a curvature is poorly manifest.

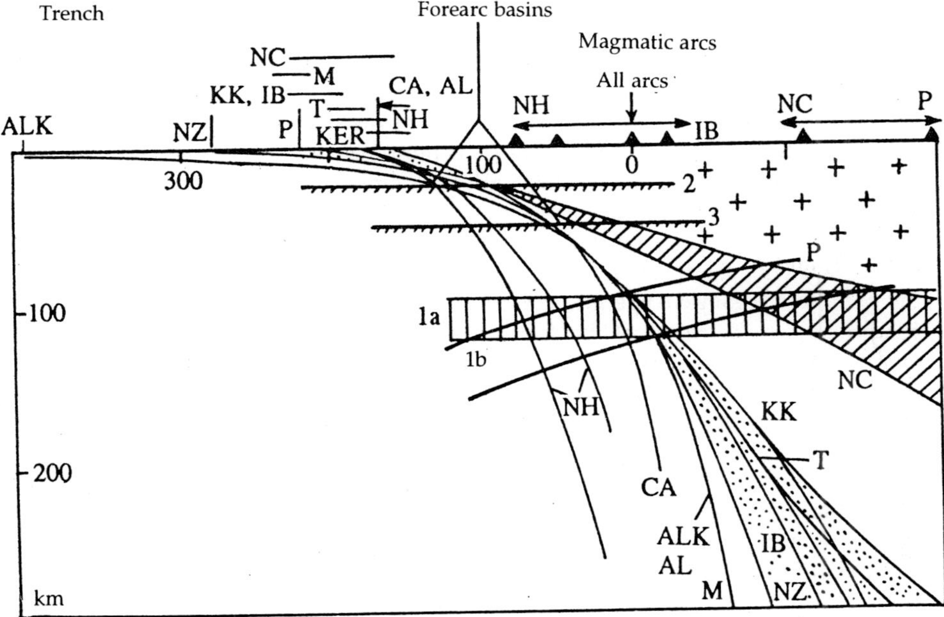

Fig. 5.44. Configuration of the upper surface of the subducting plate for different subduction zones (modified from Isacks et al., 1974; Cloos, 1993).

1a and 1b—Zones of magma formation of constant depth (1a) and depending on the rate of subduction (1b); 2 and 3—minimum depth of transition of amphibolite-eclogite for 'hot' (2) and 'cold' (3) subduction zones.

AL—Aleutian arc; ALK—Alaska; CA—central America; IB—Izu-Bonin, KER—Kermadec; KK—Kuril-Kamchatka; M—Mariana; NC—northern Chile; NH—northern Hebrides; NZ—New Zealand; P—Peru; T—Tonga.

242

Most earthquake foci under the active continental margins are concentrated up to depths of 200 to 300 km; further, they are localised not only in the accretionary wedge and subducting plate as under island arcs, but are dispersed in the entire wedge between the surface and subducting plate. Moreover, the velocity of share waves in this wedge increases continuously to a depth of 250 to 300 km. Both these facts suggest the absence of the asthenospheric layer, i.e., the thickness of the lithosphere under the active margin is not less than 250–300 km. Seismicity ceases at a depth of 300 km but reappears in many cases at depths of 500 to 600 km. This local zone of deep-focus seismicity perhaps reflects the presence of a segment of lithospheric plate at depth (or its poorly melted remnants) displaced into the mantle and deep within the continent relative to the upper part of the seismofocal zone. These features are also characteristic of some collision zones of continents (Himalayan type).

Seismofocal zones and deep-water trenches along the active continental margins are devoid of the arcuate form characteristic of island arcs and adjacent trenches. In many cases they are not oriented perpendicular to the direction of movement of the oceanic plate (although the direction of movement of the Pacific plate in island arcs is also not always perpendicular to the trench). These characteristics can be explained by two factors: a) the thick continental lithosphere cannot adapt itself to the ideal form of an arc and b) active continental margins are overthrust on the subduction zone (island arcs are also partly overthrust during 'oblique' subduction).

2) Arched uplifts of the Andes, thickening of the crust (up to 70 km) and lithosphere (up to 250 km) under the Andes, continuous seismicity beneath them and earthquake mechanisms show that, under conditions of rapid convergent movements of plates, a significant part of the continental lithosphere under the Andes is in a state of compression. Local conditions of extension occur only at the apex of the arch with the formation of rifts, while in the frontal part and at the rear of the arch, they again change into compression. At the rear of the arch, these local conditions are the result of the formation of overthrusts and nappes which are overthrust on molassic troughs (Zonenshain and Kuz'min, 1993a), and in the frontal portion, as a result of deformations of the continental slope (flexures, overthrusts and displacements) leading to repeated regeneration of avalanche olistostromes (Ross et al., 1994).

Thus, the origin of rifts of active margins at the apex of the arch but not at the rear of the structure as in the case of back-arc seas, points to a significant difference in the origin of these two structure types with respect to tectonic conditions. They differ significantly also in the nature of magmatism. In the rifts of active margins (Altiplano in South America, Rio Grande in North America and others), alkaline basalt or alkaline bimodal series predominate and bring these two types of structures close to typical continental rifts.

3) The Andean magmatic arc itself (Fig. 5.45) contains, apart from andesite series, huge bodies of dacite-rhyolite series and granite batholiths. The alternation in space and time of volcanites and major plutons suggests the alternation of compression and extension (or reduction of compression) epochs while the composition of volcanites and plutons, including such markers as Be_{10}, Nd isotopes points to a significant participation of sediments of continental origin or slices of continental

crust in magma formation. Series of different ages (see Fig. 5.45) show that the magmatic front in the active margins is often displaced toward the continent and not the ocean as in the case of island arcs. Finally, gaps or zones in which magmatism and seismicity are absent are often seen in the magmatic arc. For example, in South America the ocean ridge Nazca approaches the subduction zone in the region of 15° S. lat. and causes a seismic and volcanic gap in the Andean arc. Such examples are known in other places also where the colliding ridges are pushed towards the zone of subduction (Ben Abraham et al., 1984; Zonenshain and Kuz'min, 1993a). Other characteristics of the magmatism of subduction zones and possible model structures are examined in the next section. It is important to emphasise here, however, that, unlike island arcs for which fluctuations and a relatively short duration of existence (30 to 50 million years) are characteristic features, active margins of the Andean type exist on much of their extent commencing from the Cretaceous, i.e., for over 150 million years. Significant reorganisations in their development are noticed only on the southern (Chile) and northern (north of California) flanks.

This leads also to a significant change in the deep-level structure. Apart from the gentle slope of the subduction zone (see Fig. 5.44), the presence of large dense masses, possibly restites from remelted lithospheric plates, under the Andean arc and absence of such masses under the island arcs are characteristic features. These peculiarities for young arcs and prolonged subduction zones of Andean type are compared in Figures 5.46 and 5.47. In young arcs dense masses are of small volume and spread along the boundaries of upper and lower mantles. In the case of arcs that have existed much longer (Indonesian), dense masses increase and are subducted into the lower mantle. This effect is manifest even more distinctly for the Andean belt of North America while, under South America, dense masses (probably restites) are spread in the form of a horizontal plate up to the Mid-Atlantic ridge (MAR).

5.7. Model of Accretionary Wedge as Regulator of Subduction Zone

The subduction process in many models (Peacock, 1993; Sharapov, 1994) is simplified, assuming a rather thin zone of displacement at the plate boundaries without taking into account the driving forces. Let us now consider a more realistic model which takes into account the presence of an accretionary wedge in the subduction zones (see Fig. 5.42) and evaluate the motive forces.

Subduction, or the submergence of an oceanic lithospheric plate under a continent or island arc, occurs under the action of a massive force arising as a result of solid phase transition (eclogitisation) in the downsinking plate and due to the force of viscous friction as a result of thermogravitational flows in the asthenosphere under ocean. For example, for extension of the asthenosphere under the ocean $x_0 = 2100$ km, the average value of the coefficient of friction at the lithosphere-asthenosphere boundary has been estimated as $\tau = 4.43 \times 10^5$ N/m^2 and friction along the length of the asthenosphere (for one running metre along the ridge axis) equal to $F = x_0 \tau = 9.32 \times 10^{11}$ N/m (Kirdyashkin, 1989).

During the relative displacement of lithospheric oceanic plate and continent, either sliding according to Coulomb's law is possible, which is accompanied by very large resistance forces, or interaction is possible through a layer of less viscous

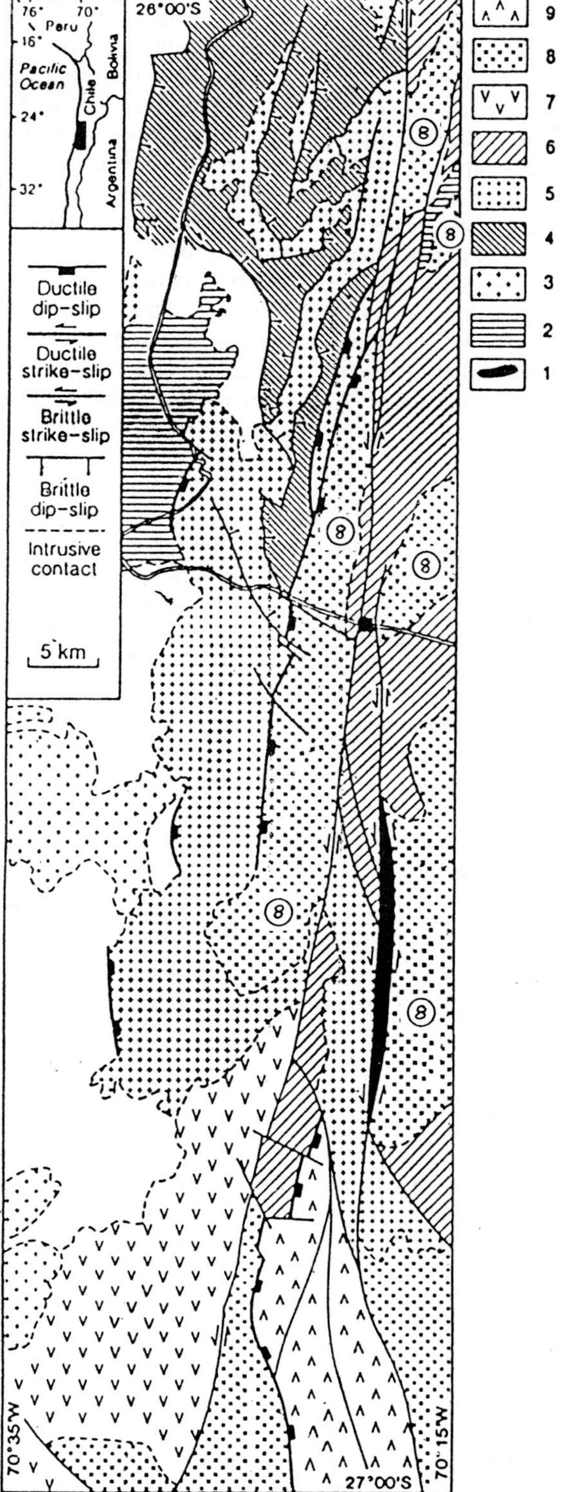

Fig. 5.45. Geological structure of the Mesozoic part of the Andes and northern Chile (Grocott et al., 1994). 1—Upper Palaeozoic accretionary complex including palaeovolcanites; 2, 3—Late Triassic and Early Jurassic plutonic complexes; 4—Early-Middle Jurassic volcanic and sedimentary rocks; 5—Late Jurassic plutonic complex; 6—Lower Cretaceous volcanic rocks; 7 to 9—three phases of Early Cretaceous plutonic complex. Inset on left depicts the location of the mapped section and different types of ductile and brittle overthrust and displacement deformations.

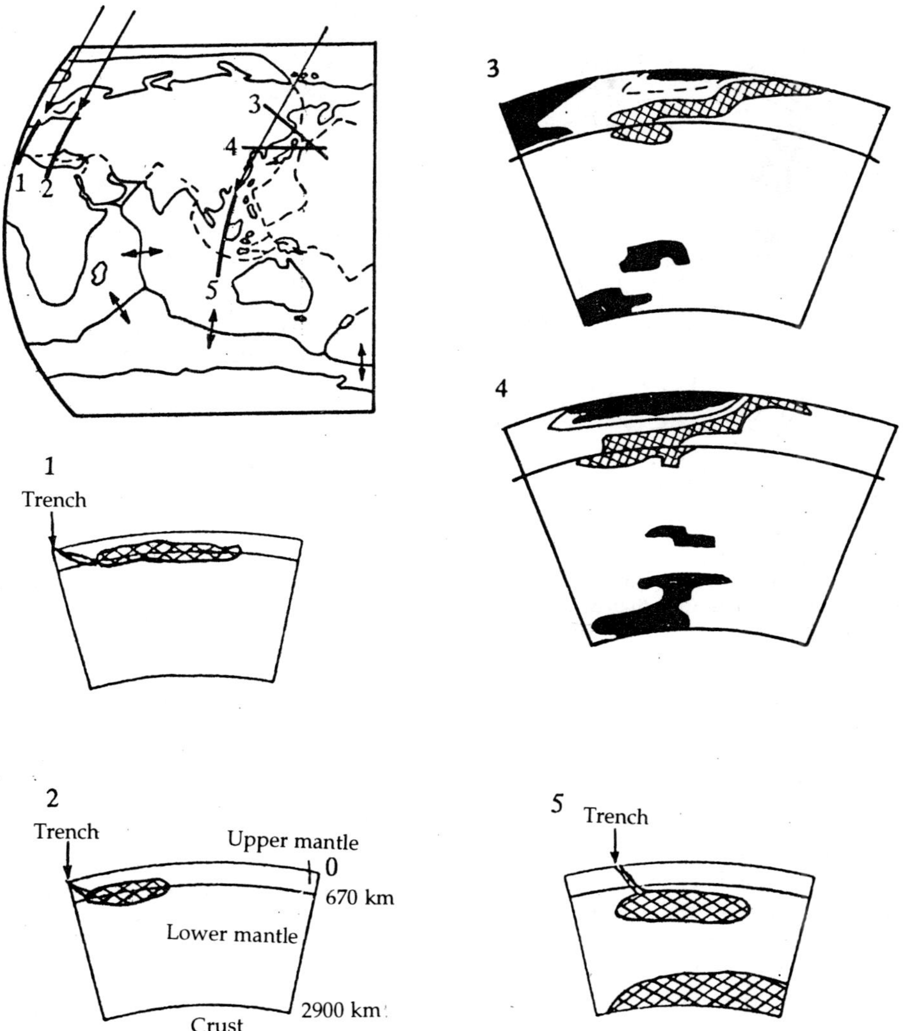

Fig. 5.46. Schematic sections of upper and lower mantle across active and extinct island arcs of the Alpine region and Pacific Ocean.
Hatched sections—major heavy masses (restites) identified from anomalies $+ \Delta V$ (Fukao et al., 1994, modified); black sections—heated zones. 1—Western Alps; 2—central Mediterranean; 3 to 5—arcs: 3—Japan; 4—Mariana; 5—Indonesia.

liquid compared to the viscosity of the two mutually displacing plates. Sliding according to Coulomb's law adopted in the model of the local displacement zone (Peacock, 1993) calls for overcoming a friction of the order of 10^{14} N/m which

246

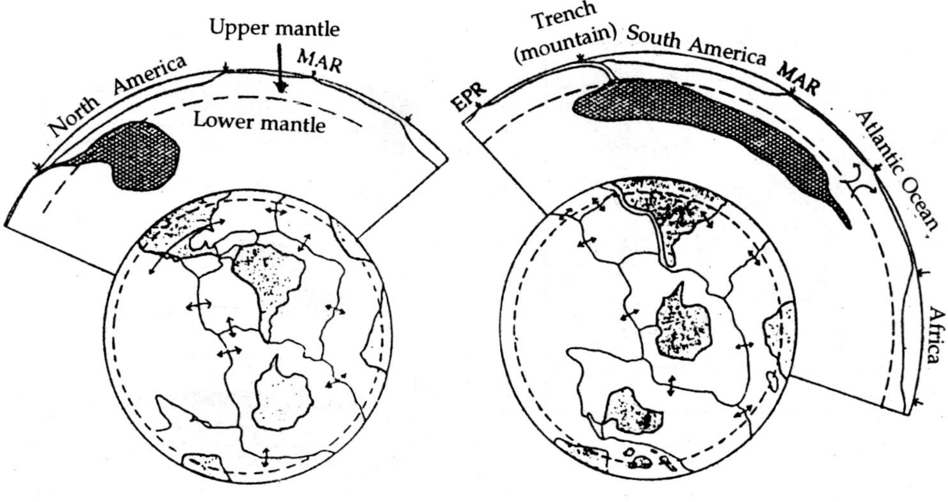

Fig. 5.47. Global sections through North and South America showing the plates subducted and accumulated restites (hatched) (Fukao et al., 1994). EPR: East Pacific Rise; MAR: Mid-Atlantic Ridge.

might considerably exceed the motive forces of subduction: friction of asthenospheric flows 10^{12} N/m (see above) and Archimedean force associated with eclogitization of the order of 10^{12} to 10^{13} N/m (Ueda, 1982). The presence of a 'lubricant' between the plates is therefore very important.

When there is a difference of a few orders in the coefficient of dynamic viscosity between the lithosphere and intermediate layer, the presence of viscous friction can be assumed according to Newton's law $\tau = \eta\,(\partial U/\partial y)$. This may also be promoted by a possible temperature rise in the intermediate layer due to dissipative heating during subduction of the plate.

The margin of the continent or island arc is exposed to the action of the oceanic plate during the entire period of subduction. Any fixed region of oceanic lithosphere interacts with the continent only for the period L/U (10^6 years), where L is the length determined by the thickness of the lithosphere and U the rate of subduction. Therefore, the margin of the continent may be subjected more to the processes of damage (tectonic erosion) if a layer of less viscous 'liquid' is absent. Such a layer may consist of sediments and products of 'abrasion', to which reference has been made above in the case of active margins of the Andean type. The presence of a layer of sediments drawn in the zone of underthrusting of the plate is confirmed by geological facts and numerical modelling (*Oceanology...*, 1979; Zonenshain and Kuz'min, 1993a) (see Fig. 5.42).

Experimental investigations point to the increasing strength of rocks as all-round pressure increases. It can therefore be assumed that thickness of a viscous layer between the oceanic lithosphere and the continent $h(x)$ decreases with depth (x), in the first approximation, according to the linear dependence $h(x) = \delta\,(a - x)$ and is in the form of a flat wedge, where a and δ are constants (Fig. 5.48).

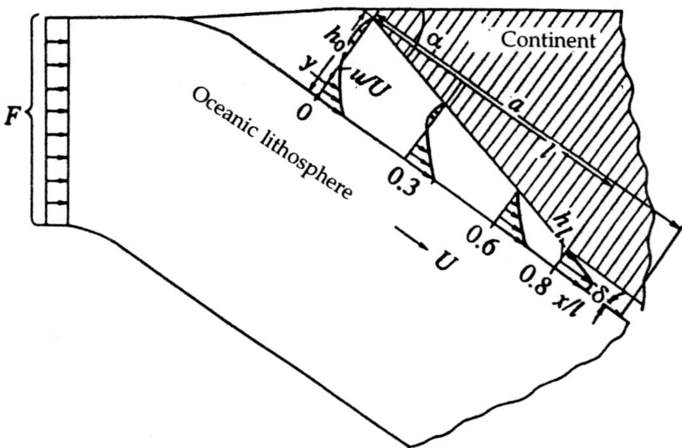

Fig. 5.48. Model of accretionary wedge and principal dimensions. Velocity profiles correspond to $\bar{l} = 0.8$, $P_0 = P_l$ and Fig. 5.49a.

According to the theory of lubrication, in the relative displacement of two planes in a wedge-like viscous layer, a high pressure is generated between them while the friction between the surfaces corresponds to the friction in the viscous layer between them (Shlikhting, 1969). If it is assumed that the viscous zone between the lithospheric plates of the ocean and continent (any lithospheric plate in general) is in the form of a wedge, a high pressure is generated in the wedge during the displacement of the oceanic plate relative to the continent. This high pressure will prevent the proximation of the plates and Coulomb's sliding friction cannot arise between them. On random change of wedge configuration, there is decrease or increase in average pressure in it and correspondingly divergence or convergence of plates to the dimensions of the wedge when the total force of pressure in the viscous wedge becomes equal to the force of the oceanic lithosphere on the continent (island arc). Pressure generated due to convective flow in the asthenosphere has been estimated as $F_1 = 10^{12}$ N/m. This is a minimal value without effect of the eclogitisation of a subducted slab.

Let us consider a case when the continent is stationary while the oceanic plate is displaced at velocity U_0. The all-round force on the continental margin from the side of the oceanic plate corresponds to the force of viscous friction F due to asthenospheric flow. The angle of orientation of the flat wedge relative to the gravity vector depends on the ratio of the vertical and horizontal velocity components of the subducting lithosphere.

Let us assume that $h(x)/l \ll 1$ (Fig. 5.48). The model of the wedge corresponds to geophysical data (see Fig. 5.43).

In the first approximation, the convective terms in the equation of motion (creeping motion) as well as the change in pressure along thickness $\partial P/\partial y = 0$ are ignored. Here, P is the additional pressure expressed as the excess of lithostatic pressure ($P = P_{tot} - P_{lit}$). The following equation is derived:

$$\partial P/\partial x = \eta\,(\partial^2 u/\partial y^2), \tag{5.23}$$

where η is the dynamic viscosity; x and y are co-ordinates and u the rate of subduction.

The continuity equation for a thin layer $h(x)/l >> 1$ is expressed as follows:

$$Q = \int_0^{h(x)} u\,dy = \text{const}, \tag{5.24}$$

where Q is the rate along the layer thickness. For boundary conditions, $u = U$ at $y = 0$; $u = 0$ at $y = h$; $P = P_0$ at $x = 0$; and $P = P_1$ at $x = 1$, solutions to eqns (5.23) and (5.24) have the following form:

$$u = U\left(\frac{1-y}{h}\right) - \left(\frac{h^2}{2\eta}\right)\left(\frac{y}{h}\right)\left(\frac{1-y}{h}\right)\left(\frac{dP}{dx}\right), \tag{5.25}$$

$$Q = (P_0 - P_l)a^2\delta^3\,\frac{(a-l)^2}{6\eta l\,(2a-l)} + Ua\delta\,\frac{(a-l)}{(2a-l)}, \tag{5.26}$$

$$P(x) - P_0 = 6\eta U\left[\frac{x\,(l-x)}{h^2\,(2a-1)}\right] - (P_0 - P_l)\left[\frac{x\,(a-l)^2\,(2a-x)}{l\,(a-x)^2\,(2a-l)}\right], \tag{5.27}$$

$$\Delta P = \overline{P(l)} - P_0 = \frac{6\eta U}{\delta^2 a}\left[\frac{-2}{(2-\bar l)} - \frac{1}{l}\ln\,(1-\bar l)\right] + \frac{(P_l - P_0)}{(2-\bar l)}, \tag{5.28}$$

where $\Delta P = \overline{P(l)} - P_0$ is the average pressure drop across the layer relative to P_0; $\bar l = l/a$; and $P(x)$ is the local pressure.

Equations (5.27) and (5.28) represent the excess pressure in the wedge depending on its dimensions a, l and h, rate of displacement U and viscosity η.

The following equations are derived using eqns (5.25) to (5.28) when the pressure at inlet (P_0) and exit (P_1) are equal:

$$u/U = (1-\bar y)\,[1 - \bar y\,(3 - (6(1-\bar l)/(1-\bar x)\,(2-\bar l)))]$$

$$Q = Uh_0(1-\bar l)/(2-\bar l),$$

$$P(x) - P_0 = [(6\eta U a/h_0^2)\,(2-\bar l)]\,[\bar x(\bar l - \bar x)/(l - \bar x)^2], \tag{5.29}$$

$$\overline{P(l)} - P_0 = (6\eta U/\delta^2 a)\,[-2/(2-\bar l) - \ln\,(1-\bar l)/\bar l],$$

$$(P(x) - P_0)/\overline{(P(l)} - P_0) = [\bar x(\bar l - \bar x)/(1-\bar x)^2]\,\{(\bar l - 2)\ln(1-\bar l)/\bar l] - 2\}^{-1},$$

where $\bar{y} = y/h$; $\bar{l} = l/a$; and $\bar{x} = x/a$ are dimensionless co-ordinates. The results of calculations using eqn (5.29) are depicted in Figure 5.49a in which the dimensionless values are given. This Figure shows that the maximum pressure is shifted towards the exit section and increases with increase in slope of the wedge and decreasing cross-section of the exit. The pressure rise generated in the wedge counterbalances the pressure of the oceanic plate at the continental margin ($F_2 = x_0 \tau_{av}$) which exists due to friction between asthenospheric convective flows and lithsophere. For $\bar{l} \to 1$, a sharp increase in pressure in observed around $x = l$ and $P(l) - P_0 \to \infty$ (Fig. 5.49a, curve 3). This points out that as the exit section is enclosed, pressure (and friction) may increase to the level of ultimate strength and destruction occurs in the region $x = l$ leading to tectonic erosion of the continental plate in the region of pinching since, as pointed out before, the continent is subjected to the action of the oceanic plate all during subduction. Calculations also show that depending on the increase of angle (δ) of the slope (increase of h_0/a), the flow of viscous liquid decreases and a reverse flow arises in the region of inlet section h_0. It may cause tectonic upheaval of subducted wedges, i.e., obduction. The maximum coefficient of friction on the continental side of the wedge is in the vicinity of exist section h_1, i.e., the carryover of the eroded rock deep into the mantle as well as obduction is effected by the capture of wedges of continental lithosphere and subducted sediments near h_l.

The characteristic observed dimension of the wedge or the layer of dragged sediments $l = (1 - 2) \times 10^5 \text{m}$ (see Figs. 5.42 and 5.43) while the rate of subduction is 3 to 10 cm/yr. The viscosity of dragged sediments has been estimated as $\eta = 10^{18}$ to $10^{19} \text{N} \cdot \text{s/m}^2$ (*Oceanology...*, 1979; Fig. 1.9). For a case when $P_0 = P_l$ and maximum parameters $l = 10^5$ m; $\bar{l} = 0.8$; $U = 9.45$ cm/yr $= 3 \times 10^{-9}$ m/s; $\eta = 3 \times 10^{18} \text{N} \cdot \text{s/m}^2$; and $\overline{P(l) - P_0} = F/l = 10^7 \text{N/m}^2$, we find $h_0 = 1.52 \times 10^4$ m; $h_1 = 3.05 \times 10^3$ m; $a = 1.25 \times 10^5$ m; and $\delta = 0.122$. The resultant estimates are close to the site and form of the accretionary wedge or deformed rocks in the lower part of the island arc slope in many trenches (see above). For $\bar{x} = 0.67$, $P_{max} = 1.93 \times 10^7 \text{N/m}^2$ (197 kg/cm^2) but for $x \geq 0.8$, P_{max} rises perceptibly. Nevertheless, these estimates of pressure are low compared to the estimates of pressure in glaucophane schists and eclogites originating in the subduction zones (see below).

The average value of the coefficient of friction on the oceanic plate can be determined using eqn (5.29):

$$\tau = (\eta/l) \int_0^l (du/dy)_{y=0}\, dx = (U\eta/\delta l) \, [\ln(1 - \bar{l}) + 6/(2 - \bar{l})]. \tag{5.30}$$

The maximum value of τ at the continental margin is: $\tau = 9 \times 10^6 \text{N/m}^2$ (92 kg/cm^2). For $\bar{x} = 0.3$, the friction on the submerging lithospheric plate $\tau = 2.16 \times 10^6 \text{N/m}^2$ (22 kg/cm^2). The total friction per running metre of the oceanic plate $F_{fr} = \tau_{av} l = 2.16 \times 10^6 \times 10^5 = 2.16 \times 10^{11}$ N/m which is less than the friction applied to the lithosphere as a result of thermogravitational forces in the oceanic asthenosphere and has been estimated above at 9.3×10^{11} N/m.

The change in length of interaction between continental and oceanic lithospheres through the wedge-shaped viscous layer (l) at constant values of F and $\bar{l}$ negligibly influences the dimensions h_0 and h_l. The liquid flow rate can be reduced

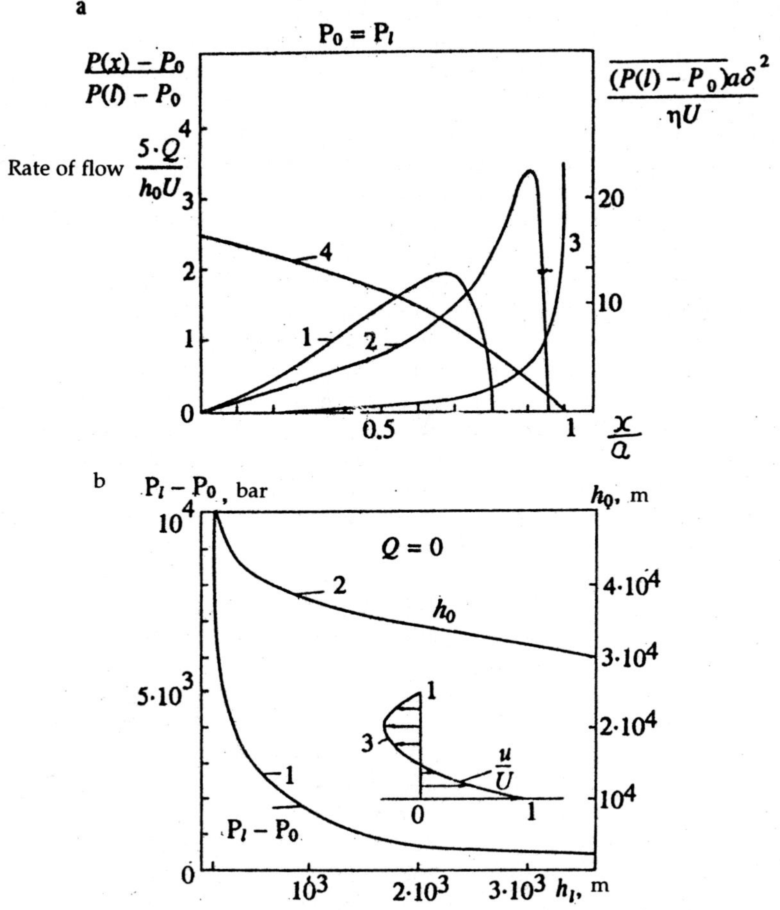

Fig. 5.49. Model of viscous wedge between the interacting plates.
a—results of calculation using eqns (5.29): change of local pressures $(P(x) - P_0)/(P(l) - P_0)$ along the length of the wedge: 1—at $\bar{l} = 0.8$; 2 —at $\bar{l} = 0.95$ and also the average pressure $\overline{P(l) - P_0}$ (3) and flow (4) at different values of $\bar{x}$ and $\bar{l} = l/a$; b—results of calculation using eqn (5.31): 1—change in pressure for $Q = 0$; $\eta = 3 \times 10^{18}\,\text{N} \cdot \text{s/m}^2$, $l = 10^5$ m, $U = 3 \times 10^{-9}$ m/s, $P(l) - P_0 = 10^7$ N/m^2; 2—change of h_0 for different values of the exit section of wedge h_1; 3—velocity profile.

by pinching the exit section h_l but it cannot attain the zero value due to a sharp increase in pressure and damage of rocks in the vicinity of $x = l$ (Fig. 5.49a, curve 3).

As pressure (P_l) in the exit section increases $(P_l - P_0 \geq 0)$, instances of $Q > 0$ and $Q < 0$ are possible. Let us study a case when $Q = 0$. From eqns (5.25) and (5.28), it follows that

$$\left.\begin{array}{l} u/U = 1 + \bar{y}\,(3\bar{y} - 4); \\[4pt] P(x) - P_0 = 6\eta\,U\bar{x}/a\,\delta^2\,(1 - \bar{x}); \\[4pt] P_1' - P_0 = 6\eta\,\overline{Ul}/a\delta^2\,(1 - \bar{l}\,); \\[4pt] \overline{P(l)} - P_0 = (-6\eta U/a\delta^2)\,[\ln\,(1 - \bar{l}\,)/\bar{l} + 1]. \end{array}\right\} \qquad (5.31)$$

Only in this case the velocity profile (Fig. 5.49b, curve 3) is self-modelling and the coefficients of friction depend linearly on x. The calculated pressure drops $P_l - P_0$ against the size of the exit section (h_l) are shown in Fig. 5.49b. For the above-stated parameters and $h_l = 10^2$ to 3.6×10^3, pressure drop $P_l - P_0$ at which $Q = 0$ varies in the range $P_l - P_0 = 0.5$; to 5 kb and attains 10 kb at $h_l \leq 100$ m. The latter values (5 to 10 kb) are comparable to pressures operating due to the metamorphism of glaucophane schists but are possible only when the exit section is clogged.

If it is assumed that the erosion of interacting surfaces of continent and oceanic plate depends on the increase in rock strength with increasing depth and value of viscous friction on these surfaces, the thickness of the viscous layer may vary non-linearly as x increases, since the surface friction for $Q \neq 0$ changes non-linearly according to the above equations.

Under real conditions, the surface of the oceanic plate with sediments has uplifts and unevennesses with an amplitude of hundreds of metres or even larger inhomogeneities of the type seamounts (Cloos, 1993). During subduction of an oceanic plate, they can generate significant fluctuations in local pressures and coefficients of friction and influence the shape of the viscous layer. Formation of the latter and its configuration may also depend on the depth-wise distribution of forces applied by the oceanic plate on the continental plate. These perturbations may cause instability of the viscous flow region.

The values of h_0 and h_l for which a stable state is observed, i.e., the force of the plate acting on the viscous wedge is the same as the total force of the viscous liquid in the wedge acting on the continent, are shown in Figure 5.50. Calculations were made for a constant value of the pressure of oceanic plate on the continent $F = 5 \times 10^{11}$ N/m (per running metre along the ridge axis) and average value of $U = 3 \times 10^{-9}$ m/s at $\eta = 10^{17}$ to 5×10^{18} N $\cdot$ s/m and $l = 50$ to 200 km.

For values of $h_0 < h_l$ and constant velocity of convergence of lithosphere and continent along x-axis, equilibrium conditions will never be achieved and convergence will continue until there is contact interaction between the oceanic plate and the continent, i.e., in the zone left of the broken line $h_0 = h_l$, a viscous intermediate layer cannot exist. The region in which $h_0 > h_l$, corresponds to the zone of stable existence of a wedge-shaped intermediate viscous layer between the margin of continents and oceanic lithospheric plate for a self-regulating wedge configuration under conditions of generation of disturbances of wedge shape, as well as variations in liquid properties in the wedge, and primarily its viscosity. The region under the curves (Fig. 5.50) represents such a ratio of wedge sizes for which pressure in the wedge exceeds the force exerted by the oceanic lithosphere on the continent (correspondingly also on the viscous liquid in the wedge). For example, for parameters corresponding to curve 2 (Fig. 5.50), $h_0 = 12$ km and $h_l = 1.65$ km (point a), pressure in the wedge exceeds the force of the plate acting on the wedge,

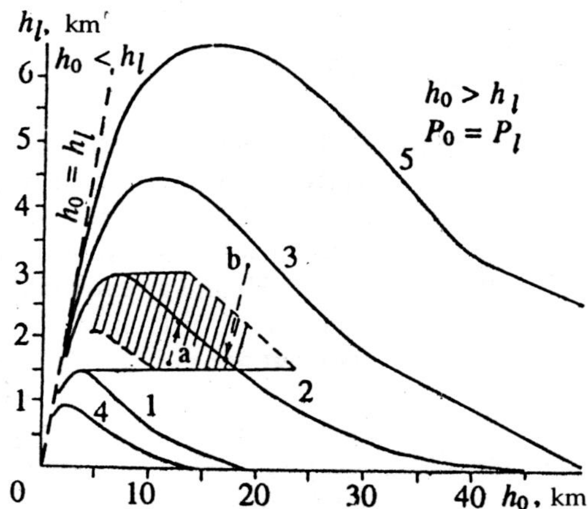

Fig. 5.50. Dependence of the thickness of the viscous wedge (h_0 and h_1) between the lithospheric plates of the ocean and continent according to eqns (5.29) when $P_0 = P_1$, $F = 5 \times 10^{11}$ N/m and $U_0 = 9.3$ cm/yr under conditions of stable equilibrium: at $\eta = 10^{18}$ N $\cdot$ s/m^2: 1—$l = 50$ km; 2—100 km; 3—150 km; 4—at $\eta = 10^{17}$ N $\cdot$ s/m^2, $l = 100$ km; and 5—at $\eta = 5 \times 10^{18}$ N $\cdot$ s/m^2, $l = 100$ km. Hatched zone depicts the most probable wedge dimensions.

leading to divergence of oceanic and continental plates along the broken straight line corresponding to the equidistant displacement of the continent and the oceanic plate to such dimensions when the force of pressure in the wedge becomes equal to the force acting on the wedge (to the point of intersection of the broken line with curve 2). If, however, the resultant form of the wedge is such that it corresponds to point b, the pressure in the viscous wedge is less than the pressure of the plate on the wedge and oceanic and continental plates move close to each other. For the same rate of convergence of lithospheric plates of ocean and continent along the x-axis, the change in wedge dimensions will correspond to the broken curve until these dimensions become equal to the point of intersection of the broken line with curve 2, i.e., until equilibrium conditions are reached.

From the data shown in Figure 5.50, the limiting dimensions of wedge thickness h_0 and h_1 can be determined for the corresponding liquid parameters and wedge length. For example, for parameters corresponding to curve 3, the size of h_0 does not exceed 50 km. For $h_0 \geq 50$ km, $h_l \rightarrow 0$ and the continent comes into contact with the oceanic plate. In this case there is either disruption and increase in h_l or the Coulomb forces of friction during the contact of the oceanic and continental lithospheres becomes more than the motive forces of the lithosphere, movement ceases and the oceanic lithosphere may undergo rupture at a new site.

Probably, h_l cannot be less than the irregularities of the lithosphere. For parameters of viscous wedge corresponding to curve 2 (Fig. 5.50), the maximum dimension is h_1 = 3.0 km. For $h_1 > 3.0$ km, the pressure in the layer of viscous liquid is less than the pressure of oceanic and continental lithosphere and, as a result, the viscous liquid is squeezed out from the wedge and the lithosphere and the continent come close to dimensions corresponding to curve 2 (Fig. 5.50). The probable wedge sizes correspond to h_1 = 1.5 to 3 km (Fig. 5.50).

For fixed values of F and η, an increase in maximum values of h_0 and h_1 is observed as l increases. As the viscosity of the liquid in the wedge decreases, the values of h_0 and h_l at which stable equilibrium is observed decrease perceptibly (Fig. 5.50, curves 2 and 4). Thus, for each value of h_l there are two of h_0 at which equilibrium of pressure in the wedge and pressure of lithospheric plates on the viscous wedge in seen. As the value of h_0 varies in this range, the pressure in the wedge increases relative to that of the lithospheric plates. As the pressure in the wedge decreases, the plates close together to wedge dimensions corresponding to the establishment of equilibrium. In this case, the viscous liquid is squeezed out from the wedge and the velocity of return flow increases to values greater than the velocity of plate descent (see below).

The possible relations in mutual displacement of oceanic and continental plates in the subduction zone are depicted in Figure 5.51. In the presence of an intermediate viscous zone in the form of a wedge, the pressure of oceanic lithosphere on the continent is counterbalanced by this region and for $x > l_1$, the interaction between the oceanic plate and continent may be represented as friction in the flow of Couette liquid in a flat channel under conditions of mutual displacement of plates.

Depending on the values and ratio of velocities u_l and u_c and also the variation pattern of subduction velocity u_l in the subduction of lithospheric plate, the wedge structure may be varied (Fig. 5.51). At constant rate of subduction v = const, the slope of the oceanic plate is constant, and a wedge of length l_1 and a crevice zone of constant thickness l_2 are delineated (Fig. 5.51a). As subduction rate varies $v = f(x)$, which may be caused by a change of gravity with depth, eclogitisation or reduction of friction along the contact, a monotonic change in angle of slope of the lithospheric plate is possible (Fig. 5.51b). Such an angle variation is characteristic of Aleutian and some other island arcs. (see fig. 5.44)

Two stages can be distinguished during the interaction of two lithospheric plates of comparable thickness (Fig. 5.51c). In the first stage, under pressure, when creep flow and crumpling of plates occur, there is thickening and the height of uplift of the crumpled plate is determined to the first approximation by pressure $P_{pl} = h_1 \rho_c g$. The depth of downsinking of the lithosphere into the asthenosphere h_2 in the zone of crumpling, ignoring the resistance forces in the movement of lithosphere into the asthenosphere, is determined by the equilibrium between pressure P_{pl} and buoyancy due to the difference in density between asthensophere ρ_a and lithosphere ρ_l.

$$h_2 = P_{pl}/(\rho_a - \rho_l) \cdot g \text{ and } h_2/h_1 = \rho_c/(\rho_a - \rho_l). \tag{5.32}$$

For example, at ρ_c = 2.8 g/cm^3, $\rho_a - \rho_l$ = 0.15 g/cm^3, P_{pl} = 1 kb, h_1 = 3.572 km and h_2 = 67 km.

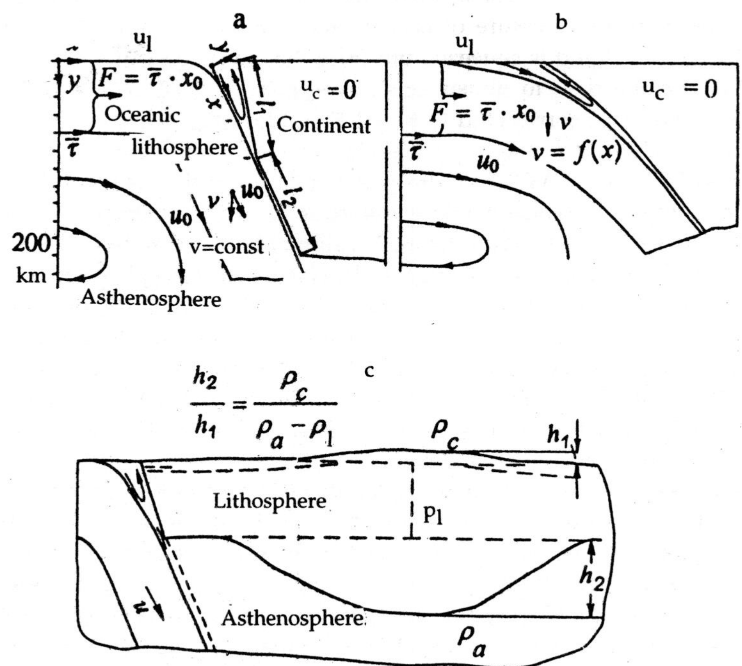

Fig. 5.51. Model of flow in the interaction of continental and oceanic lithosphere in the process of submergence.
a—thickness of continental lithosphere more than the oceanic lithosphere and $u_c = 0$; b—thickness of oceanic lithosphere less than the thickness of the continent, $u_c = 0$ and $v = f(x)$; and c—interaction between lithospheric plates of comparable thickness. Comments given in the text.

Fracture in the weakest section of the lithosphere and origin of subduction are possible in the second stage. Stable subduction is possible only when a 'lubricant' is available in the form of a viscous accretionary wedge.

From our model of the interaction of oceanic lithosphere with the continent, it follows that the origin of a viscous wedge-like zone between the subducted lithospheric plate and continent is highly probable and this zone is stable and capable of 'self-regulation' during disturbances of the form of wedge and viscosity of the intermediate layer. Such a region of viscous interaction between the continent (or island arc) and oceanic lithosphere considerably reduces the friction between them and promotes a stable plate subduction. From a solution of equations for stationary plate subduction and absence of velocities normal to the surfaces of oceanic and continental lithosphere, the rate of ascending movement cannot be greater than the rate of subduction at $\eta = $ const.

Apart from ours, many other models of accretionary wedge have been suggested (Ernst, 1975; Cloos, 1982, 1986; Wentworth et al., 1984; Platt, 1986; Avigad, 1992; Liou et al., 1994; Ring and Brandon, 1994). Closest to ours are the empirical models of Ernst (1975) and Wentworth et al. (1984) developed for Franciscan formations of California and related to the problems of glaucophane schists (see below). Other models consider the non-stationary subduction process and the special situations of collisions and cessations (or variations) of the subduction process. These situations are examined in the following sections by comparing our model with other examples of subduction modelling.

5.8. Models of Magmatism and Metamorphism in Subduction Zones

Models of an elastoviscous accretionary wedge with an exit crevice, which ensures a stable subduction process, were studied in the preceding section. Here we continue with the modelling of subduction processes and examine the processes in the crevice below the subducted wedge, i.e., the role of friction and dissipative heat and nature of melting in it and in the underlying oceanic crust as well as the upwelling mechanism of ascending melts. This melting, as discussed in Sections 5.5 and 5.6, ensures active andesite volcanism of island arcs and active margins.

Figure 5.52 depicts the general model of the zone of magmatism and distribution of metamorphic facies in the subduction zone and above it in the lithosphere of an island arc (Ernst, 1974). The magmatic column in the form of infiltrating melts and a series of intermediate chambers is formed in the melting zone of the upper part of the subducted plate at a depth of 100 to 140 km (see Fig. 5.44). Another model in the form of diapirs of diameter 30 to 50 km and a fine feeding channel are studied below on the basis of experimental results (Whitehead and Luther, 1975; Olson and Singer, 1985). A high heat flux occurs above the magmatic column in the island arc. According to experimental results, this flux is 1.5 to 3 times greater than the average steady flux (Sharapov, 1994). As a result, the isotherms serving as the boundaries of facies rise upward: isotherm 500°C representing the boundary of greenschist (5) and epidote-amphibolite (6) facies to a depth of 5 km and isotherm 750°C representing the boundary of amphibolite (7) and granulite (8) facies to a depth of 15 to 20 km.

In the upper part of the plate subducted, isogrades on the contrary are turned downwards due to loading by 'cold' material and the general succession of metamorphic facies to a depth of 70 km (predominantly within the accretionary wedge) is as follows: 1) zeolites; 2) pumpellyite-actinolite; 3) glaucophane schist; 4) eclogite or 6) epidote-amphibolite facies. Facies 5, 7 and 8 appear in the island arc section (Fig. 5.52). The zone of partial melting commences from 100 to 120 km and total melting of rocks of basalts (eclogite) composition from depth 140 to 150 km (Fig. 5.52).

The foregoing general model varies widely depending on the rate and duration of the subduction process and presence or absence of friction in the zone of the main crevice. In fact, the latter was adopted as the main factor in Peacock's model (1993). Stress in the zone of the main crevice varies from 33 to 100 MPa and, correspondingly, the temperature at a depth of 50 km (1.7 GPa), according to the calculations of Peacock (1993), varies from 230°C in the absence of stress, 400°C at stress 33 MPa, 640°C at stress 67 MPa and 940°C at stress 100 MPa. The value of

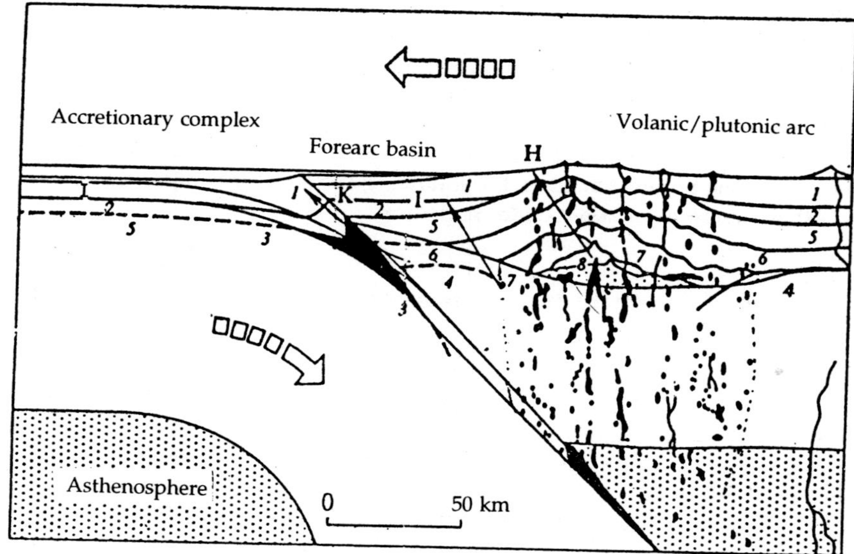

Fig. 5.52. Schematic section of subduction zone and island are reflecting the distribution of metamorphic facies and magmatic front of the island arc (according to the data of Ernst (1974), modified).
Facies 1 to 8 are shown in the text and in Figure 5.53. Lines with arrows H, I and K show the conditions of metamorphic zones of Hidaca, Idonappu and Kamuikotan on Hokkaido and their analogues. For comments, see text.

$\Delta P(\tau)$ is constant along the zone of displacement. The effect of subduction rate in this model was reflected in two ways. In the absence of shearing friction, the increase in subduction rate reduces the temperature by 100 to 150°C for the same depth of 50 km. Commencing from a shearing stress of 20 to 25 MPa, the increase in subduction rate raises the temperature due to increase in frictional heat.

In these calculations, it is difficult to regard the stable conditions along the shear zone and the accepted stress value $\Delta P(\tau) \geq 33$ MPa as realistic. In the absence of stress, temperatures at depths of 30 to 60 km (100 to 300°C) are distinctly conservative due evidently to the unrealistic heat flux. More correct (agreeing better with geological and mineralogical data) is the diagram shown in Figure 5.53 (Cloos, 1993). Here, on the background of the facies scheme (Liou et al., 1985; Evans, 1990), four different P-T curves corresponding to different geodynamic conditions are shown.

1) The high-temperature curve corresponds in the extreme case to the ascending flows of magma in Figure 5.52. It is close to our theoretical curve for ascending flow in the mantle (see Fig. 4.14) and conditions of metamorphism in the mid-oceanic ridges (Silant'ev, 1995) as well as typical P-T estimates in the interior of

island arcs, especially the conditions of metamorphism M_1, M_2 and M_3 in Hidaca zone, Hokkaido (Komatsu et al., 1992).

2) The averaged continental curve corresponds also to the 'hot' subduction conditions of the Andean type.

3a) The curve of 'hot' subduction corresponds to the estimates of Cloos (1993) for slow subduction zones (3 to 5 cm/yr) and/or young 'hot' plates in the new subduction zones.

3b) The 'cold' subduction curve is perhaps more realistic in the rapid (8 to 10 cm/yr) subduction zones where the 'cold' plate is submerged in subduction zones that have long been effective. The actual estimates of P-T conditions in eclogite-glaucophane belts which are regarded as products of metamorphism of the subducted plate correspond well either to curve 3b (Fr—Franciscan complex and Ma—Maksyutov complex in the Urals) or 3a (P—Penzhina belt of Kamchatka, SL—Sesia-Lanzo zone in the Alps and Sp—Motalfiela complex in Spitsbergen (Dobretsov et al., 1974, 1989). These examples are discussed in detail in the next section.

The upper part of Figure 5.53 shows the variations of melting conditions which determine the genesis of island arc magmas. They correspond to the continuation of curve 2 (points 2a and 2b), curve 3a (points 3a and 3b) and curve 3b (point 3c). For subduction of the Andean type, points 2a and 2b correspond to the dry, highest temperature (1300 to 1400°C) (melts generated at depths of 90 to 120 km). Points 3a and 3b correspond to the water-bearing melts at $T = 1170$ to 1250°C at depths of 92 to 110 km. Finally, point 3c, for 'cold' subduction, corresponds to the low-temperature (900°C) melts richest in water that are possible at depths of ≥ 100 km.

This model is significantly simplified, however. In Sections 5.5 and 5.6 it has already been pointed out that mixing of two sources takes place in the suprasubduction zones: 1) rise of andesite (dacite-andesite) melts from the melting zone in the upper part of the subsiding plate and 2) direct melting of suprasubduction mantle in the zones of ascending flow (see Figs. 4.15 and 5.52) in which the conditions are similar to those of mid-oceanic ridges and MORB-type basalts or those enriched with non-coherent elements of back-arc basalts are formed. Variations in the nature and degree of interaction of these two types of melts as well as differentiation in the intermediate chambers lead to the manifestation of different types of island arc volcanic series, among which the following are delineated: 1) boninites; 2) tholeiitic island arc; 3) calc-alkaline andesite-basaltic including high-alumina basalts; 4) calc-alkaline continuous basalt-andesite-rhyolitic; 5) dacite-rhyolitic acidic series characteristic of Andean-type conditions; and 6) subalkaline and alkaline (shoshonitic and picrite-shoshonitic). With the advancing development of an island arc, a regular change in different types of sequences is often observed, the boninitic series being early and the concomitant 'frontal' sequence (most proximate to the trench) and acidic and alkaline sequences representing the late series and/or those far removed from the magmatic front.

Boninitic series have been established in the island arcs that have advanced far into the ocean and are relatively young ('primitive'); Bonin and western Mariana (Sharaskin et al., 1980; Geology..., 1980), Tonga arc (Vysotskii, 1989), Bismarck arc and its continuation on cape Vogel, Papua where lavas containing clinoenstatite were described for the first time (Dallwitz et al., 1966), New Caledonia (Campiglio et al., 1986) and other arcs (*Geology and Petrology*, 1991).

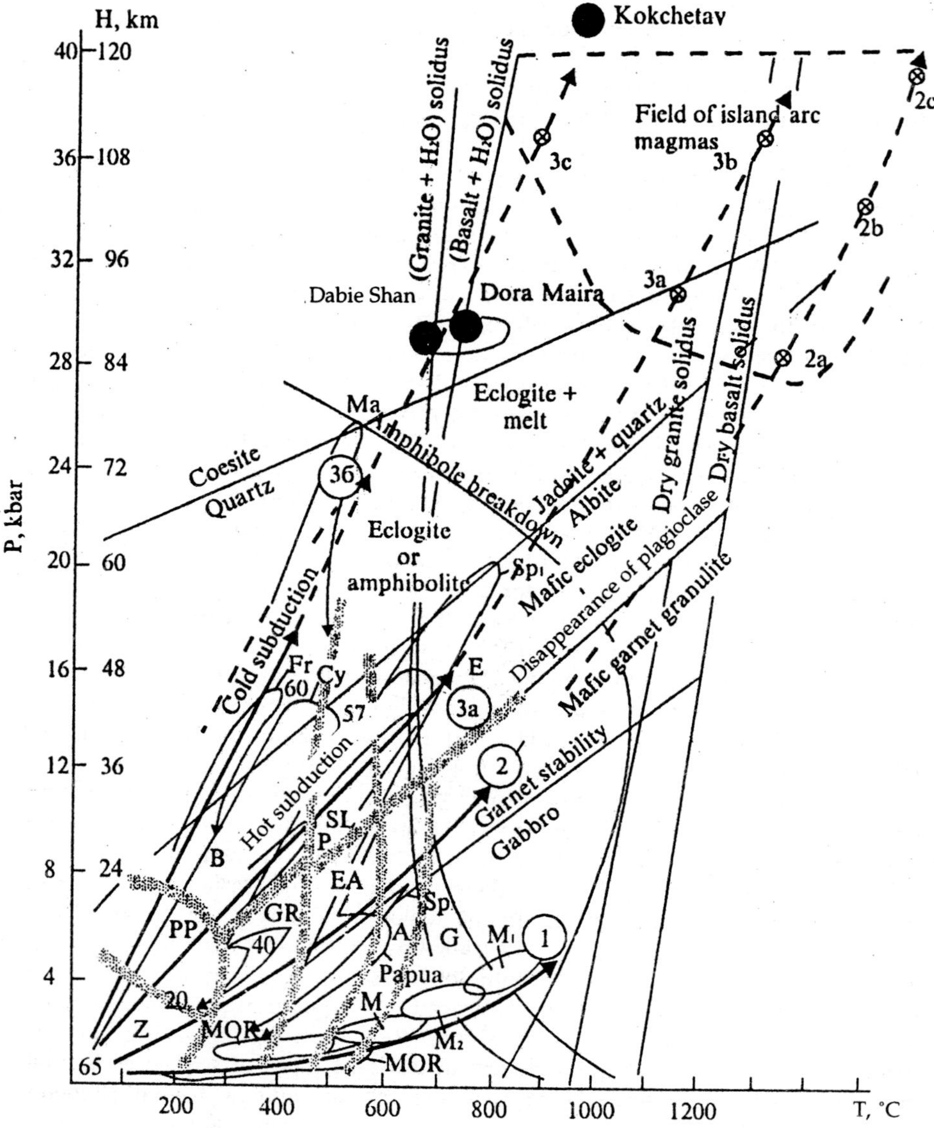

Fig. 5.53. Petrological grid, faces of metamorphism, zone of magma generation and typical
P-T curves of the evolution of metamorphic conditions (Cloos, 1993, supplemented).
Metamorphic facies are marked with alphabets and correspond to numbers used in Figure
5.52.

Faces: Z = 1—zeolitic, PP = 2—prehnite-pumpellyitic, B = 3—blueschist (glaucophane-law-
sonitic), E = 4—eclogitic, GR = 5—greenschist, EA = 6—epidote-amphibolitic and garnet-
glaucophane, A = 7—amphibolitic and G = 8—granulitic. Thin lines—mineral reactions and

Before an analysis of the petrological features of island arc volcanic series, let us look at thermophysical modelling of the melting zone, Figure 5.54 depicts the structure of the subduction zone including the melting subzone. To a first approximation, only two layers are considered: 1) viscous layer h, assuming for simplicity that it is flat throughout its length (this is also possible because the subducting lower part of the wedge corresponds roughly to the exit crevice, see Fig. 5.48) and 2) oceanic crust as layer k. For thermophysical modelling, suffice it to assume that the heat flux at the bottom of layer k in the subduction zone is equal to zero, i.e., at $y = 0$, $\partial T/\partial y = 0$. This condition is not exceptional since, in the subduction of a cold plate, the upper part of its crust is heated while the lower part is cooled, i.e., the minimum temperature depression lies initially in layer k and later below layer k (Anderson et al., 1978). Therefore, on average it may be adopted that $\partial T/\partial y = 0$ at the lower boundary of layer k.

For the crust layer $k \geq y \geq 0$, the heat transfer conditions are described as follows:

$$\partial(U_0 T)/\partial x = a\partial^2 T/\partial y^2; \; \partial/\partial x \int_0^k U_0 T dy = a(\partial T/\partial y)_k, \tag{5.33}$$

where $a = \lambda/c\rho$ is the thermal diffusion coefficient, λ the thermal conduction coefficient, U_0 the rate of subduction adopted as constant for the downsinking lithosphere and layer k.

Let $T = T_0(x)$ represent the average temperature along the thickness of the crust. Then, we find from eqn (5.33) that

$$T_0(x) = (a/U_0 k) \int_0^x (\partial T/\partial y)_k \, dx. \tag{5.34}$$

Let us examine a viscous layer $(k + h) \geq y \geq k$. For small thickness $h << k$ (for example, $h = 3$ km and $k = 10$ km), $U\partial T/\partial x$ is ignored and the heat generating due to friction (viscous dissipation) is considered:

$$a\partial^2 T/\partial y^2 = -(\eta/c\rho)(\partial U/\partial y)^2, \tag{5.35}$$

where $U = U_0(y/h + (k/h + 1))$ and $\partial U/\partial y = U_0/h$.

The boundary conditions for eqn (5.35) are as follows: at $y = k$, $T_0(x)$ is determined by eqn (5.34); at $y = k + h$, $T_k = \bar{q}x \sin\gamma/\lambda$, where $\bar{q}$ is the average heat flux

---

lines of melting. Thick and broken lines (1, 2, 3a, b)—typical trajectories of P and T variations corresponding to conditions: 1—60°C/km in typical island arcs and spreading ridges; 2—25°C/km, normal curve for 'old' lithospheric plate (> 25 million years); 3a—10°C/km, 'hot' and 3b—5 to 6°C/km 'cold' subduction zones. Thick dots (Dabie Shan, Dora Maira and Kokchetav ranges)—conditions of metamorphism of coesite and diamond-bearing rocks. Curves with letters Fr, Cy Ma, Sp$_1$, SP$_3$, P and SL depict the evolution of P and T conditions in typical metamorphic complexes: Fr—Franciscan; Cy—Cyclades, Greece (numbers 65 to 25 refer to age in million years); Ma—Maksyutov, southern Urals; Sp—Spitsbergen (Sp$_1$ and Sp$_3$—phases of metamorphism); SL—Sesia-Lanzo in the Alps; P—Penzhina (north-western Kamchatka); M$_1$, M$_2$ and M$_3$—phases of high temperature metamorphism in the zone of Hidaca (Japan); MOR—oceanic metamorphism.

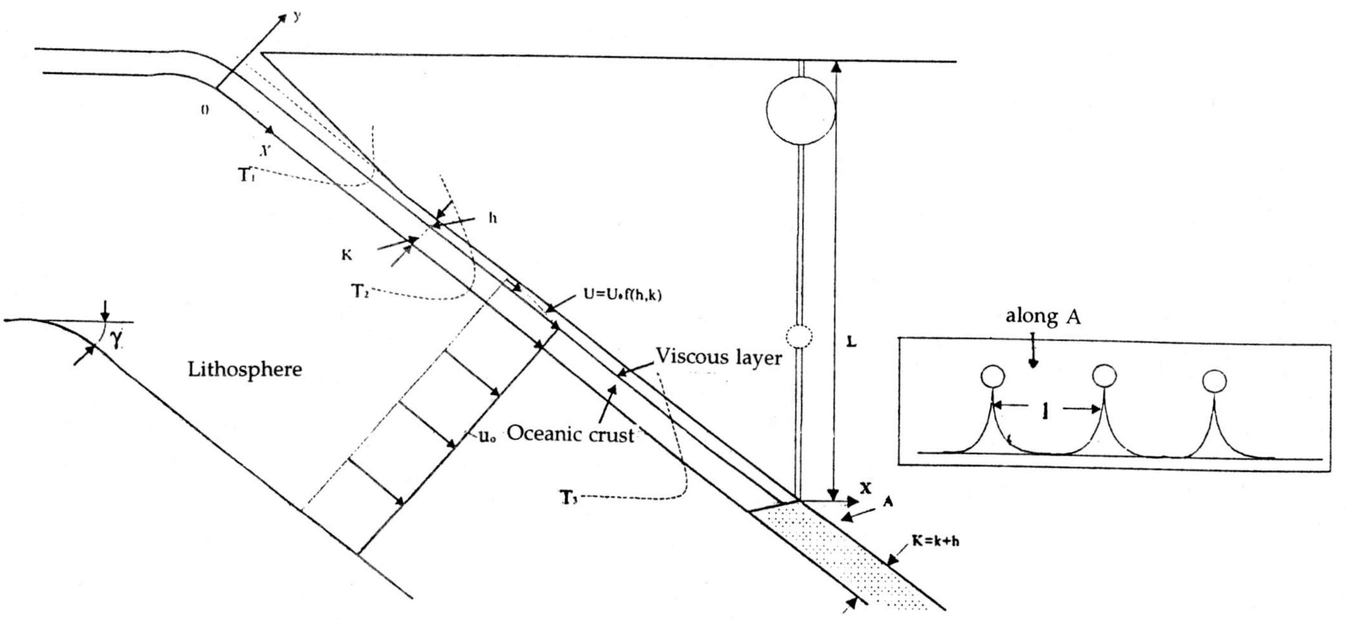

Fig. 5.54. Model of subduction zone structure showing viscous flow in layer *h*, boundary conditions, melting zone and diapir origin in the melting zone.

through the lithosphere in the subduction zone and γ the angle of subduction relative to the horizontal. For these boundary conditions, solution to eqn (5.35) has this form (Dobretsov and Kirdyashkin, 1997):

$$T_0\lambda/k \sin \gamma = \bar{x} - \text{Pe} \left[1 - vU_0\lambda/(2chk\,\bar{q}\sin\gamma)\right] (1 - e^{-x/\text{Pe}}), \tag{5.36}$$

where $\bar{x} = x/k$; $\text{Pe} = hU_0/a$ is the Peclet number. For $\lambda = 3$ W/m · °C, $k = 10^4$ m, $h = 2 \times 10^3$ and values of $\eta = (10^{18}$ to $10^{19})$ N · s/m^2; the value of dissipation term $vU_0\lambda/(2chk\,\bar{q}\sin\gamma)$ is 0.001 to 0.01 and in eqn (5.36), it can be ignored. We then find:

$$T_0\,\lambda/k\,\bar{q}\sin\gamma = \bar{x} - \text{Pe}\,(1 - e^{-\bar{x}/\text{Pe}}). \tag{5.37}$$

This estimate differs from models in which friction in the cleavage zone and energy of viscous dissipation play a significant role (Peacock, 1993).

Using eqn (5.37), let us estimate distance x and depth L corresponding to initial point of melting (i.e., $T_0(x) = 1200°C$). For $k = 10^4$ m, $\lambda = 3.5$ W/m · °C, $\bar{q} = 0.06$ W/m^2, dimensionless temperature $He = T_0\lambda/(\bar{q}k\sin\gamma)$ is equal to $He = 7/\sin\gamma$. It can be seen from Figure 5.55 that for $\gamma = 90°$, $He = 7$; as the rate of subduction increases from 1 cm/yr to 10 cm/yr, distance x corresponding to the depth at which melting commences ($T_0 = 1200°C$) varies form 80 to 150 km. At low $\gamma = 15°$, $He = 26.9$ and T_0 is reached at depths of 73 to 95 km at $U_0 = 1$ to 10 cm/yr. The distance from the trench to the zone of melting for $\gamma = 15°$ and $U_0 = 10$ cm/yr is $x = 380$ km. It must be remembered that the melting temperature attained does not strictly conform to the zone of origin of the ascending diapir.

These values can be compared with those of the section for all the main subduction zones (see Fig. 5.44). In all the arcs, the depth of initial stage of melting is roughly identical at $L = 100$ to 120 km and the distance from the trench to the zone of volcanism varied depending on the angle on inclination and curvature of the subduction surface; it is 300 to 450 km in rapid subduction zones (P and NC) and 150 to 300 km in slow zones. We find that the theoretical estimates, (Fig. 5.55) even for the simplified modelling of subduction are close to the real observed values.

Commencing from depth 100 to 120 km at $x = 200$ to 300 km (Fig. 5.44), melting in the subduction zone first covers the crust and much of the viscous layer, i.e., in the first approximation, the thickness of the melting zone $K = k + h = 10$ to 13 km (Fig. 5.54).

If it is assumed that melting occurs to great depths, i.e., ≥ 400 km, where the first phase transition of olivine can be in the ultrabasic lithosphere below layer K, total melting of layer K and partial melting of the ultrabasic layer can be anticipated at depths of ≥ 400 km. At lesser depths, melting is evidently incomplete and rhyolite-dacite or andesite melts originate specifically from basic crust. In this case the specific quantity of melt W_0 formed per running metre (r.m.) of the subduction zone in unit time along line A (melt separation, Fig. 5.54) is:

$$W_0 = U_0 k \times 1 \text{ r.m.} \times f = U_0 K f \ (\text{m}^3/\text{s}), \tag{5.38}$$

where f is the degree of melting, which varies from 0.2 to 1.

In the ascent of light melt at boundary A (melt separation from the subduction plate), local plumes or diapirs may be formed at intervals of l (Fig. 5.56) which,

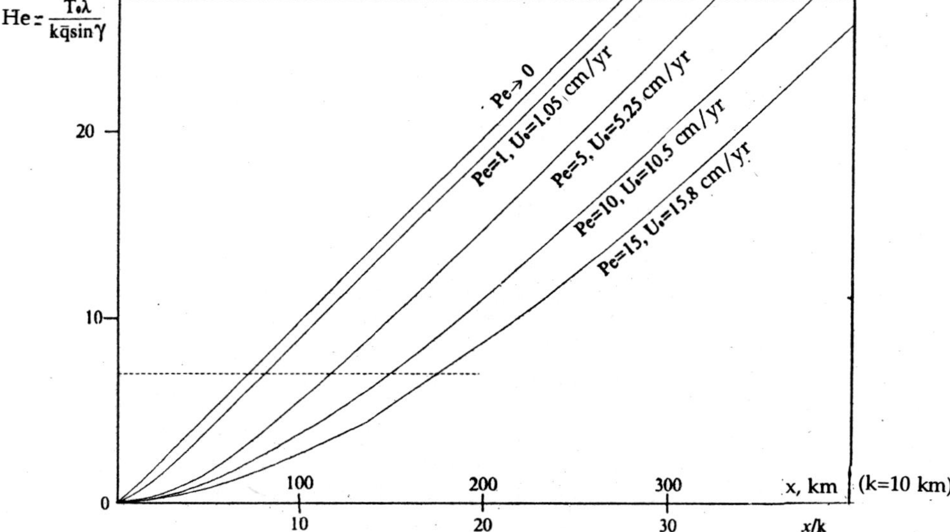

Fig. 5.55. Dimensionless temperature (*He*) versus distance (*x/k*) (*x* at *k* = 10 km) for different values of Pe and U_0 (Dobretsov and Kirdyashkin, 1997). Dotted line corresponds to subduction angle $\gamma = 90°$ and upper limit of figures corresponds to $\gamma = 15°$.

judging from the experimental estimates (Whitehead and Luther, 1975), are proportional to *K*, i.e.,

$$l = nK; n = l/K. \tag{5.39}$$

In this case, the quantity of melt entering from a single source of volcanism (one diapir) at $f = 1$ is

$$W = U_0 Kl = U_0 K^2 n, \tag{5.40}$$

where $n \geq 2.5$ on the basis of the following estimates: according to experiments (Whitehead and Luther, 1975), $2\pi H/l = 2.5$, where *H* is the thickness of the lighter (molten) layer. Since the melt flows along the crevice and the crevice is inclined, $H \geq K$, i.e.,

$$2\pi K/l \leq 2.5; \; l \geq 2\pi K/2.5 \geq 2.5K; \; n \geq 2.5.$$

At $K = 13$ km, $l \geq 32$ km. The actual distance between major volcanoes (or groups of volcanoes) $l = 35$ to 100 km or an average of 70 km and $n = l/K = 2.5$ to 6.

In this case the quantity of heat *Q* released from the zone of melting of a single source (diapir) is:

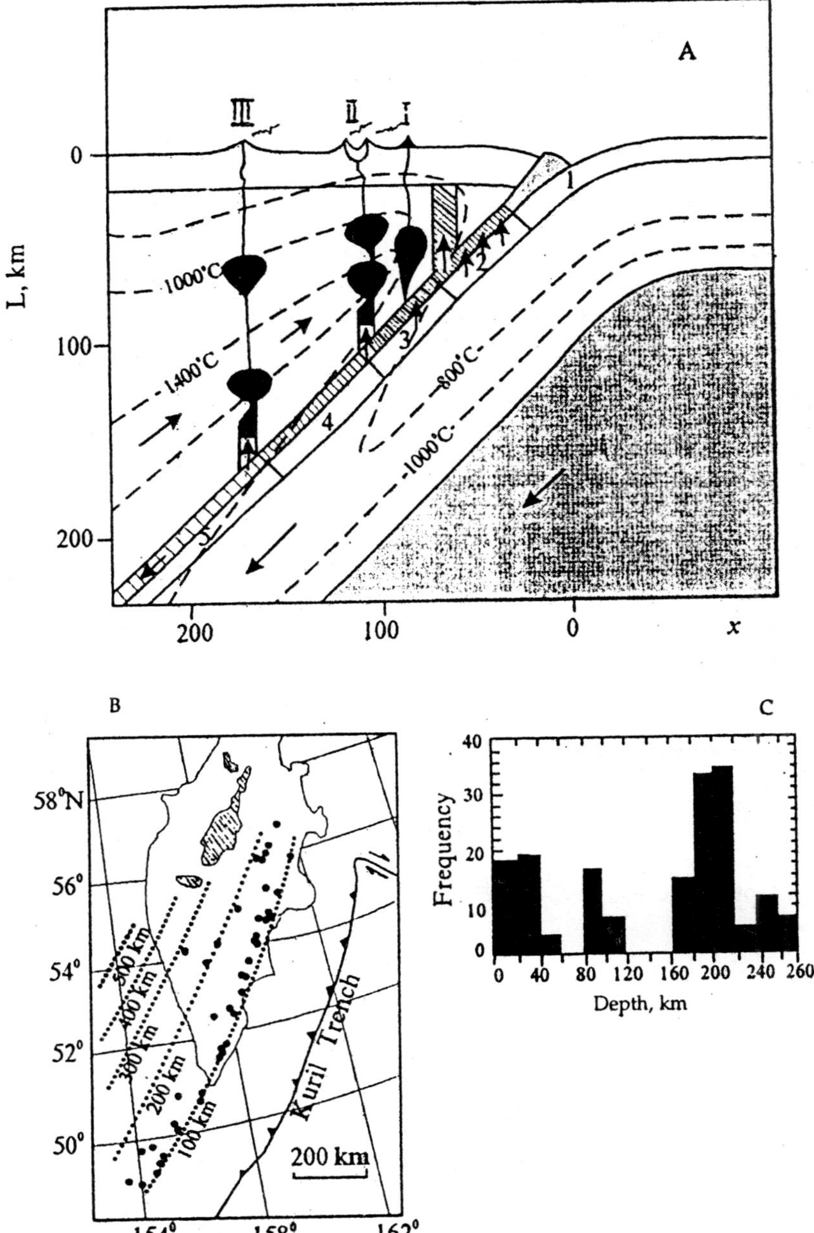

Fig. 5.56. A—Geological scheme of subduction zone with sectors 1 to 5 and three types of volcanic systems, I to III (see text); B—positions of volcanoes of northern Kuril and Kamchatka areas (dots and shaded areas in eastern Kamchatka) relative to the projection of subduction plane (dotted lines 100 to 500 km); C—frequency of earthquakes in this region at different depth intervals (Tatsumi, 1989; Hasegawa, 1994, modified).

$$Q = W\rho(B + C\,\Delta T), \tag{5.41}$$

where ρ is the density of the melt, B the heat of melting, C the heat capacity and ΔT the degree of overheating of the melt relative to solidus ($\Delta T = T_{melt} - T_{solidus} \sim$ 50 to 100°C). The heat of basalt melting (on total melting) $B \sim 210$ kJ/kg and $C = 1.2$ kJ/kg (Zharkov, 1983). At $\Delta T = 100°C$, $B + C\Delta T = B_0 \approx 320$ kJ/kg. We derive:

$$Q = W\rho B_0 = U_0 k^2 n\rho B_0. \tag{5.42}$$

According to the experimental data (Whitehead and Luther, 1975; Olson and Singer, 1985), the head of the diapir has a spherical or pear-shaped form while the feeding channel of the diaper is cylindrical in shape with radius r_0 ($D = 2\,r_0$). According to eqn (4.14), heat transfer from the cylindrical source placed in a large volume of solid mass is characterised by intense heat transfer at constant value of Nusselt number Nu: Nu $= \alpha D/\lambda = 0.5$, where, $\alpha = q/\Delta T_1$ is the coefficient of heat transfer, $\Delta T_1 = T_g - T_c$, where T_g is the temperature along the generatrix of the cylinder and T_c the temperature of the solid body away from the heater (continent). In this case

$$q(y) = 0.5\lambda\Delta T_1(y)/D(y), \tag{5.43}$$

where $q(y)$ is the specific heat flux from the cylinder at height y. The amount of heat transmitted along the channel perimeter at height y is:

$$q_t = \pi D(y)q(y) = 0.5\pi\,\lambda\,\Delta T_1(y). \tag{5.44}$$

Let us study a limiting case wherein the diapir rises to the surface and is fed by a channel throughout length L from the melting centre:

$$\Sigma q_t = \int_0^L 0.5\pi\lambda\Delta T_1\,(y)dy = 0.5\pi\lambda \int_0^L \Delta T_1\,(y)dy. \tag{5.45}$$

Let us adopt the average temperature drop:

$$\overline{\Delta T_1} = (1/L) \int_0^L \Delta T_1(y)dy.$$

Then,

$$\Sigma q_t = 0.5\,\pi\lambda L\,\overline{\Delta T_1}. \tag{5.46}$$

A steady supply to the volcanic centre at maximum height of the diapir will be possible if $\Sigma q_t \leq Q$, i.e., according to eqns (5.42) and (5.46):

$$U_0 K^2 n\rho B_0/L\lambda\overline{\Delta T_1} \geq 0.5\,\pi. \tag{5.47}$$

The ratio $B_0/\overline{\Delta T_1} = (B + C\Delta T)/\overline{\Delta T_1} = C_0$ is the analogue of heat capacity and $\lambda/\rho C_0 = a_0$ is the analogue of thermal diffusivity. Taking this aspect into account, eqn (5.47) can be rewritten as:

$$(U_0 K/a_0)\,(Kn/L) \geq 1.57.$$

By substituting $U_0 K / a_0 = \text{Pe}_0$ into the above equation, we derive

$$\text{Pe}_0 K / L \geq 1.57 / n. \qquad (5.48)$$

A solution to eqn (5.48) was derived for a case wherein all the parameters of a_0 (i.e., λ, ρ and ΔT_1), K/L and n are known. Let us make a preliminary assessment of these parameters. It was derived above that $n \geq 2.5$, $K \sim 10$ to 13 km and $L = 100$ to 120 km. Then, $K/l \sim 1/10$ and $n = 2.5$ to 6. Let us evaluate $a_0 = (\Delta T_1 / B_0)\,(\lambda / \rho)$ for $\Delta T = 100°C$, $B_0 = 330$ kJ/kg, $\rho = 3 \times 10^3$ kg/m^3 and the above value $\lambda = 2$ to 4 W/m $\cdot$ °C. We find $a_0 = (2 \text{ to } 4) \times 10^{-6} \Delta T_1$. Temperature difference ΔT_1 is difficult to estimate since it is necessary to know the temperature variation with depth in the island arc above the melting zone. When there are internal sources of heat in the island arc crust, after making some simplifications, we derive (Dobretsov and Kirdyashkin, 1997):

$$\overline{\Delta T_1} = (1/\lambda)\,\{T_{\text{melt}}\lambda - [q_M\,(0.5L - 0.9y_0) + 0.9y_0 q_s]\}, \qquad (5.49)$$

where T_{melt} is the melt temperature, q_M the mantle heat flux, q_s the heat flux at the surface (let us assume $q_M = (0.7 \text{ to } 0.8)q_s$), $y_0 \sim 12$ km the characteristic value at which specific heat release decreases by 'e' times.

For $\overline{\Delta T_1} < 0$, the temperature of the surrounding mass is higher than that of the melt and the ascending diapir is heated. At present, the values of q_s, y_0 and q_M are known for different regions and thus, using eqn (5.49), the lowest value of λ_{min} can be estimated by equating the left side of eqn (5.49) to zero:

$$\lambda_{\text{min}} = (1/T_{\text{melt}})\,[q_M\,(0.5L - 0.9y_0) + 0.9y_0 q_s]. \qquad (5.50)$$

For example, for $q_s = 0.06$ mW/m^2, $q_M = 0.04$ mW/m^2, $y_0 = 12$ km, $L = 120$ km and $T = 1200°C$, using eqn (5.50), it follows that $\lambda_{\text{min}} = 2.18$ W/m$^2 \cdot$ °C.

For $\lambda = 3$ W/m $\cdot$ deg, $\Delta T_1 = T_{\text{melt}} - T = 1200 - 872 = 328°C$. For $\lambda = 4$ W/m $\cdot$ deg, $\Delta T_1 = T_{\text{melt}} - T = 1200 - 654 = 546°C$. For values $\lambda = 3$ to 4 W/m $\cdot$ deg, $C_0 = (0.6 \text{ to } 1) \times 10^3$ J/kg $\cdot$ deg and $a_0 = \lambda / c_0 \rho = \lambda \times 0.33 \times 10^{-6}$ m^2/°C $= (1.0 \text{ to } 1.3) \times 10^{-6}$ m^2/°C.

Then the Peclet number for $k = (10 \text{ to } 13)$ km, $\text{Pe}_0 = U_0 K / a_0 = (1.0 \text{ to } 1.2) \times 10^{10}\,U_0$. Criterial eqn (5.48) assumes the following form:

$$\text{Pe}_0 K / L \geq 1.57 / n \text{ or } 1.1 \times 10^{10}\,U_0 \geq 1.57\,n, \qquad (5.51)$$

at $U_0 = 3 \times 10^{-9}$ m/s and $n = 5$, we find that criterion (5.51) is satisfied (3.3 > 0.314). The reduction or increase of $n = 2.5$ to 8 exerts little influence on the result. For a small value of $K = 4$ km, $K/L = 1/30$ and $U_0 = 10^{-9}$ m/s, $n = 5$, criterion (5.51) is not satisfied (0.12 < 0.314), i.e., feeding of the volcanic chamber is not possible.

Let us estimate the radius of the cylindrical channel r_0 feeding the diapir and the duration of the rise of the diapir. Taking into account experimental results (Shlikhting, 1969; Whitehead and Luther, 1975), we derive:

$$r_0 = \left(\frac{8\eta W}{\pi g \Delta \rho} \right)^{1/4}, \qquad (5.52)$$

where $\eta = \rho v$ is dynamic viscosity, $W = U_0 K^2 n$ (see eqn 5.40), $\Delta \rho = \rho_{\text{melt}} - \rho$ is the difference is density of the melt and surrounding rocks. According to experimental data (Olson and Singer, 1985), at constant flow at melt and $\eta_{melt} << \eta_{sol}$, the rate of uplift of the diaper dL/dt and the duration of the uplift t are equal:

$$\frac{dL}{dt} = 0.225 \, (g'/v)^{4/5} \, W^{3/5} \, t^{2/5} \tag{5.53}$$

and

$$t = (L/0.16)^{5/7} \, (v/g')^{4/7} \, W^{3/7}, \tag{5.54}$$

where $g' = g(\rho - \rho_{melt})/\rho$, L the height of uplift of the diapir and $v = \eta/\rho$ the viscosity in the solid volume around the channel.

At maximum rate of subduction $U_0 = 3.2 \times 10^{-9}$ m/s (10 cm/yr) and average values of $K = 1.2 \times 10^4$ m and $n = 3.5$, we find $W = 1.6 \, \text{m}^3/\text{s}$. At low value of $K = 4 \times 10^3$ m, $W = 0.06 \, \text{m}^3/\text{s}$, i.e., one-twnety-fifth. An ascent period close to the minimum results for maximum flux ($W = 1.6 \, \text{m}^3/\text{s}$ at $U_0 = 3.2 \times 10^9$ m/s) and low viscosity $\overline{v} = 3.3 \times 10^{17} \, \text{m}^2/\text{s}$ ($\eta = 10^{21} \, \text{N} \cdot \text{s}/\text{m}^2$). Then, for $L = 120$ km and $t = 6.6$ Ma the average rate of uplift of the diapir is about 20 km/Ma. This function and rate are 1 to 2 orders lower than the values obtained for a purely thermal plume model (see Sec. 4.10).

Let us determine the radius of the feeding channel using eqn (5.52) under the condition that the viscosity of the melt within the channel $\overline{\eta} = 10^6 \, \text{N} \cdot \text{s}/\text{m}^2$, i.e., much less than that of the surrounding rocks (decreases towards the channel wall from $\eta \sim 10^3 \, \text{N} \cdot \text{s}/\text{m}^2$ to $\eta \sim 10^{21} \, \text{N} \cdot \text{s}/\text{m}^2$). For the same flow of 1.6 m^3/s and $\Delta\rho = 200 \, \text{kg}/\text{m}^3$, we find $r_0 = 5$ km and $D = 10$ km. For a very low average viscosity of $\eta = 10^5 \, \text{N} \cdot \text{s}/\text{m}^2$, $D = 1$ km; at a very high average viscosity of the melt in the channel $\eta = 10^7 \, \text{N} \cdot \text{s}/\text{m}^2$, $D = 100$ km. The realistic values of r_0 is much less than the radius of the diapir (30 to 80 km according to the diameter of the group of volcanoes) and is $r_0 = 1$ to 10 km, which corresponds to the average viscosity of the melt in the channel $\eta = 10^5$ to $10^6 \, \text{N} \cdot \text{s}/\text{m}^2$.

The above method of modelling the subduction processes involved several simplifications and assumptions. We did not study the phase transitions in layers k and h before the initial state of melting and assumed flat layer h and melting zone $K = k + h$ without taking into account the convection in the melting zone and possible melt concentration in the upper part of the melting zone. Nevertheless, assuming that the amount of the melt corresponds to the amount of subducted crust material, thermal and hydrodynamic problems could be distinguished and conditions of manifestation of volcanism in the zones of subduction analysed.

The results obtained were close to the observed parameters of the subduction zone and enabled accurate determination of a number of physical parameters. We determined the accurate height of the viscous layer $h = 2$ to 5 km (average 3 km) and height of the melt layer $K = 10$ to 13 km. For these parameters, the distance ahead of the level of initial melt in the slow subduction zones was $x = 100$ to 250 km, in the rapid subduction zones $x = 250$ to 500 km and the average height above the level of initial melting zone $L = 100$ to 120 km while the distance between the centres of volcanoes (or groups of volcanoes fed from a single diapir) $1 = nk = 35$ to 100 km (at $n = 2.5$ to 6, average 5). Our modelling also improved the accuracy of values which had not been accurately determined before: thermal conduction

$\lambda = 3$ to $4\ \mathrm{W/m \cdot deg}$ (not 2 to 2.5 as usually adopted) and viscosity in the melt channel $\eta = 10^5$ to $10^6\ \mathrm{N \cdot s/m^2}$.

Figure 5.56 depicts a schematic section of the subduction zone with zonal distribution of segments that differ thermophysically and petrologically (Marsh, 1982; Tatsumi, 1989; Hasegawa, 1994):

1) Elastoviscous subduction wedge controlling the stability of the subduction zones and characterised by dimensions $l = 50$ to 100 km and $h_0 = 7$ to 20 km.

2) Zone of dehydration and metamorphic reactions of glaucophane schist and epidote-garnet-amphibolite facies (see Fig. 5.52).

3) Zone of eclogitisation where early melting of amphibolites with the formation of boninite magmas may occur (Sharaskin et al., 1980; Vysotskii, 1989; Sobolev, 1996).

4) Zone of partial melting and uplfit of 'frontal' diapirs with melts of dacite-andesite composition (Marsh and Carmichael, 1974; Fedotov, 1976).

5) Zone of total melting of layer K and partial melting of ultrabasic lithosphere giving rise to 'rear' magmatic melts of shoshonite-latite or picrite-tholeiite composition. Finally, diapirs rising in zones 4 and 5 will interact with the suprasubduction mantle up to isolines 1000 to 1100°C and undergo a change of composition from andesites to high-alumina basalts and also give rise to more complicated lava composition trends in zone 5.

The equilibrium of the mantle with the ascending andesite magma may be attained finally at a depth of 25 to 30 km at $T = 1175$ to 1200°C (Marsh and Carmichael, 1974) where diapirs cease moving or intermediate chambers are formed. This coincides with the estimates of Fedotov (1976), about a depth of 30 to 35 km of the intermediate chambers under Kuril and Kamchatka although there are estimates of much deeper (60 to 70 km) location of diapirs or intermediate chambers adjacent to the bottom of the lithosphere (Marsh, 1982). According to the estimates of Sharapov (1992, 1994), the heat flux q_a above the ascending diapir (= magmatic column) is 1.5 to 2 times more than the stationary value q_0. At $q_a/q_0 = 1.5$ and time of ascent 20 to 25 million years, the diapir comes to rest and the zone of magma generation occurs at a depth of 60 km. At $q_a/q_0 = 2$ and duration of ascent 10 to 20 million years, the zone of magma generation occurs at depths of 30 to 50 km.

A specific feature of island arc magmas is the high content of water, reaching up to 2.5% by weight in the primary magmas of ordinary volcanoes, e.g., Klyuchevskoi. This value has been judged from a study of melt inclusions (Sobolev, 1996). This is a direct consequence of the melting of water-bearing substratum of layer h and upper part of layer K. The maximum content of water (up to 4%) at high temperatures (up to 1500°C) is characteristic of boninite magmas (Sobolev and Chaussidon, 1996; Sobolev, 1996). This contradicts the assumption that boninite magmas are formed by the melting of amphibolites in the initial and frontal section of the melting zones (Fig. 5.56). They are more readily formed due to the melting of hot suprasubduction mantle which has not been depleted (i.e., in the initial stage of subduction in the region of 1400°C isotherm, Fig. 5.56) but under the action of fluid flows evolved from the subduction zone ahead of the melting front (Dobretsov, 1991; Sobolev, 1996).

Appearance of alkaline magmas at the rear of island arcs represent a specific problem. They cannot be a consequence of the more complete melting of basalt (eclogite) substratum feeding volcanic systems of type III (Fig. 5.56). The result of more complete melting with the participation of mantle material is the picrite series characteristic of most type III volcanoes. However, the shoshonite-latite alkaline series occur only locally and may be associated with the manifestation of lower mantle plumes intersecting the subduction zone (Edwards et al., 1991).

As an illustration, Figure 5.56C depicts the distribution of volcanoes in Kamchatka and northern Kuril showing the position of 'frontal' volcanoes with a feeding depth of 100 to 120 km (zone 4) and 'rear' volcanoes with a feeding depth of about 200, less frequently up to 300 km (zone 5) and also a histogram of the depth of earthquakes (Fig. 5.56C; see also Fig. 4.1) showing three maxima which correspond to zones 1, 3 and 5. In zones of intense dehydration (2) and initial stage of melting (4), almost no earthquakes occur.

5.9. Collision in Subduction Zones and Models of Formation of Eclogites and Glaucophane Schists

The model of accretionary wedge as the most important regulator of a stable (stationary) subduction process has been examined above and the hydrodynamics in this process has been studied in approximation of the theory of lubrication. A different situation arises in the unsteady process of subduction and, as a result, the possible convergence of the oceanic plate and continent (or island arc). The process of convergence of plates is possible under conditions of the variation of the following layer parameters: 'instant' tectonic generation (or increase) in pressure between plates; reduction in rate of subduction, pressure remaining constant; significant increase in h_c to a level comparable to h_0 due to uneven interacting surfaces; and reduction of fluid viscosity in the wedge.

The most important phenomena for such an 'unsteady' condition leading to cessation (or jump, migration) of the subduction zone is a collision in this zone and generation of collision-subduction complexes. The change in time of subduction-accretionary complexes by collision-accretionary complexes is a typical case of the evolution of folded belts (Natal'in, 1991; Borukaev and Natal'in, 1994; Dobretsov et al., 1994a).

Collision-accretionary wedges arise on the convergent boundaries of plates as a result of collision in the subduction zones as well as in the Himalayan type collision zones and ultimately determine the main fabric of the folded belts. The important elements of accretionary complexes are the lenses and slices of ophiolites, glaucophane schists and eclogites as well as zones of melange and olistostrome complexes. It is well known that ophiolites represent the remnants of oceanic crust. Glaucophane schists and eclogites are presumed to have formed in the subduction zones. Melange-olistostrome complexes indicate the different stages of overthrusting, i.e., the different stages of accretion. Quite often, all these three elements are regularly combined. Thus, glaucophane schists and eclogites are present at the base of ophiolite nappes in the form of independent slices or melange inclusions in olistostrome; olistostrome per se and part of the ophiolites may be metamorphosed in the facies of glaucophane schists (Dobretsov, 1974, 1981).

The most amazing features in this triad are the glaucophane schists and eclogites formed under high pressures (usually over 7 to 12 kb) and low temperatures (300 to 500°C) (Fig. 5.53) according to the experimental and thermodynamic estimates of mineral equilibria in these rocks (Ernst, 1975; Dobretsov et al., 1987, 1989; Miyashiro, 1994). Highest pressure complexes have been established recently in Dora Maira in the Alps, Maksyutov in the Urals, Kokchetav in Kazakhstan and Dabie in China, in which the pressure has been estimated at 30 to 40 kb, i.e., they could have formed in the subduction zones at depths of 90 to 120 km (Chopin, 1984; Dobretsov, 1991; Liou et al., 1994; Dobretsov et al., 1995b, c). The need for their submergence to 30 to 90 km and their rapid rise later to the surface was long doubted. Alternative hypotheses were therefore formulated involving tectonic overpressure (Letnikov, 1983; Parfenov, 1984) or fluid superpressure (Dobretsov, 1974; Dobretsov and Sobolev, 1975; Rubie, 1986) or sodium metasomatism (Marakushev, 1973). Tectonic overpressure is limited by mechanical strength and cannot exceed 1 kb in the presence of a fluid ($P_{lit} \approx P_{fl}$) (Brace and Kohlestedt, 1980; Etheridge et al., 1984). A reference was made before (see Fig. 5.49) to the possibility of a tectonic overpressure of up to 5 to 10 kb in the vicinity of outlets but the rock strength is much below this value in the presence of a fluid. Fluid overpressure may serve as an additional factor greatly influencing in particular the rheology (Rubie, 1986) but can hardly generate a pressure exceeding the lithostatic pressure by several times. Sodium metasomatism is observed in many glaucophane schists and associated rocks (jadeitites etc.) but association of high pressures (jadeite with quartz, quartzites with coesite, gneisses with diamond etc.) cannot be explained. Metasomatism can only provide evidence of special geochemical conditions before or during glaucophane metamorphism (Dobretsov, 1974; Dobretsov et al., 1989).

The main problem therefore is: how were the slices that submerged to a depth far exceeding the thickness of the continental crust brought back to the surface and how were their associations of high pressures and low temperatures preserved in spite of changing temperature and pressure during the ascent? On the basis of the kinetics of reverse reactions, the latter phenomenon can be explained by the high rate of ascent equal to or exceeding the rate of subduction (Dobretsov, 1981, 1991) but this high rate itself represents a complex tectonic problem. Even in the 70s, Dobretsov (1974, 1979) attempted to solve this problem and formulated a qualitative model of multistage obduction within the framework of periodically accelerating accretion. Almost simultaneously, qualitative or semiqualitative models based on experiments or approximate theoretical solutions appeared. These included: a) floating of subducted low-density material or an exotic block, as for example, pieces of a microcontinent (Ernst, 1974; Platt, 1986); b) model of floating in the form of a diapir (England and Holland, 1979); c) regional uplift caused by the underplating (injection) of subducted material at the base of an accretionary wedge together with erosion of the uplifted block (Platt, 1986); and d) plastic reverse flow into the accretionary prism on the basis of a 'clay model' (Ernst, 1975; Cloos, 1982, 1986; Shreve and Cloos, 1986; Miyashiro, 1994).

None of these models can explain the whole set of phenomena, however, and hence two directions have been suggested for tackling the problem:

a) combined model of accretion covering frontal accretion, underplating along the surface of the subducted plate, extension and uplift, controlled by the criterion of the stability of accretionary prism (Platt, 1986) and

b) combining the various solutions concerning different types of observed *P-T* trajectories in eclogite-glaucophane schist complexes (Ernst, 1988; Dobretsov, 1991).

At attempt is made in this section to develop a model of dynamic accretionary wedge in subduction zones with reverse flows and local increase in pressure and to improve the accuracy of its accord with different types of eclogite-glaucophane schist complexes.

Let us study in greater detail a limiting case when the pressure in the accretionary wedge (or intermediate viscous layer) is considerably less than the pressure between oceanic plates and continent acting to the viscous layer. This condition is satisfied by a ratio of dimensions h_0 and h_c much higher than the equilibrium curve of a stable wedge state, including also the limiting unstable state (see Fig. 5.50). For this case, let us examine the equidistant convergence of two parallel rigid slabs $(h_0 = h_1)$ and $(h/l << 1$ and $U = 0)$ between which the viscous liquid is present (Fig. 5.57).

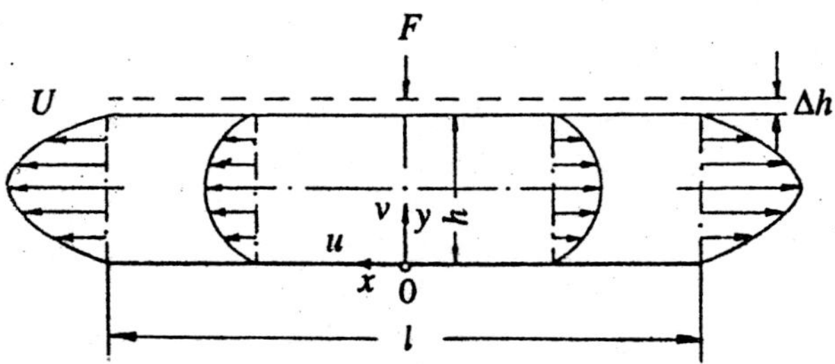

Fig. 5.57. Flow of viscous liquid squeezed between rigid plates.

In a thin layer, $h/l << 1$ and the following approximation holds at high viscosity values:

$$\sigma_x = \sigma_y = P; \quad \int_0^l P\,dx = F. \tag{5.55}$$

Convective terms in the equation of momentum conservation for viscous liquid can be ignored (Slezkin, 1955, Shlikhting, 1969):

$$\rho\,\partial u/\partial t = -\,\partial P/\partial x + \eta\partial^2 u/\partial y^2, \tag{5.56}$$

$$\partial u/\partial x + \partial v/\partial y = 0,$$

where t is the time, ρ the density and u and v the longitudinal and transverse components of velocity respectively (Fig. 5.57). It follows from a solution to the time of initiation of steady Couette flow for a plate 'instantly' set in motion that the maximum time of initiation of steady-state flow is determined using the equation $(vt)^{1/2}/h = 2$ (Batchelor, 1967), where v is the coefficient of kinematic viscosity. In our case, the time of steady state flow $t = 4h^2/v = 4 \times 10^8/10^{18} = 4 \times 10^{-10}$ s, which value is much less than the process time of the order of $\geq 10^4$ years. It would therefore be correct to solve the problem in a quasi-stationary approximation. From eqn (5.56) at $\partial u/\partial t = 0$ and boundary conditions $U = 0$ at $y = 0$ and $y = h$ (Fig. 5.57), the following values are derived:

$$u = -(h^2/2\eta)(\partial P/\partial x)(y/h - y^2/h^2), \tag{5.57}$$

where $\partial P/\partial x$ depends on the longitudinal co-ordinate x. The average value of velocity $\bar{u} = (h^2/12\eta)(\partial P/\partial x)$. This value calculated using the continuity equation (at $v_0 = $ const along the length of the layer) is equal to:

$$\bar{u} = v_0 x/h, \text{ where } v_0 = -\partial h/\partial t. \tag{5.58}$$

In this case, we find from eqns (5.57) and (5.58) that

$$\partial P/\partial x = -12\eta v_0 x/h^3. \tag{5.59}$$

After integrating eqn (5.59), taking into account that $P = 0$ for $x = l/2$ (Fig. 5.57), we find

$$P = (3\eta v_0/h_0^3)(l^2/2 - 2x^2). \tag{5.60}$$

Equation (5.60) should satisfy condition (5.55), i.e.,

$$\int_0^{l/2} (3\eta\, v_0/h^3)(l^2/2 - 2x^2)\, dx = F/2. \tag{5.61}$$

By integrating eqn (5.61), we find that

$$F = v_0\eta l^3/h^3 \text{ and } v_0 = h^3 F/\eta l^3. \tag{5.62}$$

Since $v_0 = -\partial h/\partial t = h^3 F/\eta l^3$, after integrating this equation for the condition $h = h_0$ (at $t = 0$), we find

$$h/h_0 = [\eta l^3/(2Fh_0^2 t + \eta l^3)]^{1/2}; t = (\eta l^3/2h_0^2 F)[(h_0/h)^2 - 1]. \tag{563}$$

The dependence of velocity on x, y and t as the plates move closer together is as follows:

$$u = (6xFh^2/\eta l^3)(y/h - y^2/h^2), \tag{5.64}$$

where h is determined using eqn (5.63). The average value of velocity is derived according to eqn (5.64):

$$\bar{u} = v_0 x/h = Fxh_0^2/(2Fth_0^2 + \eta l^3). \tag{5.65}$$

The maximum value of $\bar{u}$ for $t = 0$ and $x = l/2$ is as follows:

$$\bar{u}_{\max} = Fh_0^2/2\eta l^2 = Ph_0^2/2\eta l. \tag{5.66}$$

The results of calculations using eqn (5.66) are depicted in Fig. 5.58 (Dobretsov and Kirdyashkin, 1993b). The velocity averaged over length $l/2$ is equal to 0.5 $\bar{u}_{\max}$. Fig. 5.58 shows the length-average velocity. For the same geometry of the wedge, the reduction of viscosity by 1 or 2 orders (variant I) or increase of internal pressure P_{add} to 3 to 5 kb as a result of clogging of the subduction zone by the microcontinent or seamount (variant II) results in an increase in maximum rate of reverse flow to 3 to 5 m/year. Such a velocity leads to rapid decrease in pressure and temperature in the ascending slice or block to less than the kinetic limit. This ensures the preservation of such high-pressure minerals as coesite, aragonite, jadeite etc. (Dobretsov, 1991). The time taken for the thickness of the layer to decrease one-half is calculated using eqn (5.63): at $\eta = 10^{18}\,N \cdot s/m^2$, $t = 2.38 \times 10^5$ years; at $\eta = 10^{17}\,N \cdot s/m^2$, $t = 2.38 \times 10^4$ years. Assuming that a symmetrical problem was investigated, whereas in real conditions the flow at depth l would be considerably less than at the surface due to resistance at the channel exit, the velocity values obtained were underestimated somewhat.

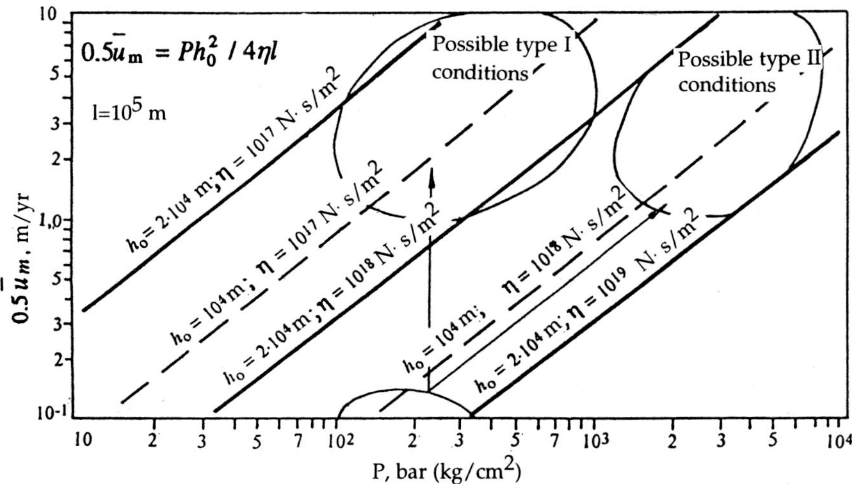

Fig. 5.58. Average flow velocity in an accretionary wedge versus average value of external pressure (P), viscosity (η) and wedge height (h_0).

The isotope data recorded in the recent years confirmed the rapid exhumation of eclogites from the mantle depths (Brocker et al., 1993; Staltenpohl et al., 1993; Liou et al., 1995). The calculations given above for the limiting case $h_1 = h_0$ are close to the estimates of the rate of squeezing of rocks (slice) from the wedge allowing for a significant change of form or dimensions of the wedge from the state of stable equilibrium (i.e., under conditions of significant disturbances). The upper one-third of the wedge (i.e., 1 to 2 km for $h_1 = 3$ to 6 km) is ejected at maximum velocity. In horizontal flow over 50 to 100 km, the height of the affected upper part may be shorter by several times and comprise about 300 to 400 m. This is characteristic of many zones of melange while the less plastic lenses separated by such zones are preserved in a more complete form.

This can also be seen in the schematic section through the Polar Urals and the Alps (Fig. 5.59), deep-level rocks constituting part of the subducted complex and

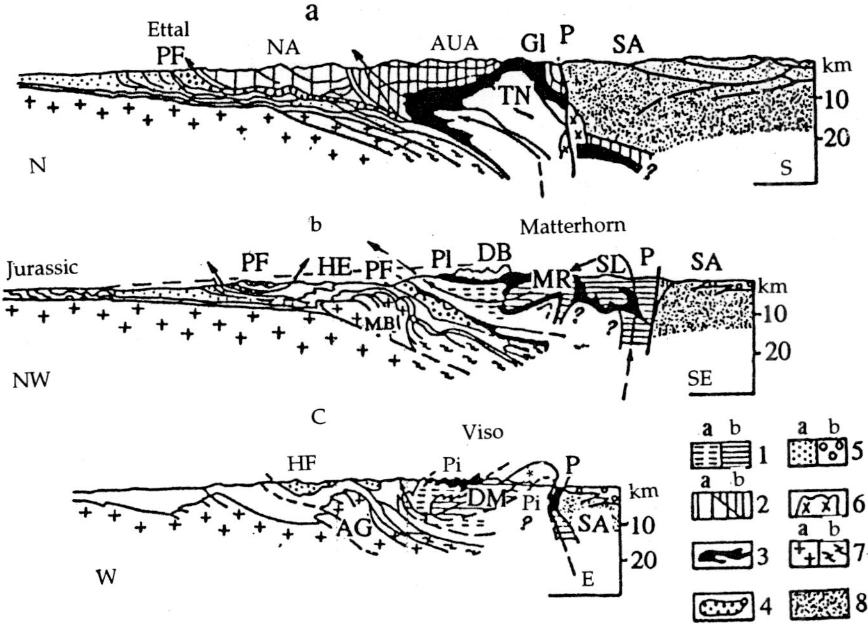

Fig. 5.59. Schematic section through the Alpine accretionary wedge in Taverne-Ettal (a) Pennine nappes (b) and Grain Alps with Mount Viso (c) (after Dal Piaz et al. (1988), abridged). 1—Eoalpine glaucophane schist (a) and eclogite-schist (b) metamorphism (the star marks the pyrope-coesite association of Dora Maira); 2—Eoalpine low-temperature facies (a) to amphibolite facies (b) metamorphism of low pressures; 3—ophiolites; often predominantly with HP-metamorphism and associated melange; 4—predominantly Cretaceous flysch; 5—Tertiary molasse–European (a) and Po valley (b); 6—Oligocene granites; 7—slightly altered (a) and remobilised (b) basement of the European plate; 8—Apennine plate. Geographic names and formations: AUA—Austro-Alpine nappe; DB—Dent Blanche nappe; HE—Helvetic; HF—Helmington flysch; NA—northern Alps; Gl, Pi and Pl—Glockner, Piemant and Platta-Aroz ophiolites; DM, MR, TN, Dora Maira, Monte Rosa, and Taverne massifs, SA—southern Alps; AG and MB—Argentera and Mont Blanc salients of the basement; PF—Pennine overthrust front; SL—Sesia-Lanzo zone; and P—Peri-Adriatic fault. Other explanations in text.

part of the hanging wall of the subjection zone are preserved in the full composition. In the Polar Urals, the deepest level rocks are drawn into the zone of blastomylonites alternating with zones of serpentinite melange. Above this zone is the hanging wall of the subduction palaeozone representing the uncovered section of the island arc (Dobretsov, 1991).

Below the zone of melange and blastomylonites is an almost complete section of subduction palaeozone although with slices brought closer together.

Our model is essentially similar to that of Cloos (Shreve and Cloos, 1986; Cloos, 1982, 1986, 1993), which emerged as a continuation of the laboratory modelling of the circulation of clay material in a wedge above a subducting plate. The model was refined by numerical modelling. According to M. Cloos, the material in the

274

wedge does not flow out (is clogged) as in the case of a collision or some other specific cases. The material was regarded as an incompressible Newtonian liquid with constant viscosity and the walls of the wedge as rigid. Numerical solutions were made for wedge angle $2\theta = 20°$, slope of subduction $\alpha = 25°$ and $l = 100$ km, with the rate of subduction varying from 2.5 to 10 cm/yr.

The calculations of M. Cloos show that a slow reverse flow arises in the subduction wedge above the subducting plate. In the zone adjacent to the subducting plate and constituting one-third of the wedge thickness, a downsinking at a rate equal to or less than the rate of subduction to the zero value is observable. In the upper zone (two-thirds of the wedge) where reserve flow can be seen, the maximum rate (about one-third the rate of subduction) is recorded in the central part of this zone. The particles, initially 0.5 km apart in that part of the wedge opening at the bottom will be separated 15 km apart after 0.5 million years. Although they will move closer together in the reverse flow, they nevertheless will remain 10 km apart. The situation with eclogite blocks is somewhat different. For the parameters given above $(2\theta = 20°$, $\alpha = 25°$ and $u = 5$ cm/yr) and additional conditions $\Delta\rho = 0.6$ g/cm^3, large blocks of eclogites (≥ 25 m) cannot ascend, medium-sized blocks (about 10 m) will ascend the fastest of all while small blocks (≤ 5 m) will lag behind the medium-size ones of 10 to 15 km from the surface. However, the data on the exhumation of real eclogite complexes are far more complicated (Krogh et al., 1994; Staltenpohl et al., 1993).

Cloos' model does not take into consideration the conditions of formation and stability of the accretionary wedge; yet we saw earlier that this is a very important aspect.

Mineralogical and experimental data on the preservation of such easily transformable minerals as aragonite, lawsonite and coesite (Dobretsov, 1991) in glaucophane schists and eclogites show that the rate of reverse movement of rock slices or melange with inclusions of glaucophane schists and eclogites containing the aforesaid minerals should be equal to or higher than the rate of subduction. Yet the Cloos' model emphasises the possibility of matching direct and reverse curves of P-T evolution during glaucophane schist metamorphism with matching rates of direct and reverse flows in the wedge, which may be less than the rate of subduction. However, it follows from kinetic data that such a relationship is hardly feasible since during slow movements the temperature decreases slowly and the 'quenching' of metastable minerals does not occur.

It can be seen from eqns (5.65) and (5.66) that the rate of ascending flow, at a low viscosity of the order of 10^{17} N $\cdot$ s/m^2, may considerably exceed the subduction rate and reach 300 cm/yr under conditions of convergence of the plate and the continent. At a more realistic viscosity of 10^{18} N $\cdot$ s/m^2, the maximum rate of ascending flow is of the order of 30 cm/yr and is observed in the initial period ($t = 0$) even in the midportion of the aperture of the wedge. It will decrease with time even in the upper part of the wedge.

Later, Cloos (1993) estimated the possibility of collision or clogging of the subduction zone when microcontinents, island arcs or seamounts enter it. According to his estimates, only small seamounts (less than 1 to 2 km high), small uplifts with crustal thickness 15 to 17 km and young arcs (younger than 20 million years) may not cause collision or interrupt subduction. Further, the entry of seamounts into the subduction zone causes only a short-run interruption of the subduction process and its migration toward the ocean. In this case the curve of $P–T$ evolution corresponds to that of the Franciscan or Pennine type (see 3a in Fig. 5.53) while the metamorphosed rocks contain Ti basalts and seamount remnants. In the event of a collision with a microcontinent or old island arc, the collision is more significant and subduction ceases altogether. As a result of rapid reverse flow, the deeper level rocks comprising metamorphosed slices of microcontinent or products of metamorphism of old island arc are brought up to the surface while HP metamorphism changes into low-pressure as in the Alpine or Sambogawa type (Dobretsov, 1991; Dobretsov and Kirdyashkin, 1993b).

Manifestation of dome structures in glaucophane schist belts with a reverse metamorphic zoning is of special interest. Such imbricate-dome structures have been described for Cuba in Escambray complex (Somin and Mil'yan, 1981), for Papua New Guinea (Hill and Baldwin, 1993) and for Tinos island in the Cyclades belt in Greece (Brocker et al., 1993). Relics of dome structures have also been detected in other glaucophane schist belts. They arise in the case of thrusts of high-pressure slices from subduction-accretion zones to a sialic bottom of the microcontinent (continent) together with the island complex covering the slices. In this case, gravitational instability gives rise to slow upward float of the sialic dome and repeated metamorphism of the high-pressure complex under conditions of low pressures and high temperatures. The complex $P–T–t$ curves reveal flexures corresponding to intraplate (average) temperature distribution as shown for the Cyclades (Cy) and Spitsbergen complex (Sp_1 and Sp_3) (see Fig. 5.53).

Thus the tectonic history of glaucophane schist complexes combines in itself their rapid initial transport from the mantle in the form of slices or melange zones, followed by interruptions and deformations, including the growth of domes.

The real complex structure of the accretionary wedge formed during the collision of two continental plates (Adriatic and European) can be seen on the example of the Alps (Fig. 5.59). This structure can be interpreted as the extrusion of deep-level rocks along the rigid basement of the Adriatic plate and later the flow of extruded rocks in the direction of the foreland of the European plate. Further, as in California and at many other places, glaucophane schists and eclogites already metamorphosed at great depths (≥ 50 km) have been thrust on unmetamorphosed rocks of the same age (or even much younger ones). This process of the combination of different facies and rocks of different ages occurred, as noticed before, very rapidly. In the section (Fig. 5.59a) only metamorphic rocks of moderate pressures (5 to 6 kb) elevated to the surface from a depth of 10 to 15 km are established. The geologic structure and motion pattern conforming to the structure in this section are far simpler than in the other sections (Fig. 5.59b, c) in which high-pressure rocks have been established. Rocks with maximum estimates of pressures containing coesite and identified by a star in the section (Fig. 5.59c) have been uplifted from depth exceeding 100 km.

Other important examples are the Dabie complex in China and Kokchetav massif in Kazakhstan. These complexes also contain ultrahigh-pressure (UHP) rocks containing coesite and diamond and exhumed from depths of over 150 km. The model of collision of continental blocks is suggested for these complexes (Coleman and Wang, 1995; Ernst et al., 1995). The tectonic units constituting these complexes are compared in Table 5.1. Taking into consideration the mineralogical and experimental data systematised in Figure 5.60, the following conclusions have been drawn (Dobretsov et al., 1995b, c).

Table 5.1. Comparison of tectonic units in Kokchetav and Dabie Shan UHP complexes

Kokchetav	Dabie Shan
Unit 1: Diamond-bearing metasediments (metacarbonate, metapelites) + eclogite and ultramafic bodies in orthogneisses	Coesite-bearing unit: eclogites, ultramafics, marble, calc-silicate metasediments + orthogneisses
Unit 2: Melange and mica schist with inclusions of eclogites, Zo-Ky-Gt and Gt-Ta-Phen rocks, including 'cold eclogite' subunit (Sulu-Tube)	Sujiahe melange zone to the NW from unit 1, separate melange zone (?) inside 'cold eclogite' terrane of South Dabie
Unit 3: Gt-Bi-Ky-Sill schists, corona dolerites, Gt-amphibolites	?
Unit 4: Cord-And-Bi gneisses and schists	?
Unit 5 + 7: Precambrian gneisses with Upper Precambrian sediments (quartzite, graphite schists + marble)	Possible metamorphic basement of North Dabie zone (?). Partly, the Susong complex
Unit 6: Epidote-amphibolite schists and quartzites with rare crossite (dominating metavolcanics)	EA, Gs + Bs zones (metavolcanics and metasediments) in the Hongan block
Unit 6b: Vendian-Cambrian IA-volcanics and gabbro	Devonian and older volcanics and gabbro in Beihuaiyang and Susong zones
North zone: Upper Precambrian rocks (dolomite, black shale, phosphorite, volcanics), tectonically intercalating with Ordovician rocks	South framing: Susong complex + Palaeozoic sediments (including phosphorite) of the Yangtze craton
Zerenda dome with younger (than UHPM) granites + alkaline ultramafic bodies	North Dabie block: younger granites and sheeted dykes, alkaline (?) ultramafics

1) UHP complexes are present either as megamelanges (Kokchetav massif) or as a series of tectonic slices (Dora Maira and partly Dabie). In all three cases the thickness of slices made up of UHP rocks does not exceed 1.5 km. At many places UHP rocks are in the form of small blocks (1 to 500 m in diameter) within the tectonic melange (Kulet in Kokchetav massif, Shuang He in Dabie and Maksyutov complex melange in the Urals). Such dimensions of slices and blocks correspond to the above-discussed model of rapid reverse flow but contradict the model of ascent of large blocks (Ernst et al., 1995; Liou et al., 1995; and others). As a result, these rock slices and blocks may undergo rapid cooling during exhumation along the tectonic contact with the upper lithospheric plate.

2) The preservation of coesite and the presence of thin reaction rims around coesite, magnesite and diopside from Dabie metacarbonate rocks provide proof of a very small duration of retrograde reactions and correspondingly very rapid uplift of these blocks. Fig. 5.60a shows different curves (1, 2, 3a and 3c) of P-T-t evolution characteristic of the different units of Dabie complex (Table 5.1). Three stages of exhumation can be suggested for the most important cases (2a–2c and 3a–3c):

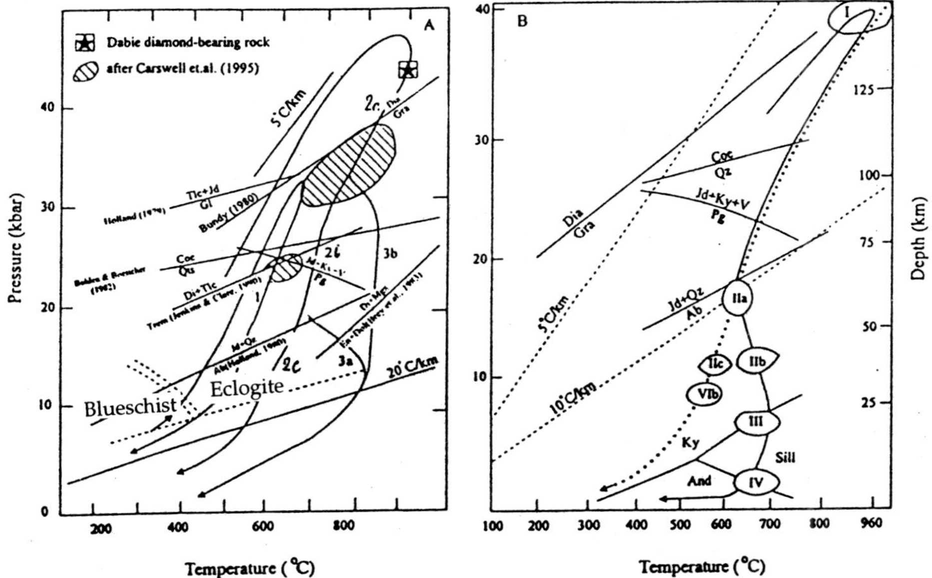

Fig. 5.60. Comparison of *P-T* evolution of Dabie Shan-Sulu UHP metamorphic rocks (Liou et al., 1995; supplemented from Carswell et al., 1995; Dobretsov et al., 1995c) and Kokchetav metamorphic units I–IV and VIb (Dobretsov et al., 1995b).

1) 2a—superfast uplift (from 900 to 750°C and from 140 to 80 km in 0.1 million years); 2) 2b or 3b—rapid uplift up to the bottom of the continental crust (750 to 650°C, from 80 to 40 km over 1 to 2 million years); depending on block size and rate of uplift, nearly isothermal decompression (curve 2b) or with heating during interruption (curve 3b; and 3) slow rise within the continental crust with the formation of domes (curve 3a) or very rapid uplift and cooling of the tectonic slices (curve 2c).

The history of the uplift of UHP rocks of Kokchetav massif is somewhat different taking into consideration the less complete preservation of UHP associations and much higher temperature origin of most of the rocks and their partial melting during the uplift. Only two stages can be recognised here: 1) I–IIa, relatively rapid uplift to a depth of 60 km over 0.1 to 1 million years. This rapid rise ensured the preservation of diamonds and prevention of the dissociation of K-pyroxene, both of them being present as inclusions in garnet and zircon; and 2) stage IIa–IIb (+ possible continuation up to III and IV) is characterised by heating during slow decompression (Fig. 5.60b).

3) In Kokchetav massif, there are units (VIa, c, possibly III and IV) pertaining to the island arc and its basement which were formed before UHP metamorphism and exhumation of UHP rocks. Such rocks are absent in Dabie complex or are significantly interrupted in time (Table 5.1; Dobretsov et al., 1995c). Nevertheless,

a comparison of the rocks of Kokchetav massif and Dabie clearly reveals that UHP metamorphism cannot be the result of a collision; rather a collision, especially an oblique collision of the microcontinent with the island arc may have ensured rapid exhumation of subducted rocks from a depth of over 150 km.

Many alternative models have been proposed recently to explain the formation and exhumation of intriguing UHP rocks (Harley and Carswell, 1995). Apart from the above-discussed model of rapid reverse flow or 'squeezing out' of the subduction wedge as a result of collision with the island arc, as judged in a set of investigations (Dobretsov, 1991; Dobretsov and Kirdyashkin, 1991, 1994; Dobretsov et al., 1994b), many scientists have suggested that the formation and uplift of UHP rocks may be the result of direct collision of two continental masses. The variants of this model are the 'rigid' collision and uplift of large blocks as a result of the doubling of the thickness of the crust (Zonenshain and Kuz'min, 1983; Coleman and Wang, 1995; Ernst et al., 1995); subduction of the continental margin, resulting in curvature of the subduction surface, excess pressure at a depth ~ 100 cm and push-up of UHP rock wedge in the opposite direction (Hynes et al., 1996); and oblique collision and/or arc parallel strains along the subduction zone leading to rapid exhumation in the local extension zones (McCaffrey, 1996). In the latter study it was estimated that, at strain rate 2×10^{-8}/year, uplift from a depth of 30 km will occur at an average rate of 0.1 to 1 cm/yr. The main contradiction to such models is not the theoretical concept but geological facts: in Kokchetav massif (see above) and Maksyutov complex in southern Urals (Dobretsov, 1974, 1991; Dobretsov et al., 1989), there is no proof of continental collisions and, on the contrary, there are no UHP complexes in the more vast zones of continental collisions such as the Himalaya (see below). HP and UHP rocks have also not been detected in island arcs with maximum strain rate as in the Aleutian islands.

Platt's model (Platt, 1986) merits attention. In this model, frontal accretion at first and deep-level underplating later as inertial continuation of subduction (underplating) have been suggested. As a result, a large accretionary thickening arises and rapidly flows like a huge thick glacial sheet causing rapid uplift of blocks at the centre of such a thickened accretionary wedge. This model is proximate to the model of formation of Cordillera type metamorphic cores (Coney and Harms, 1989) but rocks of low and moderate pressures are known to ascend at such places.

Platt's model (Platt, 1986) is based on the example of the Alps (Fig. 5.59) and California (Fig. 5.61). Structural and kinematic data for California (Ring and Brandon, 1994) do not, however, confirm the presence of deformations due to extension parallel to faults as suggested by the flow stages in the model (Platt, 1986). In the zone of the Coast Range fault, along which the Franciscan formation of HP rocks contacts with ophiolites and the sedimentary section of the Great Valley (Fig. 5.61), only brittle compressive deformations are established (Ring and Brandon, 1994). This corresponds well with the wedge model (Wentworth et al., 1984) or the thin-slice model, one of the earliest to be formulated (Suppe, 1979). In these models the predominant compression and 'squeezing out' of ophiolites is explained by external factors—backthrusting in the wedge model or frontal thrusts in the thin-slice model. The eastern Franciscan belt (Fig. 5.61), metamorphosed 120 Ma ago for Pickett Peak terrane and 90 Ma ago for Yolla Bolly terrane (Blake et al., 1988), occur in an upper structural position before the thrusting of ophiolites and the terrane of the Great Valley which took place about 60 Ma ago.

The closing tectonic uplift and erosion of the eastern Franciscan belt may have occurred during accumulation of young sediments of the western Franciscan belt

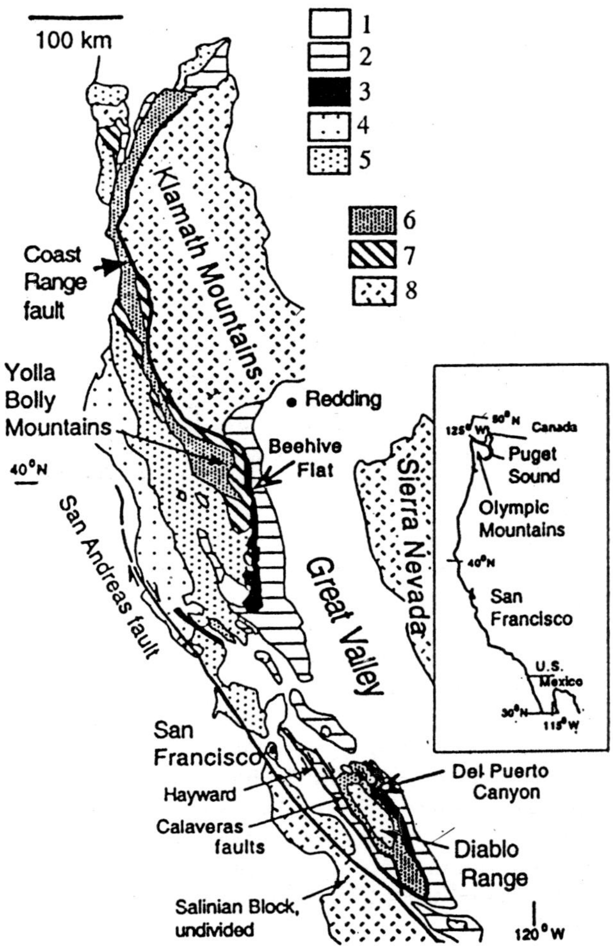

Fig. 5.61. Geological map depicting the main Mesozoic elements of western California and two regions studied along the Coast Range fault: Beehive Flat and Del Puerto Canyon. It can be seen that Diablo Range uplift is bounded by faults relative to the present-day San Andreas fault (Ring and Brandon, 1994).
1—Cenozoic sediments; 2—Great Valley series; 3—Coast Range ophiolites; 4—coastal and 5—central part of Franciscan belt; 6, 7—eastern part of Franciscan belt: Yolly Bolly terrane (6) and Pickett Peak (7); 8—Sierra-Klamath magmatic arc.

280

60 to 70 Ma ago. According to Ring and Brandon (1994), this situation is similar to the uplift of the active subduction wedge in the Olympic Mountains in northern California (Fig. 5.61) where, in the preceding 12 million years, mountains rose by 12 km as a result of accretion but most of them have been destroyed by erosion.

The proposed models represent a specific case of the model in which the normal subduction of oceanic crust is succeeded by the continuing subduction of the continental block (microcontinent) which represents the motive force (Willet et al., 1993; Beaumont et al., 1996). As a result, a wedge forms above the collision joint, which arises at a rate proportional to that of subduction and is bounded by pro- and retronappes. Ellis (1996) compared this model, called 'mantle subduction', with the 'intender' model used to describe the collision of India and Asia (see below) and proposed a combined model.

Thus, relations between the primary rapid exhumation of UHP rocks from great depths (100 to 150 km) and consequent effects of collision with the formation of domes, slow uplift and erosion are still subjects of discussion. Collision models are examined below.

5.10. Zones of Continental Collision and Related Processes of Magmatism and Metamorphism

We have just studied the transitions from subduction-accretion to accretion-collision complexes. These types of complexes constitute the basic fabric of folded belts in which fragments of old active zones are combined. However, collision does not lead only to the formation of tectonic accretion-collision complexes; it is also responsible for diverse magmatic and metamorphic processes which are localised not only within the accretionary complexes, but also extend into the adjoining blocks of the continent or microcontinent.

The main difference between the collision process and subduction is that, under collision, light lithospheric plates of continents, microcontinents and mature island arcs that are 'unsinkable' in the mantle generally collide. This leads to the amalgamation of blocks into large continents and later even into supercontinents. Collision also leads to varying degrees of thickening of the crust and the collision process may occur in different stages in the form of overthrusts, strike-slip displacements and domes. The thickening of the crust in turn leads to processes of melting and metamorphism in the thickened crust.

Orogenic belts and collision zones vary in the tectonic setting, nature of metamorphism and mode of rising of metamorphic rocks to the surface. Two contrasting types of orogens have been suggested—Scandinavian and Scottish (Dewey, 1988). In our opinion, at least three cases can be distinguished.

1) In the first case, as in the Central Himalaya, metamorphic rocks are forced out in the form of a wedge toward the plate being pushed up (Hodges et al., 1993). During the Miocene period, the metamorphic complex of the Central Himalaya moved southwards and upwards as a result of synchronous movements along the southern Tibetan detachment and the Main Thrust. The proposed kinetic model corresponds to the established P-T-t curve of medium pressures (kyanite type) and several stages of deformation (Hodges et al., 1993). Decompression close to the upper upthrow fault is almost isothermal and, adjacent to the base of the complex,

is accompanied by a decrease in temperature by 150 to 200°C and pressure by 5 to 6 kb (Fig. 5.62). This model is similar to that of a collision wedge with retronappes (Willet et al., 1993; Beaumont et al., 1996) but those models differ in rates of movement and decompression. Unfortunately, the movement rates have not been estimated in the above reports. An intender model was suggested for the Himalaya (Cobbold and Davy, 1988). For this model, the propagation rate of deformations before the collision front is much less than for the 'mantle subduction' model discussed in the preceding section (Ellis, 1996; Hynes et al., 1996). A combination of synchronised downward thrust and normal upthrust of the wedge may explain wedge extrusion conforming to the model of mantle subduction (Hynes et al., 1996).

2) Another model, typical of Scandinavian-type orogens (Dewey, 1988), assumes intense frontal compression and very large thickening amounting to 80 to 100 km, exceeding the present level in the Himalaya. Gravitational instability of a thickening crust results in a 'collapse' i.e., catastrophic expansion through rapid thrusts (of the type of gravitational sliding) and rapid rise of deep-level material in the form of diapirs accompanied by partial melting during decompression. As a result, a structure typical of Scandinavian caledonides originates (Khain, 1989; Anderson et al., 1991). This structure consists of a combination of major nappes most distinct in the frontal part and many domes and diapirs younger than the thrusts. The maximum rate and amplitude of uplift in the diapirs at the rear causes the crust-mantle mixture to rise to the surface where blocks of mantle pyrope peridotites, mantle and crustal eclogites and gneisses of the lower crust combine. A similar model was applied to the Bet Cordilleras hosting pyrope peridotites of Ronda massif (Knipper et al., 1992; Davies et al., 1993) and granite-gneiss domes characteristic for the latest stage of formation of UHP complexes with coesite and diamond (Dobretsov et al., 1995b, c; Liou et al., 1995). In the Himalaya, the aforesaid mechanism (retrogressive nappe movement) transforms in space (in the western Himalaya and the Pamirs) and probably in time into the mechanism of collapse and rapid rise of diapirs. We have already studied the combinations of these mechanisms on the example of glaucophane schists and eclogite-containing complexes (Kokchetav and Dabie). Deep-level parageneses (HP and UHP) are elevated and preserved during rapid uplift while, on slow rise, diapirs undergo remelting and low-pressure complexes arise (andalusite-sillimanite type) in the form of simple and fringed granite-gneiss domes (Miyashiro, 1961, 1994; De Yoreo et al., 1991).

3) Under less powerful or oblique compression, intense thickening does not occur and large thrusts arise together with zones of local extension. This type of orogen has been called the Scottish type (Dewey, 1988; Miyashiro, 1994). In Scotland, Dalradian zonal metamorphism of disthene-sillimanite type is succeeded by andalusite-sillimanite type in close association with granites; typical domes are rare and thrusts and upthrows as well as linear, more rarely isometric granite plutons, are widely manifest (Miyashiro, 1994).

Analogues of Scottish and Scandinavian types may be found in the Palaeozoides of southern Siberia. The Scandinavian type is characteristic of the eastern part of the Palaeozoides (Transbaikal): granite-gneiss domes and metamorphic cores together with granites (Belichenko et al., 1994). Major collision thrusts

282

(Irtysh, north-western and northern Sayan and other crumpled zones) are characteristic of the western part of the Palaeozoides (Altai-Sayan region). Andalusite-sillimanite or transitional type of metamorphism is seen along these collision thrusts. Sengör (Sengör et al., 1993, 1994) emphasised the special role of major strike-slip faults leading to end joints and mosaic structure for the Altaides and hence this type of orogen may be called the Altai-Scottish type. In the Altai type, extended domes or 'swells' of gneiss-granites arising in the thrust zones by the pull-apart mechanism play an important role. Such structures have been established in the Irtysh strike-slip zone in particular (Dobretsov et al., 1997). They are also likely for some Transbaikal gneiss-granite domes. In these cases the domes are predominantly made up of granites with the magmatic texture preserved at the centre. Formation of domes and gneiss variety has been observed even in the magmatic stage. They arise due to 'absorption' of the melt arising in the lower crust during its thickening in local 'pull-apart' type structures.

Generalising all the collision models discussed above, it can be concluded that the observed structure and, in particular metamorphic zonation, is the result of a combination of several tectonic, metamorphic and denudational processes (Willet et al., 1993; Beaumont et al., 1996). The main regulator is the rate of tectonic processes involved in uplift of the metamorphic rocks to the surface. In the most general and simplest case, the collision zone may be represented as the result of the doubling of crust thickness due to rapid thrusts, further heating of the thickened crust and equalising of temperature to a new equilibrium geotherm (Fig. 5.62; England and Thompson, 1984). The following stages can be established here: (0–1) stage of slow sedimentation studied earlier on the example of continental rifts (see Section 5.4); (1–2) stage of rapid compressive submergence; (2–3) nearly isobaric, slow heating simultaneous with continuing deformations; (3–4) slow rise of the diapir with reduction of pressure and continuation of heating; and (4–0) regressive branch: cooling along the new geotherm (Fig. 5.62). According to the estimates of England and Thompson (1984), stage (3–4) requires up to 10 million years for each 50 to 100°C of heating and maximum temperature is reached at point 4 when nearly 30% of block uplift has already occurred accompanied by a corresponding decrease in pressure.

P-T curves of the evolution of metamorphism calculated for this model are situated wholly in the field of kyanite at normal initial gradient of 20°/km or correspond to the transition kyanite → sillimanite → andalusite ± kyanite at a very high initial gradient of 30°/km. This model (Fig. 5.62) shows the main stages of metamorphism and tectonic processes. The model is simplified since it takes no account of the melting of the continental crust and heat transfer by magma, nor of the more complicated ways of uplift and relaxation arising after compression and thickening.

In the modern model (Whittington, 1996), calculated geotherms can be reconciled with geochronological and thermobarometric data, if exhumation rates lie in the range 3–4 mm/yr. Stages (2–4) and (4–0) are characteristic of many zonal complexes. In particular, the progressive and regressive stages of orogenic zonal metamorphism described by Korikovskii (1995) are proximate to stages (2–4) and (4–0) (Fig. 5.62).

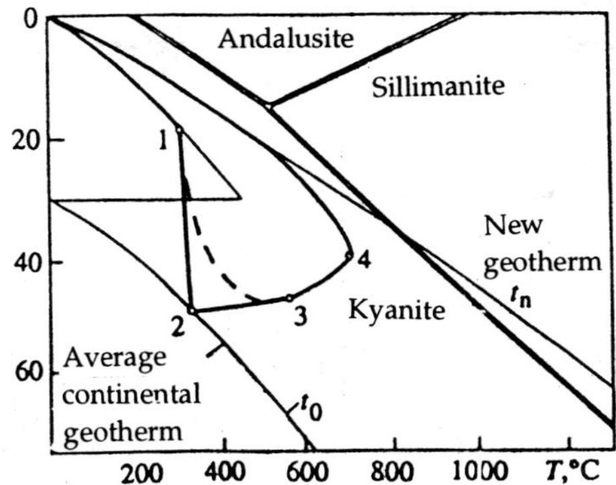

Fig. 5.62. Evolution of *P-T* conditions in the course of thickening of the continental crust and collision metamorphism (England and Thompson, 1984, modified).
1, 2—points on the initial geotherms after 'instant' thickening of crust at moment t_0; 2-3-4—lines of heating and pressure decrease during growth of domes; 4 to 0—regressive stage and adaptation to new geotherm t_n. Heavy lines—stability of kyanite, sillimanite and andalusite.

As pointed out before, local zones of extension and close association with granite complexes of dome as well as linear types arise in the second and third types of orogens. Taking also into consideration the complex history of magmatism, metamorphism and deformations, certain similarities are established with the Cordilleran type of active margins.

Metamorphic complexes or metamorphic cores originating in the active margins in close association with granite batholiths are intermediate between island arc (Hidaka type) and collision types. Metamorphic core complexes in the western states of the USA provide a typical example (Coney and Harms, 1984). They experienced a complex history of magmatism, compressive deformation and extension from the Jurassic to the Tertiary. As pointed out by De Yoreo et al. (1991), it is not quite clear which part of the zonal low-pressure rocks characteristic of these complexes originated at the end of the Nevada orogeny, which part during the Tertiary expansion, which part was uplifted during movement along the dividing low-angle fault (Coney and Harms, 1984) and which part elevated together with the melts in the form of diapirs as in Scandinavia and other collision-type belts.

A similar history has been described in the western part of the Idaho batholith (Manduca et al., 1993) and in southern California (Karlström et al., 1993). In western Idaho collision of the continent with the island arc occurred in the Middle Cretaceous (about 125 Ma). Deformations due to displacement and upthrow faults occurred simultaneously (125 to 120 Ma). The root portions of the structure are associated with the early mafic dykes and plutons and the first stage of deformation (Hudman and Foster, 1984). Later, two more granite complexes intruded and metamorphism and D_{2-3} deformations occurred between them. A metamorphic zone similar to the Altai 'pull-apart zones' was finally formed. At about 85 to 90

Ma, the intrusion of gabbro-tonalite plutons associated with crust-mantle melts and the concluding stage of deformation occurred.

These same three concluding phases are also observed in southern California (Karlström et al., 1993): metamorphism and D_{2-3} deformation stage, intrusion of pluton (85 Ma) and final strike-slip deformations, also applied to the pluton.

A similar history of metamorphism and deformation but in the Late Carboniferous-Permian (310 to 270 Ma) has been established in Irtysh zone (Dobretsov et al., 1997). The main stages and age of metamorphism (490 to 450 Ma) in Sangilen are similar to Scottish metamorphism (Cliff et al., 1993; Lebedev et al., 1993). The events in all these cases are covered in an interval of 30 to 40 Ma.

One of the characteristic features of major displacement zones is the intrusion of gabbro-tonalite plutons in the form of a belt of bedded and lenticular bodies of small dimensions such as the Irtysh shearing zone as well as elongated linear bodies confined to the main displacement zone. An example of the latter is the 'great tonalite sill' extending in south-eastern Alaska for a distance of over 800 km at a width of 5 to 20 km along a major displacement fault similar to San Andreas fault in California (Fig. 5.63). This displacement fault separates the island zone (or superterranes) and inter-montane accretionary zone of terranes into which diverse plutons intruded. They form the Coast Range plutonic belt adjoining the tonalite sill (Wood et al., 1992).

Thus, the main problems in the collision zones requiring additional study and modelling are: origin of granite-gneiss domes and prominent granitoid plutons.

Granitoid plutons are seen in a different geodynamic setting. Small gabbro-plagiogranite intrusions are characteristic of island arcs while very large diorite-granodiorite ('andesitoid') plutons are features of active margins of the Andean type (Kuz'min, 1985). The main mass of the largest granitoid batholyths is, however formed in a collision and early post-collision setting. Some of them, called S-granites (Chapell and White, 1979; Pitcher, 1979), are the result of direct melting (anatexis) of thickened continental crust. Other types (I- and A-granites) are the result of complex interactions of mantle melts (often residues of subduction melting) and crustal anatectic melts (Pitcher, 1979; Dobretsov, 1981; Dobretsov and Chupin, 1993; Litvinovskii and Dobretsov, 1993). The so-called stress-granites manifest in the early stages of collision in the displacement fault zones are of special interest (Vladimirov et al., 1992).

Based on formal but stable characteristics, Izokh (1978) classified the gabbro-granite series according to the degree of basicity, contrast (or discontinuity) and type of alkalinity (sodium, sodium-potassium or potassium series). In any case, massive manifestation of potassium granites is regarded as a characteristic of transition to mature continental crust. The potassium content of granite melts in general depends on the composition of the substratum, temperature and fluorine content of the fluid.

For a proper understanding of the identified empirical characteristic features of granitoid magmatism, further experimental and theoretical modelling is necessary. The basic principles of anatexis are well explained on the basis of well-known diagrams of granite eutectic—H_2O-CO_2 (Shkodzinskii, 1976; Dobretsov, 1981) and simple thermophysical modelling (Sharapov et al., 1977). Melt inclusions (Fig. 5.64) provide significant information on the origin of different types of granitoids. Figure 5.64 shows how the composition of fluids dissolved in the melt varies systemati-

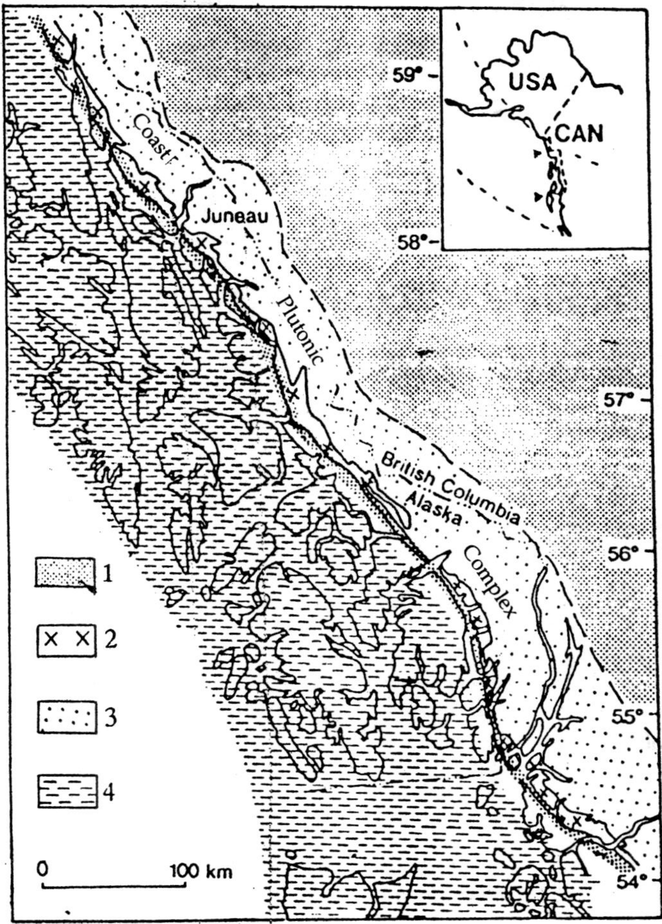

Fig. 5.63. The great tonalite sill in south-eastern Alaska.
1—shearing zone; 2—great tonalite sill; 3—intermontane and 4—island arc superterranes.

cally with the changing temperature and depth of the granitoids, from 'dry' CO_2-rich lower crust granites and migmatites to upper crust hypabyssal granites enriched with H_2O and F.

The interactions between crustal and subcrustal melts, as well as between mantle fluids and crustal melts, are very important. They have been conceptualised as syntexis and fluid syntexis; see, for example, N.L. Dobretsov (1981) and G.L. Dobretsov (1985). Far more complex problems arise in this study. Partial solutions are intended only for certain of these problems, for example in reports by Litvinovskii et al. (1993, 1994). Models related to local fluid-heat flux along fault zones

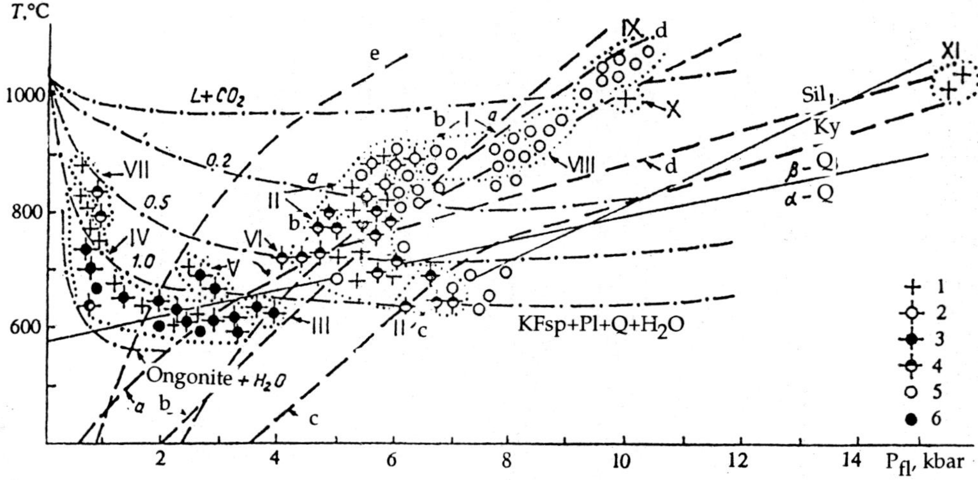

Fig. 5.64. *P-T* zones of crystallisation of granites and granulites (Dobretsov and Chupin, 1993). I—autochthonous and subautochthonous charnockites (a—Kansk series, Yenisei range and b—Aldan, Ukraine and Baltic shields); II—migmatites and anatexic—granite of granulite (a) and amphibolite (b, c) facies; III—regenerated granites (Lake Onega island and Yurchik massif in Kamchatka); IV—rapakivi granites of Korosten', Vyborg and Salmin plutons (Russian Craton); V and VI—high-alumina granites of Pamir and Shugnan complexes (south-eastern Pamir); VII—rapakivi-like granites of Ara-Litsk complex (Kola peninsula); VIII and IX— granulites of Chogar (VIII) and Sutam (IX) blocks; (Siberian Craton); X—deep-level zone of melting and crystallisation of rhyolite melts above a centre of basic rocks (Alichur pluton, southern Pamir); XI—kyanite granulites from xenoliths in a volcanic pipe of fergussite porphyries (eastern Pamir); composition of inclusion: 1—melts of granite composition; 2—melt + fluid (CO_2); 3—melt + fluid (H_2O); 4—melt + fluid (CO_2 + H_2O); 5—fluid (CO_2); and 6—fluid (H_2O); geotherms; a, b—in the Early Archaean; c—average for the continental crust (according to Dobretsov, 1981); d, e—Archaean granite-greenstone and oceanic (according to K. Condie, 1981); f—collision geotherms for doubled crust in the Pamir (obtained from the results of studying the melt inclusions). Lines of melting: granite + H_2O + CO_2 and lines Sil $\leftrightarrow$ Ky, $\beta - Q \leftrightarrow \alpha - Q$ (Dobretsov, 1981).

or associated with the intrusion of basic rock plutons (Sharapov, 1994) have yet to be correlated with real geodynamic settings, geology and geochemistry of granitoid plutons (G.L. Dobretsov, 1985).

5.11. General Model of Evolution of Orogenic Belts and Periodicity of Endogenic Processes

In the preceding sections we have studied the tectonics, petrology and the basic models of all the present-day active zones of the Earth. Analogues are known for all of them in the geological past and succeed each other regularly in the history of ancient orogenic (or folded) belts. Their location at the Earth's surface is illustrated in Fig. 1.1 in a simplified form while other illustrations in Chapters 4 and 5, represent many sections in detail.

As pointed out already, fragments of subduction-accretion and accretion-collision complexes constitute the main components of folded belts. These complexes cement the blocks of cratons, microcontinents and mature island arcs. In turn, the belts are perforated by collision and post-collision granites and are overlain by major post-collision sedimentary basins such as the West Siberian basin. In a similar manner, major cratons are covered by a sedimentary mantle in the form of large sedimentary basins as well and are divided by young and old (aulacogen) rifts.

In this remarkably complicated structure, a regular repetition of basic elements (structural-formational complexes) in space and time was recognised long ago. In the old 'geosynclinal' terminology, these complexes were regarded as initial, early, middle, late and final stages of the evolution of geosynclines or folded belts. Long ago, these stages helped regularise the status of ore deposits (Bilibin, 1955). Although the interpretation of these stages from the viewpoint of plate tectonics has undergone significant change, the division itself into stages and most complexes classified under them have retained their importance. An attempt is made here to give a new interpretation to these stages taking into consideration the data of modelling.

The **initial stage** should be understood as the stage of extension and destruction of continents and opening up of oceans. This stage is presently observed in the Red Sea, Gulf of Aden and the adjoining part of the Indian Ocean as well as in the northern Atlantic and Arctic (especially north of Iceland). The main processes occurring in this stage are the perioceanic rifting, spreading in the oceanic ridges, thinning of continental margins and formation of a passive margin with thick carbonate-terrigenous beds transiting into deepwater sediments. These features were described in Sections 5.1, 5.2 and 5.3. The relics of oceanic ophiolites (early oceanic crust and deepwater oceanic sediments) may survive in the subduction complexes. Sediments and structures of independent rifts on the continents as well as terranes made up of sediments of a passive margin are best preserved.

The **early stage** is characterised by formation of subduction zones even though only on one side of the evolving ocean. These subduction zones were initially in the form of island arcs which may have later been succeeded by active continental margin conditions. Examples of such subduction zones, their interchanges and mechanisms controlling the subduction zones were described in Sections 5.5 to 5.8. The formation of subduction zones signifies that the movement of continents slowed down or came to halt and the vector of movement changed while spreading and movement of oceanic plates continued. In this stage, fragments of ophiolites (more frequently ophiolites of marginal seas or supersubduction zones), island arcs, subduction-accretion complexes, sedimentary terranes of marginal seas and passive margins are formed and later preserved. Distinguishing these formations from the sedimentary formations of the preceding stage may be difficult. Along with the formation of subduction zones, the formation of sediments of passive margins and independent rifts may continue on another margin (*History of Development...*, 1984; *History of Tethys Ocean...*, 1987., Wilson, 1990, Unrug, 1997).

The duration of the first two stages may differ. The first stage which has not yet completed in the northern Atlantic and the northern part of the Indian Ocean has been continuing for over 100 Ma. The existence of island arcs succeeded by active margins or vice versa has been found in the setting of the Pacific Ocean and the north-eastern part of the Indian Ocean for more than 150 Ma. Thus the first two stages may add up to 250 Ma although short-lived oceanic basins (Meso-Tethys, Neo-Tethys and Mongolian-Okhotok) with the first two stages extending for 60 to 90 Ma are restored.

The **middle collision stage** (also formely known as the inversion stage) may run into many stages commencing with the collisions of island arcs, microcontinents among themselves or with a large craton, reduction in ocean area and conclude with the collision of continents and total closing of the ocean. Each stage of collision may be relatively brief (10 to 15 Ma, see below) but on the whole falls in an interval of about 60 to 90 Ma. For example, closing of the Riphean-Vendian Palaeo-Asiatic ocean (and the early Palaeo-Pacific) started in the middle of the Early Cambrian and completed in the Middle Ordovician (530 to 440 Ma). Closing of the Hercynian ocean commenced in the Late Devonian and ended in the Middle Carboniferous (360 to 300 Ma) (Berzin and Dobreisov, 1993; Berzin et al., 1994; Dobretsov et al., 1995a). These stages are marked by olistostromes, peaks of glaucophane metamorphism, marine molasses, intrusion of granites and formation of early domes (see Sections 5.9 and 5.10).

Late (post-collision) stage commences with the massive intrusion of post-collision granites and formation of granite-gneiss domes and is accompanied by the formation of continental, often molassic troughs of volcanic origin. Its demarcation from the preceding stage is not always distinct. For example, for the Hercynian stage of the Palaeo-Asiatic ocean, it encompasses the Middle Carboniferous-Permian (300 to 240 Ma), at places continuing to the Early Triassic. (Berzin et al., 1994).

Finally, the **final stage** is characterised by the absence of volcanic activity (manifestation of only dykes) or manifestation of alkaline or bimodal basalt-alkaline volcanism associated with hot spots. During this period, major post-collision basins of lacustrine or shallow marine origin are formed and overlie on the preceding molasses, volcanic troughs and rifts such as the West Siberian or Junggar and Tarim basins containing major oil and gas reservoirs (Surkov et al., 1982; Berzin et al., 1994). The main stage of formation of these basins covers Jurassic-Cretaceous, i.e., about 150 Ma, but is actually continuing even at present (for more than 200 Ma). Prolonged periods of formation of such basins are known on other ancient platforms (Belt, Adelaide in Australia, over 400 Ma; and Riphean basins of the Siberian platform, over 300 Ma) (Zonenshain and Kuz'min, 1993a; Condie and Rosen, 1994). These estimates (150 to 300 Ma) were obtained in Section 5.4 by theoretical modelling.

Such a sequence of stages is characteristic of relatively young belts commencing from the border of 1800 to 2000 Ma. The older belts reveal specific features calling for special investigations (Dobretsov et al., 1992; Condie and Rosen, 1994). The duration of individual processes estimated empirically as well as using theoretical models may differ. The prolonged process continuing for hundreds of millions of years characterises the formation of sedimentary basins, deep-level metamorphism, island arc magmatism and also the collision stages of mountain formation and metamorphism (in an erosion model of uplift). Rapid processes (covering a few to ten million years) involve normal collision and metamorphism of progressive and especially retrogressive stages associated with the possibility of tectonic transport of slices of metamorphic rocks.

Many estimates are known concerning the formation of sedimentary basins and deep-level metamorphism under a bed of accumulating sediments. Assuming an average rate of sedimentation of 0.005 to 0.01 cm/yr (Zoetemeijer et al., 1990, 1993; Reverdatto et al., 1995) and a probable thickness of sediments at 20 km, we find a submergence duration of 40 to 20 Ma (average 30 Ma) (= maximum period of metamorphism). A more prolonged period denotes interruptions in sedimentation or other complications. Contrarily, the presence of stress causing thrusts may considerably reduce this period (Zoetemeijer et al., 1993). 'Instant' (in the geological sense, < 1 Ma) submergence caused by sizable thrusts and subsequent slow heating in a period of about 15 to 20 Ma on the whole allows 15 to 20 Ma for the progressive stage of zonal collisional metamorphism (England and Thompson, 1984). A similar estimate of time would also evidently hold good for the evolutionary curve of submergence and heating (see Fig. 5.62, 1–3–4).

Estimates of the duration of metamorphism at 15 to 20 Ma are in fact characteristic of zonal progressive metamorphism and have been made in many studies. The varying duration of the different stages of metamorphism should be borne in mind, however. A typical example has been given for the classic Dalradian metamorphism in Ireland and Scotland (Yardley et al., 1987; Cliff et al., 1993). The main high-temperature metamorphism follows directly after island arc magmatism (with a duration of about 30 Ma and interruption of less than 10 Ma) and is of extremely brief duration (about 2 to 3 Ma). Metamorphism is followed by an interruption or slow uplift of diapirs in domes in a period of about 20 Ma and later by a new stage of metamorphism of reduced pressures extending for 15 Ma associated with upthrow faults and tectonic jointing of different terranes.

A total duration of 30 to 35 Ma is characteristic of many tectonometamorphic and tectonomagmatic cycles (Dobretsov, 1988, 1995). This duration is most distinctly manifest in the case of eclogite-glaucophane schist metamorphism. A correlation of geological and isotope geochronological data (Dobretsov et al., 1987, 1989, 1994b; Dobretsov and Kirdyashkin, 1993b) shows that the interval between the peaks of glaucophane schist metamorphism in the Mesozoic glaucophane schist belts is equal to 30 Ma, in the Palaeozoic about 60 Ma and in the Precambrian belts 90 to 120 Ma. Further, 2 or 3 stages of metamorphism are combined in many glaucophane schist belts.

Typical examples studied in Oman, California and the western Alps are compared in Figure 5.65. It can be clearly seen that stages of metamorphism of about 120 and 90 Ma are well correlated in these essentially differing features located far apart from each other. Older stages of 140 to 150 Ma have also been established in the Alpine-Mediterranean belt and in the Pacific folded setting (Dobretsov et al., 1994b). Two stages have been found in Oman, three in California and as many as five stages in the western Alps; the latter interpretation is debatable. In California, the various stages have been documented in different geological formations which have been combined as a result of successive accretion. Old ages (140 to

150 Ma) have been established for eclogites in coarse-grained garnet-glaucophane schist rocks in zone of melange. An age of 115 to 120 Ma has been established for the eastern zone of the Franciscan belt comprising metamorphic schists of the type Rattlesnake Creek and South Fork Mountains (Dobretsov et al., 1987; Blake et al., 1988; Fig. 5.61). The youngest dates (60 Ma) are characteristic of the western lawsonite-pumpellyite zone. These three events separated by an interval of about 30 Ma may be interpreted as three stages of accretion accompanied by obduction phenomena. This assumption was used before in a model of multistage obduction (Dobretsov, 1979).

At present, it may be suggested that the first stage of such a process is associated with the rapid return flow from the deep levels of the subduction zone as a result of its clogging by a microcontinent or seamount, as depicted in Figure 5.48. The subsequent stages of accretion are associated with analogous events in a parallel subduction zone resulting from the clogging of an earlier zone. In this new stage, the deep-level material from the new subduction zone is uplifted to the surface and blocks or slices of rocks of deep-level metamorphism in the old zone of subduction may experience an additional impulse and move to the surface. Such a model has been marked by the dotted line in Figure 5.65b for the rocks of the Sesia-Lanzo zone. Another variant emerging from the model of a continuous, multistage but prolonged uplift of deep-level rocks along the former Benioff-Zavaritskii zone, transformed into an inclined collision zone (Hsu, 1991), is shown by the continuous line.

The regular periodicity of such stages in a given belt or as a whole in similar tectonic conditions in different parts of the Earth is related to the model of pulsating, periodically repeating forces of tectonic activity. It is probably associated with the periodic detachment of mantle plumes from the core-mantle boundary (CMB) causing periodic (at intervals of 30 Ma) change of convection flows in the asthenosphere and acceleration (or change of pattern) of movements of lithospheric plates (Dobretsov and Kirdyashkin, 1993a) (see Chapter 4).

New, surprising correlations have been determined between global geological events (Dobretsov, 1988, 1994b, 1997a, b; Larson and Olson, 1991; Rampino and Caldeira, 1993). These correlations help draw the conclusion that the primary cause of all the periodic endogenic phenomena correlating with the global variations of climate may in fact be the fluctuating separation of portions of mantle plumes from the CMB. Figure 5.66 depicts the correlation of periodic variations of the frequency of inversions of the magnetic field of the Earth caused by periodic variation of flow structure in the outer liquid core and periodic variation of the intensity of mantle volcanism in the 'hot spots' and 'hot fields' of the Earth generated by ascending mantle plumes (Larson and Olson, 1991; Mazaud and Laj, 1991). The Cretaceous 'megachrone' when there was no magnetic field inversion whatsoever in the interval 124 to 84 Ma corresponding to maximum mantle magmatism is the best-known period; it correlates with the prolonged Cretaceous period of warm climate on the Earth. On the background of the general curve of variation of intensity of mantle magmatism (Fig. 5.66b), local maxima at about 120, 90 and 60 Ma and a maximum close to the present epoch are observed; the maximum at about 30 Ma is absent but well manifested in other similar graphs. Such correlations

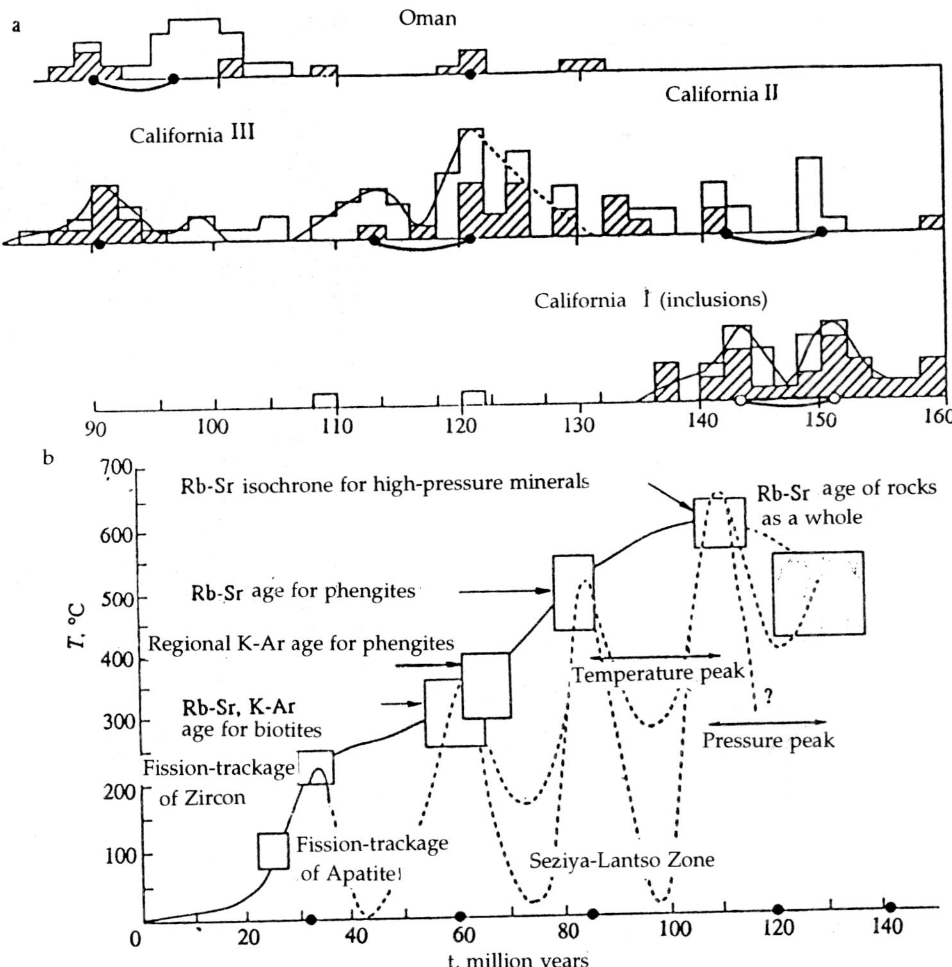

Fig. 5.65. Comparison of ages and stages of glaucophane schist metamorphism. a—in Oman (Bogdanov et al., 1991) and three stages of metamorphism (I, II and III) in the Franciscan belt in California (Ernst, 1983; Dobretsov et al., 1989); b—Sesia-Lanzo zones in the Alps (Hsu, 1991, modified). The different age datings (Rb-Sr, K-Ar, fission track) have been interpreted as a long curve of evolution of *P-T* conditions (continuous line) with estimate of the time corresponding to temperature and pressure peaks (Hsu, 1991) or as a multistage periodic obduction process (broken curve, authors' interpretation). Thick dots on the time axis represent the peaks of glaucophane schist metamorphism in Oman, California, Alps, Japan and other places (Dobretsov et al., 1987).

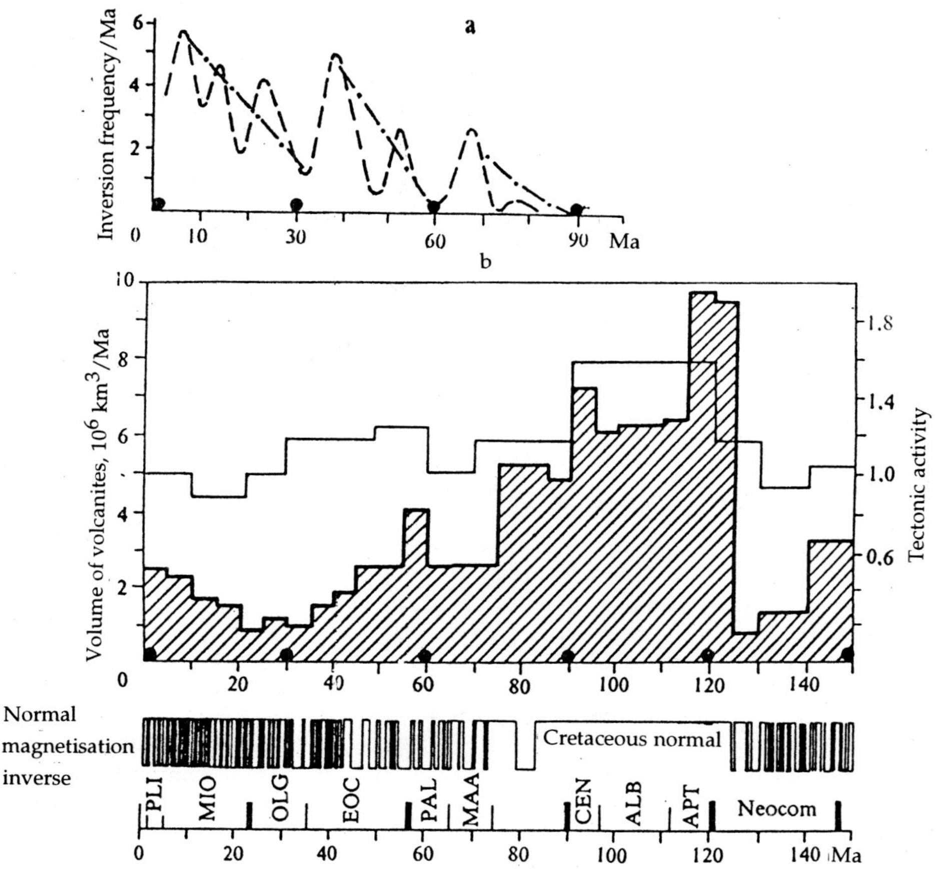

Fig. 5.66. Frequency (a) and time scale (b) of magnetic inversions compared to the intensity of mantle magmatism (in marine plateaus, 'hot spots' and continental plateau basalts) (from the data of Larson and Olson, 1991; Mazaud and Laj, 1991; Zonenshain and Kuz'min, 1993b; supplemented by the authors).

Points in Figs. a and b correspond to peaks of eclogite-glaucophane metamorphism. Each step in the hatched histogram (b) shows the volume of the oceanic crust formed over 5 Ma while the thin line represents change in tectonic activity.

are prolongated in the Palaeozoic—Early Mesozoic (Dobretsov, 1988, 1994b, 1997a, b).

Such correlations are not random phenomena and can be conclusively explained as follows. The rate of flow and vigour of heat and mass transfer in the liquid core are not compensated by the slow flows in the lower mantle. As a result, the outer core is 'heated' and the instability of flows and frequency of magnetic inversions in it increase. As soon as a major mantle plume is detached from the

CMB, it carries away the excess heat, the outer core is cooled, more stable flows are established and the frequency of magnetic inversions decreases or they disappear as in the Cretaceous and Permian. The detachment of major plumes (or groups of plumes) occurs (or is intensified) periodically at intervals of 30 Ma (like a periodic 'dropper') on the background of a periodicity of 300 to 200 Ma (see above). The ascending mantle plumes during their period of maximum intensity accompanied by intense mantle magmatism modify the asthenospheric flow structure. This in turn causes periodic variations of velocity and (or) direction of movement of lithospheric plates. This leads to the acceleration of subduction-accretion processes on the convergent plate boundaries. This explains the periodic maxima of the ages of glaucophane schists, island arc volcanism and mineralisation (Krivtsov et al., 1986; Dobretsov, 1988, 1994b). In turn, the change of relief in the oceans and on the continents leads to correlated processes of transgression-regression and change of currents in the hydrosphere and atmosphere causing climatic variations.

The effect of endogenic processes on global changes caused by periodic mantle plumes (Dobretsov, 1994b) is manifest in the form of periodic spurts of catastrophic volcanic activity in the oceans as well as in the island arcs and also periodic variations in the movements of the lithospheric plate ensemble and relief in the ocean and on land. This in turn leads to a change in the system of currents in the hydrosphere and atmosphere and to a change in climate at intervals of 30 to 34 Ma. Variations occurring over much shorter durations are associated with cosmic factors, primarily with the Milankovitch cycles although the interactions between endogenic and cosmic factors exert an influence at all levels (Schwarzacher, 1992).

In conclusion, let us attempt to summarise all curves depicting various types of magmatism and metamorphism on the P-T diagram and compare them (Fig. 5.67) with the P-T curves characteristic of descending (in subduction zones) and ascending (in mid-oceanic ridges and back-arc cells) flows in the asthenosphere. These theoretical curves were calculated for models of two-layer convection in the mantle (Dobretsov and Kirdyashkin, 1993a) and discussed in Chapter 4. It can be seen from Figure 5.67 that they accord well with the limiting empirical curves of the evolution of P-T conditions in metamorphic belts and mantle zones.

The P-T curve of descending flows coincides with the line of 'cold subduction' (see Fig. 5.53, 3b) and with many curves of the progressive stage of metamorphism in eclogite-glaucophane schist complexes including the Maksyutov complex in the southern Urals and Dora Maira in the Alps containing coesite, and the Kokchetav complex in northern Kazakhstan containing microdiamonds. Values of P-T parameters of diamond-bearing xenoliths of the Siberian platform are also situated along this same descending curve. Some differences have been established for xenoliths of Africa (Sobolev et al., 1984; Dobretsov et al., 1989; Pokhilenko et al., 1993). This confirms the suggestion made in many works that diamond-bearing zones in the mantle below the platforms represent relics of the deep parts of the ancient subduction zones.

The P-T curve of ascending flows in the mantle is close to curve 1 in Figure 5.53 and to the curve of P-T evolution of metamorphism in the mid-oceanic ridges and axial parts of island arcs, examples of which were cited earlier. Estimates of P-T parameters of the formation of xenoliths of spinel and pyrope peridotites from the zone of ascent of plumes in Hawaii, Baikal rift and other regions also lie in

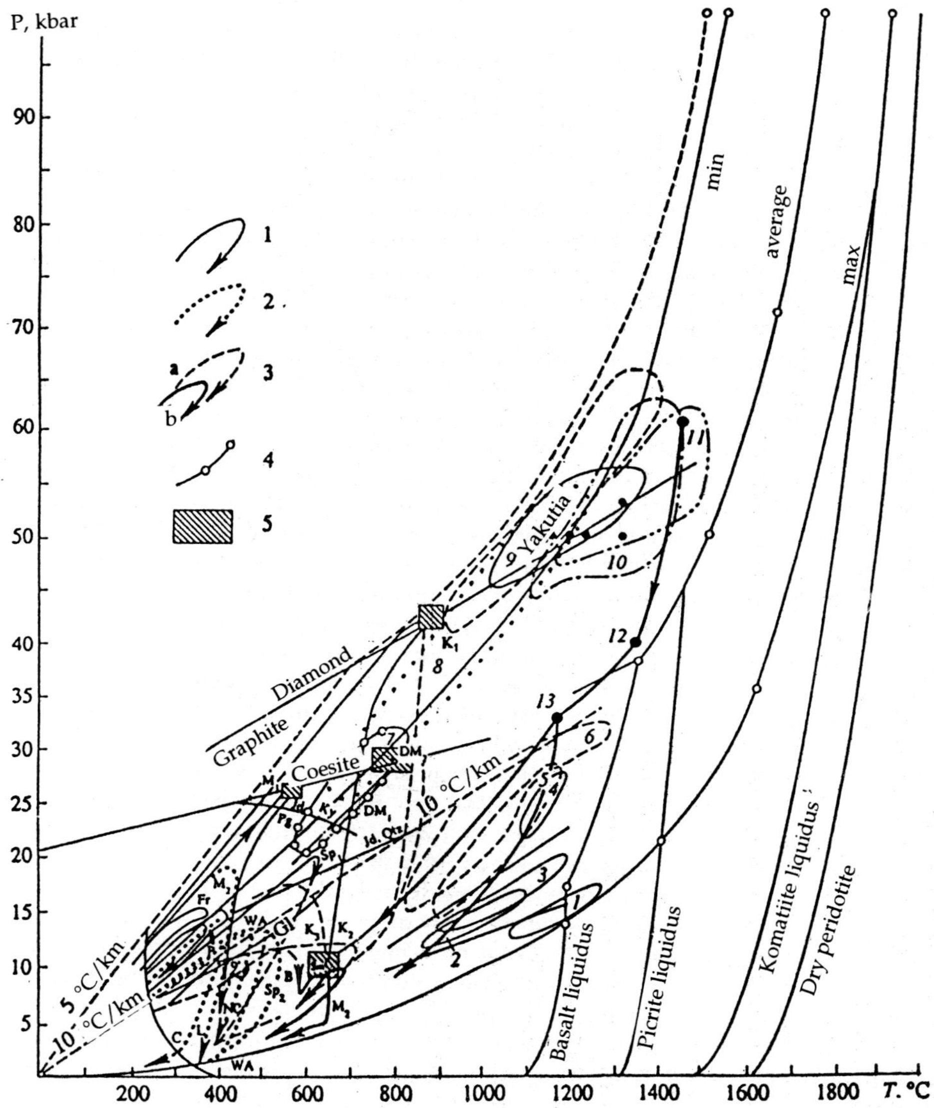

Fig. 5.67. Diagram summarising *P-T* conditions of regional metamorphism and crystallisation of mantle rocks (prepared by N.L. Dobretsov).
1—curves of evolution of Franciscan-type metamorphism (Fr—Franciscan, P—Penzhina and Sp₁ and Sp₂—Spitsbergen); 2—curves of evolution of Alpine-type metamorphism (B—Borus, western Sayans, C—Corsica, L—Liguria, NC—New Caledonia and WA—western Alps) (for more details of curves 1 and 2, see Fig. 5.53); 3—curves of evolution: a—polymetamorphic

this same region (Dobretsov and Ashchepkov, 1991; Ashchepkov, 1991; Dobretsov and Kirdyashkin, 1993a).

Between these extreme curves are situated the *P-T* parameters of most other cases of regional metamorphism, including metamorphism in the collision zones. A special case, unique for collision zones, covers the Ronda-type ultrabasic massifs in the Betic Cordilleras and Beni Bushera in Morocco. The evolution of these rocks on the example of Ronda massif (Davies et al., 1993) is also shown in Figure 5.67. The primary mantle rocks with diamonds were formed in a field comparable to the diamond-bearing peridotites of Africa. The final stages of evolution of the diapir are analogous to diapir domes of the Scandinavian type.

The diagram shown in Figure 5.67 characterises only one aspect of the wide variety of endogenic processes, i.e., various ways of evolution of *P-T* conditions. These are combined with various geodynamic conditions, structural characteristics, complex histories of deformations, diverse proportions of magmatic and metamorphic processes and varying duration of processes and their periodicity. These conditions have been considered rather schematically in this book. But though the set of the elementary processes in each type of events correlated is limited, it is nevertheless unique in any geological complex. This dictates, on the one hand, the need for thorough geological documentation of a variety of processes in each geological complex and, on the other, emphasises no less the importance of an extensive study and modelling of elementary processes, as attempted in this chapter.

belts (of the Sambogawa belt-type in Japan) and b—metamorphism of orogenic belts; 4—curves showing temperature distribution according to the model of two-layer convection in the mantle in a descending flow (minimum), at the centre (average) and ascending flow (maximum) (Dobretsov and Kirdyashkin, 1993a); 5—zone of *P-T* conditions of characteristic complexes (DM_1 and DM_2—Dora Maira; K_1, K_2 and K_3—Kokchetav; M_1 and M_2—Maksyutov, and Sm—Sambogawa). Numbered ovals—field of *P-T* conditions for mantle xenoliths: 1 to 6—Baikal rift, 7 to 9—spinel, transitional and diamond-bearing rocks of Yakutia, 10—South Africa (according to the authors' data and Sobolev et al., 1984; Ashchepkov, 1991; Pokhilenko et al., 1993; Ashchepkov et al., 1994;). Numbered lines: 11 to 13—probable evolution of mantle rocks with pseudomorphoses of diamonds of Ronda and Beni Bushera massifs (Davies et al., 1993). Stability lines of minerals are also shown: diamond, coesite, jadeite (Jd), kyanite (Ky), paragonite (Pg), jadeite and quartz (Jd, Qtz), glaucophane (Gl) and liquidus lines.

Conclusions

In summarising, we would like to emphasise the important points of the preceding discussions and to identify some of the problems that remain unsolved thus far.

1) One of the main methods of studying the deep-level processes in the Earth's interior is the **method of thermophysical modelling.** We have therefore considered it necessary to detail the fundamentals of thermophysical modelling as clearly as possible from the mathematical point of view, without simplifying the physical essence of the phenomena. Thermogravitational convection represents the main motive force of geodynamic movements. Therefore, the principles of the theory of convective heat transfer and the theory of similarity were outlined in Chapter 2 and the results of investigations on thermal and hydrodynamic structure of flow in a horizontal fluid layer presented in Chapter 3. Most of our experimental data on thermogravitational currents have been presented when solving the actual geodynamic problems in Chapters 4 and 5.

It must be pointed out that the usual level of studies in the field of heat and mass transfer is inadequate for solving the problems of geodynamic modelling. Quantitative evaluations of free convective heat transfer in the mantle and core of the Earth point to the fact that core and mantle convection proceed at Rayleigh numbers corresponding to developing or developed turbulent convection. Therefore, geodynamic modelling is necessary to study unsteady free convective and developed turbulent flows. A particularly large number of unsolved thermophysical problems is encountered when modelling thermal plumes, magmatic centres and hot spots. These problems include: free convection flows under conditions of phase transitions at the 'liquid—crystalline body' boundary in the presence of a local heat source. For this reason, our initial investigations revealed only the general structure of convection flows and stability characteristics of the phase boundary.

Considering that the geodynamic processes often proceed in complex (multiphase) media, it can be visualised how broad a class of problems will have to be solved when modelling geodynamic processes. It can be said that geodynamic modelling has already opened up entirely new formulations of thermophysical problems.

2) **From the geological point of view,** an attempt was made to understand the mechanisms of processes and develop some models of the most important

active zones and centres of the Earth. Their surface manifestation is described by the following two independent series:

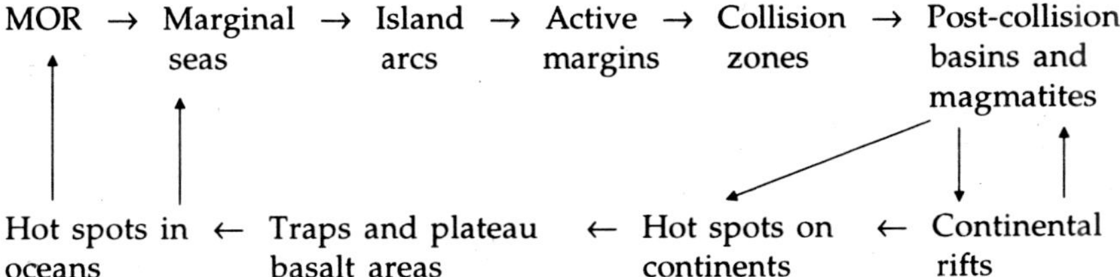

The upper series is the result of the interaction of asthenosphere and lithospheric plates as well as interaction between plates. It can be described by the convection model in the asthenospheric layer not associated (directly) with the lower mantle. The model for this series includes: ascending flows beneath the mid-oceanic ridges and marginal seas; descending flows in the subduction zones beneath the island arcs and active margins; and the 'unsinkability' of continental crust and lithosphere, experiencing thickening in the collision zones and thinning in the zones of basins and rifts. All of the main parameters (thickness of the lithosphere and asthenosphere, magmatism, metamorphism, tectonics and ore formation) undergo systematic changes in this series.

The lower series is formed by the action of mantle plumes (local jets) on the asthenosphere and lithosphere. The 'primary' plumes rise from the core-mantle boundary and have a thermal or combined thermal and chemical nature. The 'secondary' plumes are formed under the action of the 'primary' ones in intermediate layer C of the upper mantle or above the zone of subduction. The lower series is fairly independent of the upper one and may superimpose on any structure of the upper series but the main interaction occurs at the ends of these series: lower mantle plumes (Iceland type) may be superimposed on the mid-oceanic ridges and marginal seas; post-collision basins and magmatites are associated with continental rifts and hot spots.

In the present state of the Earth's interior the lower mantle interacts actively (and exchanges matter?) with the outer liquid core. The interaction with the upper mantle is limited: a portion of the material of the lower mantle penetrates with the plumes into the upper mantle; convection cells in the upper mantle reflect (partly or wholly) the structure of convection cells in the lower mantle. Convection in the asthenosphere and secondary plumes cause intense heat and mass transfer and mechanical interaction between the asthenosphere and lithosphere. The end-product of interaction is the formation of light 'unsinkable' continental crust and 'depleted' upper mantle which can only partly be regenerated under the action of lower mantle plumes.

To test and develop the thermophysical and geological consequences, two more aspects have to be taken into consideration: geohistorical and comparative-planetological. In the present monograph, these aspects have been dealt with only cursorily. On the geohistorical plane, the most important aspect is the repeated 'coalescence' of continents into supercontinents (Pangaeas) and their fragmentation and separation. The last Pangaea existed during the Permian (200 to 250 Ma ago)

(Zonenshain and Kuz'min, 1993a) while the one preceding it broke up 750 to 950 Ma ago (Condie and Rosen, 1994; Maruyama, 1994; Dobretsov et al., 1995a, Unrug, 1997). This break-up probably occurred in two stages. In the first stage of 950 to 750 Ma, the Palaeo-Asiatic Ocean and South Pacific were formed. In the second stage (after 700 to 750 Ma), the North Pacific and Palaeo-Atlantic opened up and the Palaeo-Asiatic Ocean underwent modification (Dalziel et al., 1993; Young, 1994; Dobretsov et al., 1995a; Unrug, 1997). Perhaps, one more Pangaea existed 1650 to 1700 Ma ago and was partly or totally fragmented during 1450 to 1100 Ma, probably also in two stages (Duncan and Turcotte, 1994; Khain, 1995). In the evolution of each newly formed ocean, there were repeated collisions between island arcs and microcontinents culminating in the formation of supercontinents (Kumazawa and Maruyama, 1994; Osazawa, 1994; Dobretsov et al., 1995; Khain, 1995). What factors are responsible for fragmentation and coalescence of supercontinents giving rise to an extremely long periodicity of 'Pangaea to Pangaea' called the Wilson cycle with a duration of about 500 Ma? Is it related to the period of convection in the lower mantle approaching 500 Ma (according to our estimates)? When did the processes of fragmentation and movement of major lithospheric plates, i.e., plate tectonics in full force, commence? What are the reasons for the 'main' periodicity of global processes on the Earth with a period of about 30 Ma? Only some preliminary answers have been given to these aspects in the preceding chapters as well as in other published sources (Dobretsov, 1981, 1994a, b; Larson, 1991; Dobretsov and Kirdyashkin, 1992, 1993b, 1994; Duncan and Turcotte, 1994; Khain, 1994; Kumazawa and Maruyama, 1994; Janssen et al., 1995; Kazansky, 1995; Larson and Kinkaid, 1996; Maruyama, 1994, 1997).

Our knowledge of geodynamics older than 1.7 to 2.0 billion years is even more uncertain. In the history of the Earth, the border of the Archaean and Proterozoic at about 2.6 billion years is regarded as the most important one. This boundary is assumed as the cessation of intense formation of the core (Dobretsov, 1980) and/or the catastrophic collapse of the stable density stratification in the core (Kumazawa et al., 1994). The Early Archaean stage (older than 3.1 billion years) is called the nuclear stage (Dobretsov, 1980; Bogatikov et al., 1987) since the first of large plates or 'nuclei' of continental crust of adequate thickness (≥ 30 km) probably originated in this period. The extensive distribution of komatiites, i.e., the high-temperature magnesial lavas in the greenstone belts, as well as other features of Archaean volcanism and sedimentation suggest a high heat flow in the Archaean (Ringwood, 1977; Condie, 1981; Dobretsov, 1981, 1994a; Ringwood et al., 1992; McCulloch, 1993).

Assuming that the heat flow in the Archaean was 5 to 10 times higher than the present level and the same value of superadiabatic temperature drop in the mantle of ~ 1200°C, we determined that the viscosity of the lower mantle in the Archaean was close to that of the upper mantle asthenosphere (i.e., $\eta = 10^{18}$ to 10^{19} poises) (Dobretsov and Kirdyashkin, 1995). In this case the velocity of convection in the lower mantle was the same as in the upper mantle (10 to 20 cm/yr) and the convection covered the entire mantle and was unstable with a developed turbulence (Dobretsov and Kirdyashkin, 1995) while a three-layer convection has been assumed by other investigators for the Archaean (McCulloch, 1993; see Fig. 4.20). In both models, however, a consequence is the 'small-plate tectonics', frequent interaction of microplates with relatively short term, intense impulses of subduction which give rise to greenstone belts with komatiites. Volcanites and sediments of these belts hold many characteristics of present-day 'primitive' island arcs (Mariana type) and marginal seas while the tectonics of these belts is similar to the accretionary tectonics of Phanerozoic folded belts. On this basis, Borukaev (1996) suggested that plate tectonics prevailed even in the Archaean in a regime that is similar to that of the present. However, systematic evolution of the composition of volcanites, sediments (including ferruginous and manganiferous sediments) and mineralisation during the Archaean and Proterozoic (Kazansky, 1995; Dobretsov, 1997a) suggest a regular and objective evolution of geodynamics during the Precambrian.

There are almost no traces of the earlier, pre-geological history (older than 3.9 to 4.0 billion years) of the Earth (Froude et al., 1983). It has been suggested that this was the result of intense meteorite bombardment of the early Earth and formation of a 'magmatic' ocean 400 to 600 km in depth, that is, a thick layer of the upper mantle. This ocean cooled and crystallised at around 4.1 to 4.0 billion years. This stage has a possible analogy with the geological history of the Moon in the period 4.6 to 4.1 billion years (Galimov, 1995; Guyot, 1995).

It was suggested in Chapter 1 (Sandwell and Shubert, 1992; Kumazawa et al. 1994) that Venus illustrates the early history of the Earth. It is quite possible that the present state of Venus may foretell the future course of development of the Earth (Herrick, 1994). Here, following formation of a thick lithosphere below a major part of the surface of Venus and depletion of its upper mantle, plate tectonics nearly ceased and only plume tectonics manifested (Maruyama, 1994). In any case, the uniqueness or universality of the various geodynamic processes on the Earth can be understood only from a study of comparative planetology and astrochemistry.

Thus, a vast field of activity is open to researchers working in the fields of geodynamics, geophysics and geochemistry and outstanding discoveries can be anticipated in the very near future.

Literature Cited

*Afanas'ev, S.L. (1993). Isotopic geochronological scale of Vendian—Phanerozoic. *Geol. i Geofiz.*, 34(3): 3–9.

Akimoto, S. and Fujisawa, H. (1965). Demonstration of the electrical conductivity jump produced by the olivine-spinel transition. *J. Geophys. Res.*, 70: 443–449.

Allegre, C.J. (1982). Chemical geodynamics. *Tectonophysics*, 81: 109–132.

Allegre, C.J., Hamelin, B., Provosti, A. and Dupre, B. (1986). Topology in isotopic multispace and origin of mantle chemical heterogeneities. *Earth Plan. Sci. Lett.*, 8: 319–337.

Allegre, C.J., Poirier, J.-P., Humler, E. and Hofmann, A.W. (1995). Chemical composition of the Earth. *Earth Plan. Sci. Lett.*, 134: 515–526.

Al'mukhamedov, A.I., Kashintsev, G.L. and Matveenkov, V.V. (1985). *Evolution of Basalt Volcanism in the Red Sea Region*. Nauka, Novosibirsk, 192 pp.

Issue in English Al'mukhamedov, A.I., Medvedev, A.Ya. and Zolotukhin, V.V. (1996). Space-time interactions of riftogenic and blanket basalts of Siberian platform. In: *Geodynamics and Evolution of the Earth*. Novosibirsk, pp. 105–108.

Anders, E. and Grevasse, N. (1989). Abundance of the elements: meteoric and solar. *Geochim. Cosmochim. Acta*, 53(1): 197–214.

Anderson, D.L. (1994). The sublithospheric mantle as the source of continental flood basalts. *Earth Plan Sci. Lett.*, 123: 269–280.

Anderson, R.N., Delong, S.E. and Schwarz, W.M. (1978). Thermal model for subduction with dehydration in the downgoing slab. *J. Geol.*, 86: 731–739.

Anderson, T.B., Jamtveit, B., Dewey, J.F. and Swenson, E. (1991). Subduction and eduction of continental crust: major mechanisms during continent-continent collision and orogenic extensional collapse—a model based on the south Norwegian caledonides. *Tera Nova*, 3: 303–310.

*Aplonov, S.V. (1987). *Geodynamics of Early Mesozoic Ob' Palaeo-ocean*. Institute of Oceanography AN SSSR, Moscow, 98 pp.

*Artyushkov, E.V. (1979). *Geodynamics*. Nauka, Moscow, 279 pp.

*Artyushkov, E.V. (1993). *Physical Tectonics*. Nauka, Moscow, 450 pp.

---

*Asterisked works are in Russian. However, *Geol. i. Geophys.* is published in both Russian and English (English edition by Allerton Press, New York)—General Editor.

*Ashchepkov, I.V. (1991). *Deep-seated Xenoliths of Baikal Rift*. Nauka, Novosibirsk, 160 pp.

Ashchepkov, I.V., Litasov, Yu.D. and Dobretsov, N.L. (1994). Pyroxenites and composite garnet peridotite xenoliths from picrite-basalt, Vitim plateau (Transbaikal): implication for thermobarometry and mantle reconstruction. *Proc. V Internat. Kimberlite Conf., San Paolo*, pp. 455–466.

Aubry, M.P. (1991). Sequence stratigraphy: global eustasy or tectonic imprint. *J. Geophys. Res.*, 96: 6641–6679.

Aurnou, J.M., Battles, J.L., Neumann, G.A. and Olson, P.L. (1996). Electromagnetic core-mantle coupling and preferred geomagnetic reversal path. *EOS, AGU Spring Meeting*, Abstracts, p. 82.

Avigad, D. (1992). Exhumation of coesite-bearing rocks in the Dora Maira massif (West Alps, Italy). *Geology*, 20: 947–950.

Azbel, I.Ya. and Tolstikhin, I.N. (1993). Accretion and early degassing of the Earth: constraints from modelling of Pu-U-I-Xe isotopic systematics. *Meteoritics*, 28: 609–621.

*Balla, Z. and Dobretsov, N.L. (1984). Mineralogy and petrology of magmatic rocks of ophiolite complex of Sarvashke region (Byukk mountains, Hungary). *Geol. i Geofiz.*, 9: 11–26.

Basu, A.R., Podera, R.J., Renne, P.R., Teichmann, F., Vasiliev, Yu.R., Sobolev, N.V. and Turrin, B.D. (1995). High-He plume origin and temporal-spatial evolution of the Siberian flood basalts. *Science*, 269: 822–825.

Batchelor, G. (1967). *An Introduction to Fluid Dynamics*. Cambridge University Press, 615 pp.

*Baum, B.A. (1979). *Metallic Fluids—Problems and Hypotheses*. Nauka, Moscow, 120 pp.

Beaumont, C., Ellis, S., Hamilton, Y. and Fullsack, Pb. (1996). Mechanical model for subduction-collision tectonics of Alpine-type compressional orogens. *Geology*, 24: 675–678.

*Belichenko, V.G., Sklyarov, E.V., Dobretsov, N.L. and Tomurtogoo, T. (1994). Geodynamic map of the Palaeo-Asiatic Ocean, Eastern segment. *Geol. i Geofiz.*, 35 (7–8): 29–40.

Ben Avraham, Z., Nur, A., Jones, D. and Cox, A. (1994). Continental accretion: from oceanic plateau to allochthonous massifs. In: *Modern Problems of Geodynamics*. 'Mir', Moscow, pp. 101–121.

Benard, H. (1901). Les tourbillions cellulaires dans une nappe liquide transportant de la chaleur par convection en regime permanent. *Ann Chim. et Phys.*, 23: 62–144.

Bercovici, D. and Mahoney, J. (1994). Double flood basalts and plume head separation at the 660 km discontinuity. *Science*, 266: 1367–1369.

Bercovici, D., Schubert, G. and Glatzvair, G.R. (1989). Three-dimensional spherical models of convection in the Earth's mantle. *Science*, 244: 950–955.

*Berdnikov, V.S. and Kirdyashkin, A.G. (1978). Structure of free convective currents in a horizontal fluid bed at different boundary conditions. In: *Structure of Boundary Wall Layer*, Inst. Thermophysics, SO AN SSSR, Novosibirsk, pp. 5–48.

*Berdnikov, V.S., Malyshev, V.N. and Markov, V.A. (1987). Experimental investigations of the effect of the relative dimensions of horizontal fluid layer on some characteristics of thermogravitational convection. In: *Transport Processes in Forced and Free Convective Currents*. Inst. Thermophysics, SO AN SSSR, Novosibirsk, pp. 50–71.

*Berzin, N.A., Coleman, R.G., Dobretsov, N.L., et al. (1994). Geodynamic map of the western part of the Palaeo-Asiatic Ocean. *Geol. i Geofiz.*, 35(7–8): 8–28.

Berzin, N.A. and Dobretsov, N.L. (1993). Geodynamic evolution of Siberia in Late Precambrian-Early Palaeozoic time. In: *Reconstruction of the Palaeo-Asian (Palaeo-Asiatic) Ocean* (R.G. Coleman, ed.). VSP Intern. Sci. Publ., Netherlands, pp. 45–63.

*Bilibin, Yu.A. (1955). *Metallogenic Provinces and Metallogenic Epochs*. Gosgeoltekhizdat, Moscow, 87 pp.

*Birikh, R.V. (1968). Thermocapillary convection in a horizontal fluid bed. *PMTF*, 3: 69–72.

Blake, M.C., Jayko, A.S., McLaughlin, R.J. and Underwood, M.B. (1988). *Metamorphic and Tectonic Evolution of the Franciscan Complex, North California* (W.G. Ernst, ed.). Prentice-Hall, New Jersey, pp. 1036–1059.

Bochler, R. (1992). Melting of the Fe-FeO and Fe-FeS system at high pressure. *Earth Plan. Sci. Lett.*, 121: 247–257.

*Bogatikov, O.V., Bogdanova, S.V., Borsuk, A.M., et al. (1987). *Magmatic Rocks. Evolution of Magmatism in the History of the Earth*. Nauka, Moscow, 440 pp.

*Bogdanov, N.A., Dobretsov, N.L. and Knipper, A.L. (1991). Ophiolites and geological structure of eastern Oman. *Izv. AN SSSR, Ser. Geol.*, 9: 3–32.

*Bogdanov, N.A. and Khain, V.E. (eds.) (1996). Explanatory Note to the Tectonic Map of Barents Sea and the Northern Part of the European Russia on Scale 1:2,500,000. Inst. Lithosphere, Moscow, 40 pp.

Bohlen, S.R. and Boettcher, A.L. (1982). The quartz-coesite transformation: a pressure determination and the effect of other components. *J. Geophys. Res.*, 87: 7073–7078.

Bonatti, E. (1987). Serpentinite protrusions in the oceanic crust. *Earth Plan. Sci. Lett.*, 32: 107–113.

*Bortnikov, N.S., Fedorov, D.T. and Murav'ev, K.G. (1993). Mineral composition and conditions of formation of sulphide structures of Lau basin (south-western part of the Pacific Ocean). *Geol. Rud. Mestorozhd.*, 35(6): 528–544.

*Borukaev, Ch.B. (1996). *Plate Tectonics in the Archaean*. United Inst. Geology, Geoph., Mineralogy, Novosibirsk, 60 pp.

*Borukaev, Ch.B. and Natal'in, B.A. (1994). Accretionary tectonics in the southern part of the Russian Far East. *Geol. i Geofiz.*, 35(7–8) 89–94.

Bott, M. (1971). *The Interior of the Earth*. Edward Arnold, London, 367 pp.

Brace, W.F. and Kohlestedt, D.L. (1980). Limits on lithospheric stress imposed by laboratory experiments. *J. Geophys. Res.*, 85: 6248–6252.

Breuer, D. and Spohn, T. (1993). Cooling of the Earth, Urey ratios and the problem of potassium in the core. *Geophys. Res. Lett.*, 20: 1655–1658.

Brey, G., Brice, W.R., Ellis, D.J., Green D.H., Harris, K.L. and Ryabchikov, I.D. (1983). Pyroxene-carbonate reaction in the upper mantle. *Earth Plan. Sci. Lett.*, 65: 63–74.

Bridgeman, P.W. (1949). *The Physics of High Pressures*. London, 315 pp.

Brocker, M., Kreuzer, H., Matthews, A. and Okrusch, M. (1993). $^{40}Ar/^{39}Ar$ and oxygen isotope studies of polymetamorphism from Tinos island, Cycladic blueschist belt, Greece. *J. Metamorph. Geol.*, 11: 223–240.

Brown, K.M. and Westbrook, G.K. (1988). Mud diapirism and subduction in the Barbados Ridge accretionary complex: the role of fluids in accretionary processes. *Tectonics*, 7: 613–640.

Brun, J.-P., Sokoutis, D. and Van den Driessche, J. (1994). Analogue modelling of detachment fault systems and core complexes. *Geology*, 22: 319–322.

Bundy, F.P. (1980). The P-T phase and reaction diagram for elemental carbon. *J. Geophys. Res.*, 85: 6930–6936.

Burke, K. and Dewey, J.F. (1973). Plume-generated triple junctions: key indicators in applying plate tectonics to old rocks. *J. Geol.*, 81(4): 406–433.

Burke, K. and Wilson, J. (1977). Hot spots on the Earth's surface. *Achievements Phys. Sci.*, 123(4): 615–625.

Burov, E.V. and Diament, M. (1996). Isostasy, equivalent elastic thickness and inelastic rheology at continents and oceans. *Geology*, 24: 419–422.

Burov, E.V., Houdry, F., Diament, M. and Deverchere, J. (1994). A broken plate beneath the north Baikal rift zone revealed by gravity modelling. *Geophys. Res. Lett.*, 21: 129–132.

Busse, F.H. (1967). On the stability of two-dimensional convection in a layer heated from below. *J. Math. Phys.*, 46(2): 140–150.

Cadec, O., Kyvalova, H. and Yuen, D.A. (1995). Geodynamic implications from the correlation of surface geology and seismic tomographic structure. *Earth Plan. Sci. Lett.*, 136: 615–627.

Calderwood, A.R. (1996). Mineral physics constraints on the chemical composition and temperature of the earth mantle from density and bulk sound velocity profiles. *EOS, AGU Spring Meeting, Abstracts*, p. 261.

Cameron, A.G.W. (1978). Physics of the primitive solar nebula and giant gaseous protoplanets. In: *Protostars and Planets* (T. Gehrels, ed.). University of Arizona Press, Tuscon, pp. 453–487.

Campbell, I.H. and Griffith, R.W. (1990). Implications of mantle plume structure for the evolution of flood basalts. *Earth Plan. Sci. Lett.*, 99: 79–83.

Campiglio, C., Narion, C. and Wapnie, M. (1986). Etude d'une boninite à olivine de Nouvelle Caledonia: pétrographie et chémisme des phases. *Bull. Mineral.*, 109(4): 423–440.

Cann, J.R., Blackman, D.K., Smith, D.K., et al. (1997). Corrugated slip surfaces formed at ridge-transform intersections on the Mid-Atlantic Ridge. *Nature*, 385: 329–332.

Carswell, D.A., O'Brien, P.J., Wilson, R.N. and Zhai, M. (1995). Garnet-omphacite-phengite barometry of eclogistes in Dabie Shan, central China. *Chinese Sci. Bull.*, vol. 40, suppl., pp. 76–79.

*Carter, U.E. and Robertson, D.S. (1987). Study of the Earth using superlong range interferometry. *V Mire Nauki*, 1: 16–25.

Cazenave, B. and Thoraval, S. (1994). Mantle dynamic constrained by degree 6 surface topography, seismic tomography and geoid. *Earth Plan. Sci. Lett.*, 122: 207–219.

Cazenave, B., Dominh, K., Rabinowicz., M., et al. (1988). Geoid and depth anomalies over ocean swells and troughs. *J. Geophys. Res.*, 93: 8064–8067.

Chandrasekhar, S. (1961). *Hydrodynamic and Hydromagnetic Stability.* Clarendon Press, Oxford, 654 pp.

Chapel, B.W. and White, A.J.R. (1979). Two contrasting granite types. *Pacif. Geol.*, 8: 173–194.

Charoki, A. and Sparrow, E.M. (1987). Melting in a vertical tube rotating about a vertical colinear axis. *Intern. J. Heat Mass Transfer*, 30(4): 613–622.

Chase, C.D. and Sprawl, D.R. (1983). The geoid and ancient plate boundaries. *Earth Plan. Sci. Lett.*, 62: 314–320.

Chopin, C. (1984). Coesite and pure pyrope in high-grade blueschists of the western Alps. *Contrib. Mineral. Petrol.*, 86: 107–118.

Christensen, U.R. and Yuen, D.A. (1984). The interaction of a subducting lithospheric slab with a chemical or phase boundary. *J. Geophys. Res.*, 89: 4389–4402.

Clemens, J.D., Holloway, J.R. and White, A.J.K. (1986). Origin of A-type granite. Experimental constraints. *Amer. Miner.*, 71: 317–324.

Cliff, R.A., Yardley, B.W.D. and Bussy, F. (1993). U-Pb isotopic dating of fluid infiltration and metasomatism during Dalradian regional metamorphism in Konnemara, western Ireland. *J. Metamorph. Geol.*, 1: 185–191.

Cloetingh, S. (1992). Lithospheric dynamics and tectonics of sedimentary basins. *Proc. Kon. Ned. Akad. Wetensch.*, 95(3): 349–369.

Cloetingh, S., McQueen, H. and Lambeck, K. (1985). On a tectonic mechanism for regional sea level variations. *Earth Plan. Sci. Lett.*, 75: 157–166.

Cloetingh, S., Kooi, H. and Groenewoud, W. (1989). Intraplate stresses and sedimentary basin evolution. *Amer. Geophys. Union, Geophys. Monograph*, 48: 1–16.

Cloetingh, S., Gradstein, F., Grant, A.C., et al. (1990). Did plate reorganisation cause rapid Late Neogene subsidence around the Atlantic? *J. Geol. Soc., London*, 147: 495–506.

Cloos, M. (1982). Flow melanges: numerical modelling and geological constraints on their origin in the Franciscan subduction complex, California. *Geol. Soc. Amer. Bull.*, 93: 330–345.

Cloos, M. (1986). Blueschists in the Franciscan complex of California: petrotectonic constraints on uplift mechanisms. *Geol. Soc. Amer. Mem.*, 164: 77–93.

Cloos, M. (1993). Lithospheric buoyancy and collisional orogenesis: subduction of oceanic plateaus, continental margins, island arcs, spreading ridges and seamounts. *Geol. Soc. Amer. Bull.*, 105(6): 715–737.

Cobbold, P.R. and Davy, P. (1988). Intendation tectonics in nature and experiments: central Asia. *Bull. Geol. Inst., Uppsala*, 14(5): 143–162.

Coleman, R.G. (1997). *Ophiolites.* Springer-Verlag, 262 pp.

*Coleman, R.G. (1984). Magmatic complex of Tihama Azir: ophiolites of passive continental margins. In: *27th Intern. Geol. Cong.*, vol. 9, *Petrology*. Nauka, Moscow, pp. 104–112.

Coleman, R.G. and Wang, X. (eds.) (1995). *Ultrahigh-pressure Metamorphism.* Cambridge University Press, pp. 1–32.

Condie, K.C. (1981). *Archaean Greenstone Belts.* Elsevier, Amsterdam, 434 pp.

Condie, K.C. and Rosen, O.M. (1994). Laurentia-Siberia connection revisited. *Geology*, 22: 168–170.

Coney, P.J. and Harms, T.A. (1984). Cordilleran metamorphic core complexes. *Geol. Soc. Amer. Mem.*, 153: 7–34.

*Cox, A. and Hart, T. (1989). *Plate Tectonics*. Mir, Moscow, 427 pp.

Cserepes, L. and Rabinowicz, M. (1995). Gravity and convection in a two-layer mantle. *Earth Plant. Sci. Lett.*, 76: 193–207.

Cserepes, L., Rabinowicz, M. and Rosemberg-Borot, C. (1988). Three-dimensional infinite Prandtl number convection in one and two layers with implications for the Earth's gravity field. *J. Geophys. Res.*, 93(10): 12009–12025.

Dal Piaz, G.V., Gosso, G. and Polino, R. (1988). The Alpine Cretaceous orogeny as a precollisional wedge accreted on the Adrian margin. *Ofioliti*, 14(1/2): 51–56.

Dallwitz, W.B., Green, D.H. and Thompson, J.E. (1966). Clinoenstatite in a volcanic rock from the Cape Vogel area, Papua. *J. Petrol.*, 7: 357–403.

Dalziel, I.W., Dalla Salda, C.H. and Gahagan, L.M. (1994). Palaeozoic Laurentia-Gondwana interaction and the origin of the Appalachian-Andean mountain system. *Geol. Soc. Amer. Bull.*, 106: 243–252.

Davies, G.F. (1984). Geophysical and isotopic constraints on convection: an interim synthesis. *J. Geophys. Res.*, 89: 6017–6040.

Davies, G.F. and Richards, M.A. (1992). Mantle convection. *J. Geol.*, 100: 151–206.

Davies, G.F., Nixon, P.M., Pearson, D.G. and Obata, M. (1993). Tectonic implication of graphitised diamond from the Ronda massif, southern Spain, *Geology*, 21: 471–474.

Defant, M.J. and Drummond, M.S. (1993). Mount St. Helens: potential example of the partial melting of the subducted lithosphere in a volcanic arc. *Geology*, 21: 547–550.

Dewey, J.F. (1988). Extensional collapse of orogens. *Tectonics*, 7: 1123–1139.

De Yoreo, J.J., Lux, D.R. and Guidotti, C.V. (1991). Thermal modelling in low-pressure-high-temperature metamorphic belts. *Tectonophysics*, 188: 209–238.

Dickinson, W.R. (1970). Relation of andesites, granites and derivative sandstones to arc-trench tectonics. *Rev. Geophys. and Space Phys.*, 8: 813.

Dietz, R.S. (1961). Continent and ocean basin evolution by spreading of the sea floor. *Nature*, 190: 854–857.

*Dobretsov, G.L. (1985). *Variscan Granitoid Magmatism of Kazakhstan (Geology, Formation and Genetic Models)*. Nedra, Leningrad, 252 pp.

*Dobretsov, N.L. (1974). *Glaucophane-schist and Eclogite-glaucophane-schist Complexes of the USSR*. Nauka, Novosibirsk, 436 pp.

Dobretsov, N.L. (1979). The new overthrusting model on blueschist metamorphism with reference to the Franciscan Great Valley problems. *Ofioliti*, 5: 17–24.

*Dobretsov, N.L. (1980). *Introduction to Global Petrology*. Nauka, Novosibirsk, 200 pp.

*Dobretsov, N.L. (1981). *Global Petrological Processes*. Nedra, Moscow, 236 pp.

*Dobretsov, N.L. (1988). Correct periodicity of the formation of glaucophane schists as an index of the periodicity of geological processes. *Dokl. AN SSSR*, 300: 427–431.

Dobretsov, N.L. (1991). Blueschists and eclogites: a possible plate tectonic mechanism of their transportation from the upper mantle. *Tectonophysics*, 186: 253–268.

*Dobretsov, N.L. (1994a). Geological factors of global changes and periodicity of geological processes. *Geol. i Geofiz.*, 35(3): 3–21.

*Dobretsov, N.L. (1994b). Periodicity of geological processes and deep-level geodynamics. *Geol. i Geofiz.*, 5: 3–19.

*Dobretsov, N.L. (1995). Problems of tectonics and metamorphism. *Petrologiya*, 3(1): 5–25.

Dobretsov, N.L. (1998). Ore formation and global geological processes: evolution and the problems of periodicity. In: *Theophrastus' Contributions to Advanced Studies in Geology*, vol. 11, Athens.

*Dobretsov, N.L. (1997) Permian-Triassic magmatism and sedimentation in Eurasia as a result of a superplume. *Doklady Rus. Acad. Sci.*, 354(4): 491–500.

Dobretsov, N.L. and Sobolev, V.S. (1975). Eclogite-glaucophane schist complexes of the USSR and their bearing on the genesis of blueschist terranes. *Geol. Soc. Amer. Spec. Paper* 151. pp. 145–155.

*Dobretsov, N.L. and Ashchepkov, I.V. (1991). Evolution of the upper mantle of the Baikal rift zone. *Geol. i Geofiz.*, 1: 3–16.

*Dobretsov, N.L. and Kirdyashkin, A.G. (1991). Dynamics of subduction zones: models of the formation of accretionary wedge and uplift of glaucophane schists and eclogites. *Geol. i Geofiz.*, 3: 4–20.

Dobretsov, N.L. and Kirdyashkin, A.G. (1992). Subduction zone dynamics: models of accretionary wedge. *Ofioliti*, 17(1): 155–164.

*Dobretsov, N.L. and Chupin, V.P. (1993). Genesis of granitoids and formation of continental crust. In: *Granitoids of Folded Regions*. VSEGEI, St. Petersburg, pp. 7–26.

*Dobretsov, N.L. and Kirdyashkin, A.G. (1993a). Application of two-layer convection to the structural characteristics and geodynamics of the Earth. *Geol. i Geofiz.*, 1: 3–26.

Dobretsov, N.L. and Kirdyashkin, A.G. (1993b). Experimental modelling of two-layer mantle convection. *Ofioliti*, 18(1): 61–81.

Dobretsov, N.L. and Kirdyashkin, A.G. (1994). Blueschist belts of north Asia and models of subduction-accretion wedge. In: *Reconstruction of the Palaeoasiatic Ocean* (R.G. Coleman, ed.). VSP Intern. Sci. Publ., The Netherlands, pp. 91–106.

*Dobretsov, N.L. and Kirdyashkin, A.A. (1995). Heat exchange and rheology of the lower mantle in the early periods of the development of the Earth. *Dokl. RAN*, 345: 103–105.

*Dobretsov, N.L. and Kirdyashkin, A.G. (1997). Modelling of subduction processes. *Geol. i Geofiz.*, 37(5): 846–857.

*Dobretsov, N.L., Melamed, V.G. and Sharapov, V.N. (1970). Dynamics of regional metamorphism for a simple subsidence model of the oceanic-type crust. *Geol. i Geofiz.*, 10: 14–20.

*Dobretsov, N.L., Sobolev, N.V. and Shatsky, V.S. (1989). *Ecologites and Glaucophane Schists in Folded Regions*. Nauka, Novosibirsk, 235 pp.

Dobretsov, N.L., Konnikov, E.G. and Dobretsov, N.N. (1992). Precambrian ophiolite belts of southern Siberia, Russia, and their metallogeny. *Precambr. Res.*, 58: 427–446.

*Dobretsov, N.L., Kirdyashkin, A.G. and Gladkov, N.N. (1993). Problems of deep-level geodynamics and modelling of mantle plumes. *Geol. i Geofiz.*, 12: 5–21.

*Dobretsov, N.L., Kolobov, V.Yu. and Simonov, V.A. (1994a). Formation of oceanic lithosphere in slow-spreading ridges of the central Atlantic. *Petrologiya,* 2(4): 363–369.

Dobretsov, N.L., Berzin, N.A. and Buslov, M.M. (1995a). Opening and tectonic evolution of the Palaeo-Asiatic Ocean. *Intern. Geol. Rev.*, 37: 335–360.

Dobretsov, N.L., Shatsky, V.S. and Sobolev, N.V. (1995c). Comparison of the Kokchetav and Dabie Shan metamorphic complexes: coesite and diamond-bearing rocks and UHP-HP accretional collisional events. *Intern. Geol. Rev.*, 37: 636–656.

Dobretsov, N.L., Theunissen, K. and Berzin, N.A. (1997). Granitic gneissic domes in the strike-slip zone as a result of intrusion in the local pull-apart structures. *Abstr., VIII European Intern. Geol. Congr., Strasburg*, pp. 177–178.

*Dobretsov, N.L., Sobolev, V.S., Sobolev, N.V. and Khlestov, V.V. (1974). *Facies of High-pressure Regional Metamorphism*. Nedra, Moscow, 328 pp.

Dobretsov, N.L., Coleman, R.G., Liou, J.G. and Maruyama, S. (1987). Blueschist belts in Asia and possible periodicity of blueschist facies metamorphism. *Ofioliti*, 12(3): 445–456.

*Dobretsov, N.L., Bogdanov, N.A., Zonenshain, L.P. and Sborshchikov, I.M. (1991). Section of oceanic crust in King trough (central Atlantic). *Izv. AN SSSR, Ser. Geol.*, 8: 141–146.

Dobretsov, N.L., Watanabe, T., Miyashita, S. and Natal'in, B.A. (1994b). Comparison of ophiolites and metamorphic belts of Sakhalin and Hokkaido. *Ofioliti*, 19(2): 53–67.

Dobretsov, N.L., Sobolev, N.V., Shatsky, V.S., Coleman, R.G. and Ernst, W.G. (1995b). Geotectonic evolution of diamondiferous paragneisses, Kokchetav complex, northern Kazakhstan: the geological enigma of ultrahigh-pressure crustal rocks within a Palaeozoic folded belt. *The Island Arc*, 4: 267–279.

*Dortman, N.B. (Editor) (1984). *Handbook of Geophysics. Physical Properties of Rocks and Minerals (Petrophysics)*. Nedra, Moscow, 445 pp.

*Duchkov, A.D., Balobaev, V.T., Volod'ko, B.V. et al. (1994). *Temperature, Cryolite Zone and Radiogenic Heat Generation in the Earth's Crust in Northern Asia*. United Inst. Geol. Geophys. Mineral. SO RAN, Novosibirsk, 143 pp.

*Duchkov, A.D. and Sokolova, L.S. (1974). *Geothermal Investigations in Siberia*. Nauka, Novosibirsk, 280 pp.

Duncan, C.E. and Turcotte, D.L. (1994). On the break-up and coalescence of continents. *Geology*, 22: 103–106.

*Dzhaluriya, N. (1983). *Natural Convection*. Mir, Moscow, 399 pp.

Dziewonski, A.M. (1984). Mapping of the lower mantle: determination of lateral heterogeneity in P velocity up to degree and order 6. *J. Geophys. Res.*, B89: 5929–5952.

Dziewonski, A.M. and Anderson, D.L. (1981). Preliminary reference to Earth model. *Phys. Earth Planet Interior*, 25: 297–356.

Dziewonski, A.M. and Woodhouse, J.H. (1989). Three-dimensional Earth structure and mantle convection. *Abstr., 28th IGC, Washington,* 1: 427–428.

Edwards, C., Menzies, M. and Thirwall, M. (1991). Evidence from Muriah, Indonesia, for the interplay of suprasubduction zone and intraplate processes in the genesis of potassic alkaline magmas. *J. Petrol.,* 32: 555–592.

Ellis, S. (1996). Forces driving continental collision: reconciling intendation and mantle subduction tectonics. *Geology,* 24: 699–702.

Embry, A.F. (1990). A tectonic origin for depositional sequences in extensional basins—implications for basin modelling. In: *Quantitative Dynamic Stratigraphy* (T.A. Cross, ed.). Prentice-Hall, New York, pp. 491–501.

Encarnacion, J., Fleming, Th.H., Elliot, D.H. and Eales, H.V. (1996). Synchronous emplacement of Ferrar and Karroo dolerites and the early break-up of Gondwana. *Geology,* 24: 535–538.

Engebretson, D.C., Cox, A. and Gordon, R.G. (1990). Relative motions between oceanic and continental plates in the Pacific. *Geol. Soc. Amer. Spec. Paper* 206, pp. 1–59.

England, P.C. and Holland T.J.B. (1979). Archimedes and Tauern ecologits: the role of buoyancy in the preservation of exotic tectonic blocks. *Earth Planet. Sci. Lett.,* 44: 287–294.

England, P.C. and Thompson, A.B. (1984). Pressure-temperature-time paths of regional metamorphism: heat transfer during the evolution of regions of thickened continental crust. *J. Petrol.,* 25: 894–928.

Erickson, S.G. (1993). Sedimentary loading, lithospheric flexure and subduction initiation at passive margins. *Geology,* 21: 125–128.

Ernst, W.G. (1974). Metamorphism and ancient continental margins. In: *The Geology of Continental Margins* (K.A. Burke and C.L. Drake, eds.). Springer-Verlag, pp. 907–919.

Ernst, W.G. (1975). Systematics of large-scale tectonics and age progressions in Alpine and Circumpacific blueschist belts. *Tectonophysics,* 26: 229–246.

Ernst, W.G. (1983). Phanerozoic continental accretion and the metamorphic evolution of northern and central California. *Tectonophysics,* 100: 287–320.

Ernst, W.G. (1988). Tectonic history of subduction zones inferred from retrograde blueschist P-T Paths. *Geology,* 16: 1081–1084.

Ernst, W.G., Liou, J.C. and Coleman, R.G. (1995). Comparative petrotectonic study of five Eurasian ultrahigh-pressure metamorphic complexes. *Intern. Geol. Rev.,* 37: 191–211.

*Ershova, T.P., Fradkova, N.D. and Litvin, Yu.A. (1981). Construction of T-P-X diagrams of the fusibility of Mn-Ni system at pressure of up to 100 kb. *Izv. AN SSSR, Metally,* 5: 192–197.

Etheridge, M.A., Wall, V.I. and Cox, S.F. (1984). High fluid pressure during regional metamorphism and deformation: implication for mass transport and deformation mechanisms. *J. Geophys. Res.,* 89: 4344–4358.

Evans, B.W. (1990). Phase relations of epidote blueschists. *Lithos,* 25: 3–23.

Fairheed, J.D. and Henderson, N.B. (1977). The seismicity of southern Africa and incipient rifting. *Tectonophysics,* 41: 19–26.

*Fedotov, S.A. (1976). Uplift of basalt masses in the Earth's crust and mechanism of fissure eruptions of basalts. *Izv AN SSSR, Ser. Geol.,* 10: 5–23.

Feighner, M.A., Kellog, L.H. and Travis, B.J. (1995). Numerical modelling of chemically buoyant mantle plumes in spreading ridges. *Geophys. Res. Lett.*, 22: 715–718.

Fjeldskaar, W. (1994). Viscosity and thickness of the asthenosphere detected from the Fennoscandian uplift. *Earth Plan. Sci. Lett.*, 126: 399–410.

Forsyth, D.A., Morel-a-l'Huissieer, P., Asudsen, I. and Green, A.G. (1986). Alpha ridge and Iceland: products of the same plume? *J. Geodyn.*, 6: 197–214.

Fouquet, Y., von Stackelberg, U., Charlon, J.L. et al. (1991). Hydrothermal activity in the Lau back-arc basin: sulphides and water chemistry. *Geology*, 19: 303–306.

Francheteau, J., Needham, H.D., Choukroune, P., et al. (1979). Massive deep-sea sulphide ore deposits on the East Pacific Ridge. *Nature*, 277: 523–528.

Fridenger, P.Dzh., Reverdatto, V.V. and Polyanskii, O.P. (1991). Modelling of rift structures. *Geol. i Geofiz.*, 9: 1–19.

Froude, D.O., Ireland, T.R., Kinny, P.D., Williams, I.S., Compston, W. and Meyers, J.S. (1983). Ion microprobe identification of 4100 to 4200 Ma-old terrestrial zircons. *Nature*, 304: 616–618.

Fukao, Y., Maruyama, S., Obayashi, M. and Inoue, H. (1994). Whole mantle P-wave tomography. *J. Geol. Soc. Japan*, 100(1): 7–23.

*Galimov, E.M. (1995). Problems of the origin of the Moon. In: *Principal Directions in Geochemistry* (E.M. Galimov, ed.). Nauka, Moscow, pp. 8–43.

Gente, P. (1987). Etude morphostructural comparative de dorsales océaniques à taux d'expansion variès. These de Doc. Sci., Brest, 373 pp.

Geology and Petrology of Deep-water Trench Zones of Western Pacific Ocean (1991). Nauka, Moscow, 261 pp.

Geology of the Philippine Sea Floor (1980). Nauka, Moscow, 290 pp.

*Gershunin, P.Z., Zhukhovitskii, E.M. and Myznikov, V.M. (1974). Stability of plane-parallel convective fluid flow in a horizontal bed. *Prykl. Mat. Teoret. Phys.*. no. 5, pp. 145–187.

Giardini, D., Xiang-Dong, L. and Woodhouse, J.H. (1987). Three-dimensional structure of the Earth from splitting in free oscillation spectra. *Nature*, 325: 405–411.

*Golitsyn, G.S. (1980). *Study of Convection with Geophysical Applications and Analogies*. Gidrometeoizdat, Leningrad, 55 pp.

Golub, J.P. and Benson, S.V. (1980). Many routes to turbulent convection. *J. Fluid Mech.*, vol. 100, pt. 3, pp. 449–470.

Goodfellow, W.D. and Franclin, J.M. (1993). Geology, mineralogy and chemistry of sediment-hosted clastic massive sulphides in shallow cores, Middle Valley, northern Juan de Fuca ridge. *Econ. Geol.*, 88(8): 2037–2068.

Gossler, J. and Kind, R. (1996). Seismic evidence for very deep roots of continents. *Earth Plan. Sci. Lett.*, 138: 1–13.

Green, D.H. (1971). Crystallisation of calc-alkaline andesite under high water pressure conditions. *Contrib. Mineral. and Petrol.*, 34: 150.

Griffiths, R.W. (1986a). Thermals in external fluids, including the effects of temperature-dependent viscosity. *J. Fluid Mech.*, 166: 115–138.

Griffiths, R.W. (1986b). Particle motion induced by spherical convective elements in Stokes flow. *J. Fluid Mech.*, 166: 139–159.

Griffiths, R.W. (1986c). Dynamics of mantle thermals with constant buoyancy or anomalous internal heating. *Earth Plan. Sci. Lett.*, 78: 435–446.

Grocott, J., Brown, M., Dallmeyer, R.D. et al. (1994). Mechanisms of continental growth in extensional arcs: an example from the Andean plate-boundary zone. *Geology*, 22: 391–394.

Gurnis, M., Lithgow-Bertelloni, C. and Richards, M.A. (1996). Cenozoic dynamic topography, uplift and subsidence history of continents. *EOS, AGU Spring Meeting, Abstracts*, p. 274.

Guyot, F. (1995). Earth's innermost secrets. *Nature*, vol. 369, no. 6479.

Haddon, R.A.W. and Bullen, K.E. (1969). An Earth model incorporating free Earth oscillation data. *Phys. Earth Planet Interior*, 2(1): 39–49.

Hager, B.H. (1984). Subducted slabs and the geoid: constraints on mantle rheology and flow. *J. Geophys. Res.*, 89: 6003–6015.

Hager, B.H., Clayton, R.W., Richards, M.A. et al. (1985). Lower mantle heterogeneity, dynamic topography and geoid. *Nature*, 313: 541–545.

Haggerty, S.E. (1994). Superkimberlites: a geodynamic diamond window to the Earth's core. *Earth Plan. Sci. Lett.*, 122: 57–69.

Hamilton, W. (1970). The Uralides and the motion of the Eurasian and Siberian platforms. *Bull. Geol. Soc. Amer.*, 81: 2553–2576.

Harley, S.L. and Carswell, D.A. (1995). Ultradeep crustal metamorphism: a prospective view. *J. Geophys. Res.*, 100: 8367–8380.

Hart, S.R. and Zindler, A. (1986). In search of bulk Earth composition. *Chem. Geol.*, 57: 247–267.

Hartley, R., Watts, A.B. and Fairhead, J.D. (1996). Isostasy of Africa. *Earth Plan. Sci. Lett.*, 137: 1–18.

Hartmann, W.K. and Davis, D.R. (1975). Satellite-sized planetesimals and lunar origin. *Icarus*, 24: 504–515.

Hasegawa, A. (1994). Seismic structure of the NE Japan and Kuril arcs. *Abstr. 101st Meeting of Geol. Soc. Japan, Sapporo*, p. 13.

Hawaii Sci. Drilling Project Team (1996). Hawaii scientific drilling project: summary of preliminary results. *GSA Today*, 6(8): 1–8.

Hawkins, J. (1989). Back-arc basins evolution, geological characteristics and importance in crustal evolution. *Joint 4 ECOP/SOPAS Intern. Workshop Abstr., Canberra*, pp. 49–50.

Hayashi, C., Nakazawa, K. and Mizuno, H. (1979). Earth melting due to the blanketing effect of the primordial dense atmosphere. *Earth Plan. Sci. Lett.*, 43: 22–48.

Herrick, R.R. (1994). Resurfacing history of the Venus. *Geology*, 22: 703–706.

Hess, H.H. (1962). History of ocean basins. In: *Petrological Studies*. A.F. Baddington Volume. Geol. Soc. Amer., pp. 599–620.

Hill, E.J. and Baldwin, S.L. (1993). Exhumation of high-pressure metamorphic rocks during crustal extension in the D'Entiecasteaux region, Papua New Guinea. *J. Metamorph. Geol.*, 11: 261–277.

History of the Development of Urals Palaeo-ocean (1984). Inst. Oceanology AN SSSR, Moscow, 164 pp.

History of Tethys Ocean (1987). Inst. Oceanology AN SSSR, Moscow, 155 pp.

Ho, C.-Y. and Viskanta, R. (1984). Heat transfer during inward melting in a horizontal tube. *Intern. J. Heat Mass Transfer*, 27(5): 705–716.

Hodges, K.V., Burchfield, B.C., Royden, L.H. et al. (1993). The metamorphic signature of contemporaneous extension and shortening in the central Himalayan orogen: data from the Nialam transect, southern Tibet. *J. Metamorph. Geol.*, 11: 721–737.

Hoffman, P.F. (1991). Did the breakout of Laurentia turn Gondwanaland inside out? *Science*, 252: 1409–1412.

Holland, T.I.B. (1979). Experimental determination of the reaction paragonite = jadeite + kyanite + water; internally consistent termodynamic data for part of the system Na_2O-Al_2O_3-SiO_2-H_2O, with application to eclogites and blueschists. *Contrib. Mineral. Petrol.*, 68: 293–301.

Holland, T.I.B. (1980). The reaction albite = jadeite + quartz determined experimentally in the range 600–1200°C. *Amer. Mineral.*, 65: 129–134.

Honda, S.A. (1982). Numerical analysis of layered convection marginal stability and finite amplitude analysis. *Bull. Earthquake Res. Inst.*, Univ. of Tokyo, 57: 273–302

Honda, S.A. (1995). Simple parameterised model of Earth's thermal history with the transition from layered to whole mantle convection. *Earth Plan. Sci Lett.*, 131: 357–369.

Honda, S.A. and Yuen, D.A. (1994). Model for convective cooling of mantle with phase changes: effects of aspect ratios and initial conditions. *J. Phys. Earth*, 42: 165–181.

Hsu, K.J. (1991). Exhumation of high-pressure metamorphic rocks. *Geology*, 19: 107–110.

Hsui, A.T., Turcotte, D.L. and Torrance, K.E. (1972). Finite-amplitude thermal convection within a self-gravitating sphere. *Geophys. Fluid Dyn.*, 3: 35–44.

Hudman, D.W. and Foster, D.A. (1984). The role of tonalities and mafic dikes in the generation of the Idaho batholith. *J. Geol.*, 94: 31–46.

*Hutchinson, D.R., Gol'mshtok, A.D., Zonenshain, L.P. et al. (1994). Preliminary results of multichannel seismic sounding on Lake Baikal in 1969. *Geol. i Geofiz.*, 34(10–11): 19–27.

Hynes, A., Arkani-Hamed, J. and Greiling, R. (1996). Subduction of continental margins and the uplift of high-pressure metamorphic rocks. *Earth Plan. Sci. Lett.*, 140: 13–25.

Inoue, H., Fukao, Y., Tanabe, K. and Ogata, Y. (1990). Whole mantle P-wave travel time tomography. *Phys. Earth Planet Interior*, 59: 294–328.

Irvine, T.N. (1988). A global convection framework. *Carnegie Inst. Washington Yearbook*, 87: 3–8.

Irvine, T.N. (1991). Global convection and Hawaiian upper mantle structure. *Carnegie Inst. Washington Yearbook*, 90: 3–11.

Isacks, B., Oliver, J. and Sykes, L. (1974). *Seismology and the New Global Tectonics.* Mir, Moscow, pp. 133–179.

Ito, G. and Lin, J. (1995). Oceanic spreading centre-hot spot interactions: constraints from along isochrone bathymetric and gravity anomalies. *Geology*, 23: 657–660.

*Ivanov, S.N. (1978). Metamorphism of the extension of continents. *Dokl. AN SSSR*, 238: 908–912.

Ivanov, S.N. and Ivanov, K.S. (1993). Hydrodynamic zoning of the Earth's crust and its significance. *J. Geodyn.*, 17: 155–180.

*Izokh, E.P. (1978). *Evaluating the Ore Content in Granitoid Formations for Forecasting Purposes.* Nedra, Moscow, 137 pp.

Jagoutz, E., Palme, H., Baddenhausen, K., Blum, M., Cendales, G., Dreibus, B., Spettel, V., Lorenz, V. and Wanke, H. (1979). The abundance of major, minor and trace elements in the Earth's mantle as derived from primitive ultramafic nodules. *Proc. Lunar Plan. Sci. Conf.*, 10: 2031–2050.

Janssen, M.E., Stephenson, R.A. and Cloetingh, S. (1995). Temporal and spatial correlations between changes in plate motions and the evolution of rift basins in Africa. *Geol. Soc. Amer. Bull.*, 107: 1317–1332.

Jarvis, G.T. (1994). The unifying role of aspect ratio in cylindrical convection with varying degrees of curvature. *Geophys. J. Intern.*, 117: 419–426.

Jarvis, G.T. and McKenzie, D.P. (1980). Sedimentary basin formation with finite extension rates. *Earth Plan. Sci. Lett.*, 48: 42–52.

Jenkins, D.M. and Clarc, A.K. (1990). Comparison of the high-temperature and high-pressure stability limits of synthetic and natural tremolite. *Amer. Mineralogist*, 75: 358–366.

Jurdi, D.M. and Stefanic, M. (1990). Models for the hot spot distribution. *Geophys. Res. Lett.*, 17: 1965–1968.

Karlström, K.E., Miller, C.F., Kingslury, J.A. and Wooden, J.L. (1993). Pluton emplacement along an active ductile thrust zone, Pimte Mountains, southern California: interaction between deformational and solidification processes. *Geol. Soc. Amer. Bull.*, 105: 213–230.

Karson, J.K. (1988). Sea-floor spreading on the Mid-Atlantic Ridge: implications for the structure of ophiolites and oceanic lithosphere produced in slow-spreading ridges. *Tectonophysics*, 155: 49–71.

Karson, J.K. and Dick, H.K. (1983). Tectonics on ridge-transform intersections of the Kane fracture zone. *Mag. Geol. Res.*, 6: 51–98.

Kaula, W.M. (1968). *An Introduction to Planetary Physics.* John Wiley and Sons, New York, 235 pp.

Kawamoto, T., Hervig, R.L. and Holloway, J.R. (1996). Experimental evidence for a hydrous transition zone in the early Earth's mantle. *Earth Plan. Sci. Lett.*, 142: 587–592.

Kazansky, V.I. (1995). *Evolution of Ore-bearing Precambrian Structures.* Oxford & IBH Pub. Co. Pvt. Ltd., New Delhi, 307 pp.

*Kaz'min, V.G. (1987). *Rift Structures of Eastern Africa—Continental Break-up and Birth of Ocean.* Nauka, Moscow, 206 pp.

Kellog, L.H. and Turcotte, D.L. (1990). Mixing and the distribution of heterogeneities in a chaotically convecting mantle. *J. Geophys. Res.*, 95: 421–432.

*Keondzhyan, V.N. (1980). Model of chemical-density differentiation of the Earth's mantle. *Izv. AN SSSR, Fizika Zemli*, 8: 3–15.

Kerr, A.C., Saunders, A.D., Tarney, J., Berry, N.H. and Hards, V.L. (1995). Depleted mantle plume geochemical signatures: no paradox for plume theories. *Geology*, 23(9): 843–846.

*Khain, E.V. (1989). Granite-gneiss domes and ultrabasite-basite intrusions in ophiolite subduction zones. *Geotektonika*, 5: 38–51.

*Khain, V.E. (1994). Plate tectonics: analysis of present status. Vestnik MGU, Ser. 4, *Geologiya*, 1: 3–10.

314

*Khain, V.E. (1995). *Fundamental Problems in Contemporary Geology (Geology at the Threshold of the 21st Century)*. Nauka, Moscow, 190 pp.

King, S.D. and Masters, G. (1992). An inversion for radial viscosity structure using seismic tomography. *J. Phys. Lett.*, 19: 1551–1554.

*Kirdyashkin, A.G. (1966). Friction and Heat Transfer during Thermogravitational Convection and in the Field of Centrifugal Forces. Cand. Diss. Tech. Sci., Novosibirsk, 153 pp.

'Kirdyashkin, A.G. (1979). Structure of thermogravitational flows on the heat-transfer surface. In: *Models in the Mechanics of Continuous Media*. Novosibirsk, pp. 69–90.

*Kirdyashkin, A.G. (1985). Periodic thermocapillary flows. Novosibirsk, 35 pp. (*Prepr. Inst. Geol. Geophys SO AN SSSR, no. 8*).

*Kirdyashkin, A.G. (1989). *Thermogravitational Flows and Heat Transfer in the Asthenosphere*. Nauka, Novosibirsk, 81 pp.

*Kirdyashkin, A.G. and Mukhina, N.V. (1971). Free convection in inclined fluid layers and during step-wise changes of temperature of heat transfer surfaces. *Prykl. Mat. Teoret. Phys.*, 6: 115–121.

*Kirdyashkin, A.G. and Dobretsov, N.L. (1991). Modelling of two-layer mantle convection. *Dokl. AN SSSR*, 318: 946–949.

*Kirdyashkin, A.G., Leont'ev, A.I. and Mukhina, N.V. (1971). Stability of laminar fluid flow in vertical layers under conditions of natural convection. *Izv. AN SSSR, MZhG*, 5: 170–174.

*Kirdyashkin, A.G., Polezhaev, F.N. and Fedyushkin, A.I. (1983). Thermal convection in a horizontal layer with lateral heat supply. *Prykl. Mat. Teoret. Phys.*, 6: 122–128.

*Kirdyashkin, A.G., Gladkov, N.N. and Sharapov, V.N. (1987). Modelling the dynamics of magma formation above a local source of heat. *Geol. i Geofiz.*, 1: 64–67.

*Kirdyashkin, A.G., Kirdyashkin, A.A. and Dolgov, V.Yu. (1994). Experimental modelling of unsteady convective flows in the lower mantle of the Earth. *Dokl. RAN*, 338(3): 394–396.

*Kirdyashkin, A.G., Fedorov, I.I, Chepurov, A.I. and Borzdov, Yu.M. (1986). Mass transport of carbon in the process of diamond recrystallisation. *Sverkhtverdye Materialy*, 42(3): 23–29.

*Kirdyashkin, A.G. and Kirdyashkin, A.A. (1998). Onset of turbulent free convection in horizontal fluid layer and regime of the lower mantle convection. *Dokl. Rus. Acad. Nauk, Moscow*.

Klein, E.M. and Langmuir, C.H. (1987). Global correlations of ocean ridge basalt chemistry with axial depth and crustal thickness. *J. Geophys. Res.*, 92: 8089–8115.

Klitgord, K.D. and Behrent, J.C. (1979). Basin structure of the US Atlantic margin. *AAPG Memoir*, 29: 85–112.

Klitgord, K.D. and Hutchinson, D.R. (1988). US Atlantic continental margin: structural and tectonic framework. In: *The Geology of North America*, vols. 1–2. *The Atlantic Continental Margin*. Publ. Geol. Soc. Amer., pp. 19–55.

*Knipper, A.L., Dobretsov, N.L. and Bogdanov, N.A. (1992). Ophiolites and pyrope peridotites of Betic Cordilleras. *Izv. AN SSSR, Ser. Geol.*, 12: 8–24.

Komatsu, M., Shibakusa, H., Miyashita, S. et al. (1992). Subduction and collisional related high and low P–T metamorphic belt in Hokkaido. *29th IGC Field Trip CO 1, Kyoto, Japan*, 62 pp.

*Kononov, M.V. (1989). *Plate Tectonics of North-western Pacific Ocean*. Nauka, Moscow, 169 pp.

*Korikovskii, S.P. (1995). Contrast types of metamorphism in orogenic belts. *Petrologiya*, 3(1): 26–42.

Krishnamurti, R. (1970). On the transition to turbulent convection. Pts. 1–2. *J. Fluid Mech.*, vol. 42, pt. 2, pp. 295–320.

*Krivtsov, A.I., Migachev, I.F. and Popov, V.S. (1986). *Copper Porphyry Deposits of the World*. Nedra, Moscow, 238 pp.

Krogh, E.J., Oh, C.W. and Liou, J.G. (1994). Polyphase and anticlockwise P–T evolution for Franciscan eclogites and blueschists from Jenner, California, USA. *J. Metamorph. Geol.*, 12: 121–134.

*Kropotkin, P.N., Efremov, V.N. and Makeev, V.M. (1987). Stress state of the Earth's crust and geodynamics. *Geotektonika*, 1: 3–24.

Kruger, J.M. and Johnson, R.A. (1994). Raft of crustal extension evidence from seismic reflection data in south-eastern Arizona. *Geology*, 22: 351–354.

*Krylov, S.V. (1984). Combining the methods of seismic explosions and earthquakes for studying the deep-level structure of Baikal rift. In: *Regional Complex Investigations of the Earth's Crust and Upper Mantle*. Radio i Svyaz', Moscow, pp. 80–87.

Kumazawa, M. and Maruyama, Sh. (1994). Whole Earth tectonics. *J. Geol. Soc. Japan*, 100(1): 81–102.

Kumazawa, M., Yoshida, S., Ito, T. and Yoshioka, H. (1994). Archaean-Proterozoic boundary interpreted as a catastrophic collapse of the stable density stratification in the core. *J. Geol. Soc. Japan*, 100: 50–59.

*Kuskov, O.L. and Khitarov, N.I. (1992). *Thermodynamics and Geochemistry of the Earth's Core and Mantle*. Nauka, Moscow, 279 pp.

*Kuskov, O.L. and Parfenov, A.B. (1992). Thermodynamic model of the pyrolitic mantle of the Earth based on phase diagram $FeO-MgO-SiO_2$. *Geokhimiya*, 12: 1779–1789.

*Kuskov, O.L., Sidorov, Yu.I. and Shapkin, A.I. (1995). Condensation model of the Earth: solar chondrite. In: *Basic Directions in Geochemistry* (E.M. Galimov, ed.). Nauka, Moscow, pp. 58–68.

*Kutateladze, S.S. and Borishanskii, V.M. (1959). *Handbook of Heat Transfer*. Gosenergoizdat, Moscow: Leningrad, 414 pp.

*Kutateladze, S.S., Kirdyashkin, A.G. and Berdnikov, V.S. (1974). Velocity field in a convective cell of horizontal fluid layer under conditions of thermogravitational convection. *Izv. AN SSSR, Fizika Atmosfery i Okeana*, 10(2): 137–145.

*Kuz'min, M.I. (1985). *Geochemistry of Magmatic Rocks of the Phanerozoic Mobile Belts*. Nauka, Novosibirsk, 200 pp.

*Kuznetsov, V.V. (1990). *Physics of the Earth and Solar System: Models of Formation and Evolution*. Revised and Supplemented 2nd ed. Inst Geol. Geophys. SO AN SSSR, Novosibirsk, 215 pp.

Larson, R.L. (1991). Geological consequences of superplumes. *Geology*, 19: 963–966.

Larson, R.L. and Olson, P. (1991). Mantle plumes control magnetic reversal frequency. *Earth Plan. Sci. Lett.*, 107: 437–447.

Larson R.L. and Kinkaid, C. (1996). Onset of Mid-Cretaceous volcanism by elevation of the 670 km thermal boundary layer. *Geology,* 24: 551–554.

Lawer, L.A. and Muller, R.D. (1994). Iceland hot spot track. *Geology,* 22: 311–314.

Le Pichon, X. (1968). Sea-floor spreading and continental drift. *J. Geophys. Res.,* 73: 3661–3697.

Le Pichon, X. and Barbier, F. (1987). Passive margin formation by low-angle faulting within the upper crust: the northern Bay of Biscay margin. *Tectonics,* 6: 133–150.

*Lebedev, V.I., Khalilov, V.A., Kargopolov, S.A. et al. (1991). U-Pb age of high-temperature metamorphism and ultrametamorphism of Sangilen (south-eastern Tuva). *Dokl. AN SSSR,* 320(3): 682–686.

*Leontjev, A.N. and Kirdyashkin, A.G. (1995). Heat transfer during free convection in horizontal crevices and large horizontal surface. *Inzh. Fiz. Zhurn.,* 9(1): 9–14.

Leontjev A.I. and Kirdyashkin, A.G. (1968). Experimental study of flow patterns and temperature fields in a horizontal free convection liquid layer. *Intern. J. Heat Mass Transfer,* 11: 1461–1466.

*Lepezin, G.G. (1978). *Metamorphic Complexes of Altai-Sayan Region.* Nauka, Novosibirsk, 229 pp.

*Letnikov, F.A. (1983). Formation of diamonds in deep tectonic zones. *Dokl. AN SSSR,* 71(2): 433–435.

Liou, J.G., Maruyama, S. and Cho, M. (1985). Phase equilibria and mineral parageneses of metabasites in low-grade metamorphism. *Mineral. Mag.,* 49: 321–333.

Liou, J.G., Zhang, B. and Ernst, W.G. (1994). Ultrahigh-pressure metamorphism. *Island Arc,* 3(1): 2–25.

Liou, J.G., Wang, Q., Zhai, M., Zhang, R.Y. and Cong, B. (1995). Ultrahigh-pressure metamorphic rocks and their associated lithologies from the Dabie mountains, central China: a field trip guide. *Chinese Sci. Bull.,* vol. 40, suppl., pp. 1–71.

*Lisitsyn, A.P. (1978). *Processes of Oceanic Sedimentation.* Nauka, Moscow, 392 pp.

*Lisitsyn, A.P. (1991). *Process of Terrigenic Sedimentation in Seas and Oceans.* Nauka, Moscow, 271 pp.

*Lisitsyn, A.P., Bogdanov, Yu.A. et al. (1993). *Hydrothermal Systems and Sedimentary Formations of Mid-oceanic Ridges.* Nauka, Moscow, 256 pp.

Lister, G.S. and Baldwin, S.L. (1993). Plutonism and origin of metamorphic core complexes. *Geology,* 21: 607.

Lithgow-Bertelloni, C. and Richards, M.A. (1995). Cenozoic plate driving forces. *Geophys. Res. Lett.,* 22: 1317–1320.

Lithgow-Bertelloni, C., Silver, P.G. and Russo, R. (1996). Do deep mantle upwellings drive plate motions? *EOS, AGU Spring Meeting, Abstracts,* p. 274.

*Litvinovskii, B.A. and Dobretsov, N.L. (1993). Evolution of geodynamic regimes of magma formation as a major factor in the evolution of granitoid magmas. In: *Granitoids of Folded Regions.* VSEGEI, St. Petersburg, pp. 49–61.

*Litvinovskii, B.A., Zanvilevich, A.N. and Wikkhem, S.M. (1994). Angara-Vitim batholith, Transbaikal: structure, petrology and model of formation. *Geol. i Geofiz.,* 35(7–8): 217–234.

*Lobkovskii, L.I. (1988). *Geodynamics of Zones of Spreading, Subduction and Two-level Plate Tectonics.* Nauka, Moscow, 252 pp.

Logachev, N.A. (1993). History and geodynamics of the Baikal rift (east Siberia): a review. *Bull. Centre Rech. Explor. Prod. Elf Aquitaine,* 17: 353–370.

Loper, D.E. and Lay, T. (1995). The core-mantle boundary region. *J. Geophys. Res.,* 100: 6397–6420.

*Lyubimova, E.A. (1968). *Thermal conditions on the Earth and the Moon.* Nauka, Moscow, 279 pp.

*Lyubimova, E.M., Lyubshits, V.M. and Nikitina, V.N. (1988). Interpretations of the anomalies of heat currents in the axial zones of mid-oceanic ridges. In: *Geothermal Investigations on the Floor of Water Bodies.* Nauka, Moscow, pp. 11–21.

Magee, M.E. and Zobak, M.D. (1993). Evidence for a weak interplate thrust fault along the Northern Japan subduction zone and implication for the mechanism of thrust faulting and fluid expulsion. *Geology,* 21(9): 809–812.

Mahon, K.J. and Harrison, T.M. (1988). Ascent of a granitoid in a temperature varying medium. *J. Geophys. Res.,* 93(B2): 1174–1188.

Manduca, C.A., Kuntz, M.A. and Silver, L.T. (1993). Emplacement and deformation history of the western margin of the Idaho batholith near McCall, Idaho: influence of major terrane boundary. *Geol. Soc. Amer. Bull.,* 105: 749–765.

Manghnani, M.N., Andro, R.R. and Clark, S.R. (1974). Compressive and shear velocities in granulite facies rocks and eclogites to 10 kb *J. Geophys. Res.* 79: 5427–5466.

*Marakushev, A.A. (1973). *Petrology of Metamorphic Rocks.* Izd. Moscow univers, Moscow, 357 pp.

*Marakushev, A.A. (1992). *Origin and Evolution of the Earth and Other Planets of the Solar system.* Nauka, Moscow, 207 pp.

Marsh, B.D. (1982). On the mechanics of igneous diapirism, stoping and zone melting. *Amer. J. Sci.,* 282: 808–855.

Marsh, B.D. and Carmichael, I.S. (1974). Benioff zone magmatism. *J. Geophys. Res.,* 79(8): 1196–1206.

Maruyama, Sh. (1994). Plume tectonics. *J. Geol. Soc. Japan,* 100(1): 24–34.

Maruyama, Sh. (1997) Pacific-type orogeny revised: Miyashiro-type orogeny proposed. *Island Arc,* 6: 91–120.

Mazaud, A. and Laj, C. (1991). The 15 million year geomagnetic reversal periodicity: a quantitative test. *Earth Plan. Sci. Lett.,* 107: 689–697.

McCaffrey, (1996). Estimates of modern arc-parallel strain rates in forearcs. *Geology,* 24: 27–30.

McCulloch, M.T. (1993). The role of subducted slabs in an evolving Earth. *Earth Plan. Sci. Lett.,* 115: 89–100.

McElhinny, M.W. and Senanatake, W.E. (1980). Palaeomagnetic evidence for the existence of the geomagnetic field 3.5 GA ago. *J. Geophys. Res.,* 85: 3523–3528.

McKenzie, D.P. (1978). Some remarks on the development of sedimentary basins. *Earth Plan. Sci. Lett.,* 40: 25–32.

McKenzie, D.P. and Morgan, W.J. (1969). Evolution of triple junction. *Nature,* 224: 125–133.

McKenzie, D.P., Roberts, Y.M. and Weiss, N.O. (1974). Convection in Earth's mantle: towards a numerical simulation. *J. Fluid Mech.*, vol. 62, pt. 3, pp. 465–538.

McNutt, M.K. and Judge, A.V. (1990). The superswell and mantle dynamics beneath the south Pacific superswell. *Science*, 250: 969–975.

McNutt, M.K., Diament, M. and Kagan, M.G. (1988). Variations of elastic plate thickness at continent thrust belts. *J. Geophys. Res.*, 93: 8825–8838.

Meyer A.W., Davies, T.A. and Wise, S.W. (eds.) (1992). *Evolution of Mesozoic and Cenozoic Continental Margins*. Elsevier, Amsterdam-N.Y., 361 pp.

*Mikheev, M.A. (1947). *Principles of Heat Transfer*. Gosenergoizdat, Moscow: Leningrad, 415 pp.

Minster, J.B. and Jordan, T.H. (1978). Present-day plate motions. *J. Geophys. Res.*, B83: 5331–5334.

Miyashiro, A. (1961). Evolution of metamorphic belts. *J. Petrol.*, 2: 277–311.

Miyashiro, A. (1994). *Metamorphic Petrology*. UCL Press Ltd., London, 404 pp.

Molnar, P. and Atwater, T. (1973). The relative motion of hot spots in the mantle. *Nature*, 246: 288–291.

Molnar, P. and England, P.C. (1990). Temperatures, heat flux and frictional stress near major thrust faults. *J. Geophys. Res.*, 95: 4833–4856.

Molnar, P., Freedmann, D. and Shih, J.S.F. (1979). Length of intermediate and deep seismic zones and temperature in downgoing slabs of lithosphere. *J. Royal Astron. Soc.*, 56: 41–54.

*Monin, A.S. (1977). *History of the Earth*. Nauka, Leningrad, 228 pp.

*Monin, A.S. and Sorokhtin, O.G. (1982). Evolution of the Earth during volume differentiation of its interior. *Dokl. AN SSSR*, 263(3): 572–575.

Montagner, J.P. (1994). Can seismology tell us anything about convection in the mantle? *Rev. Geophys.*, 32: 115–137.

Montagner, J.P. and Anderson, D.L. (1989). Constrained reference mantle model. *Phys. Earth Planet Interior*, 58: 205–227.

Morel, P. and Irving, E. (1981). Palaeomagnetism and the evolution of Pangeas. *J. Geophys. Res.*, 86: 1858–1872.

Moresi, L.N. and Solomatov, V.S. (1995). Numerical investigation of 2D convection with extreme viscosity variations. *Phys. Fluids*, 7: 2154–2162.

Morgan, W.J. (1968). Rises, trenches, great faults and crustal blocks. *J. Geophys. Res.*, 73: 1959–1982.

Morgan, W.J. (1971). Convective plumes in the lower mantle. *Nature*, 230: 42–43.

*Muratov, M.V., Belousov, V.V., Reisner, G.I. et al. (1978). *Earth's Tectonosphere*. Nauka, Moscow, 531 pp.

Nakamura, Y. (1983). Seismic velocity structure of the Lunar mantle. *J. Geophys. Res.*, 88: 677–686.

Nakayama, M. (1994). Spatial and temporal variations in chemical composition of Cenozoic volcanics from Hokkaido, Japan. *Abstr., 101st Meeting of Geol. Soc. Japan, Sapporo*, p. 10.

Nataf, H.C., Froidevaux, C., Levart, J.L. and Rabinowicz, M. (1981). Laboratory convection experiments: effect of lateral cooling and generation of instabilities in the horizontal boundary layers. *J. Geophys. Res.*, 86: 6143–6154.

*Natal'in, B.A. (1991). Mesozoic accretion and collision tectonics in the southern part of the Far East. *Tikhookean. Geol.*, 5: 3–23.

Newsom, H.E. and Sims, K.W.W. (1991). Core formation during early accretion of the Earth. *Science*, 252: 926–933.

Niu, Y.L. and Hekinian, R. (1997). Spreading-rate dependence of the extent of mantle melting beneath ocean ridges. *Nature*, 385: 326–329.

Nozette, S., Pleasance, L.P., Horon, D.M. et al. (1994). The Clementine mission to the Moon, scientific overview. *Science*, 266: 1835–1839.

Oceanology. Geophysics of the Ocean, Vol. 2: *Geodynamics* (1979). Nauka, Moscow, 416 pp.

Oho, S. (1992). Origin of the Earth's core from abundances of siderophile elements: pressure effects on partition coefficient. *Proc. 12th ISAS Lunar and Planet Symp., Tokyo*, p. 75.

Okada, M. (1984). Analysis of transfer during melting from a vertical wall. *Intern. J. Heat Mass Transfer*, 27(11): 2057–2066.

Olson, P. and Singer, H. (1985). Creeping plumes. *J. Fluid Mech.*, 158: 511–531.

O'Nions, R.K. and Tolstikhin, I.N. (1996). Limits on the mass flux between lower and upper mantle and stability of layering. *Earth Plan. Sci. Lett.*, 139: 213–222.

O'Nions, R.K., Hamilton, P.J. and Evensen, N.M. (1979). The chemical evolution of the Earth's mantle. *Sci. Amer.* 242: 91–101.

Osozawa, S. (1994). Plate reconstruction based upon age data or Japanese accretionary complexes. *Geology*, 22: 1135–1138.

Paolucci, S. (1990). Direct numerical simulation of two-dimensional turbulent natural convection in an enclosed cavity. *J. Fluid Mech.*, 215: 229–262.

Papanastassiou, D.A. and Wasserburg, G.J. (1976). Rb-Sr ages of troctolite 76 535. *Proc. 7th Lunar and Planet. Sci. Conf., Houston*, pp. 2035–2054.

*Parfenov, L.M. (1984). *Continental Margins and Island Arcs of Mesozoide of Northeastern Asia*. Nauka, Novosibirsk, 189 pp.

Parsons, B. and Sclater, J.G. (1972). An analysis of the variation of the ocean-floor bathymetry and heat flow with age. *J. Geophys. Res.*, 82: 803–827.

Parsons, T. and McCarthy, J. (1995). The active south-west margin of the Colorado plateau: uplift of mantle origin. *Geol. Soc. Amer. Bull.*, 107: 139–147.

Peacock, S.M. (1993). The importance of blueschist-eclogite dehydration reactions in subducting oceanic crust. *Geol. Soc. Amer. Bull.*, 105(5): 684–694.

Pearson, D.G., Shirbey, S.B., Carlson, R.W., Boyd, F.R., Pokhilenko, N.P. and Shimizu, N. (1995). Re-Os, Sm-Nd and Rb-Sr isotope evidence for thick Archaean lithospheric mantle beneath the Siberian craton by multistage metasomatism. *Geochem. Cosmochem. Acta*, pp. 953–957.

Pekeris, C.L. (1936). Thermal convection in the Earth. *Monat. Nat. Royal Astron. Soc., Geophys.* (suppl.) vol. 3, pp. 331–368.

Peltier, W.R. and Jarvis, G.T. (1982). Whole mantle convection and thermal evolution of the Earth. *Phys. Earth Planet Interior*, 29: 281–304.

*Perchuk, L.L. and Aranovich, L.Ya. (1979). Regime of 'subsidence' metamorphism. *Izv. An SSSR, Ser. Geol.*, 11: 57–70.

Perfit, M.R., Fornari, D.J., Smith, M.L. et al. (1994). Small-scale spatial and temporal variations in mid-ocean ridge crest magmatic processes. *Geology*, 22: 375–379.

320

Picque, A. and Laville, E. (1993). The Atlantic ocean opening: extensional reactivation of Palaeozoic structures? *C.R. Acad. Sci., Paris*, vol. 317, ser, II, pp. 1325–1332.

Pitcher, W.S. (1979). The nature ascent and emplacement of granitic magmas. *J. Geol. Soc.*, 136: 627–662.

Platt, J. (1986). Dynamics of orogenic wedges: the uplift of high pressure metamorphic rocks. *Geol. Soc. Amer. Bull.*, 57: 1037–1053.

*Pokhilenko, N.P., Sobolev, N.V., Boid, F.R. et al. (1993). Megacrystalline pyrope peridotites in the lithosphere of the Siberian platform: mineralogy, geochemical features and problem of origin. *Geol. i Geofiz.*, 1: 71–74.

*Puzankov, Yu.M., Duchkov, A.D., Mel'gunov, S.V. and Nozhkin, A.D. (1989). *Radioactive Elements and Generation of Radiogenic Heat in the Structural Complexes of the Altai-Sayan Region.* Inst. Geol Geoph., Novosibirsk, 158 pp.

Raleigh, J.W. (1916). On convective currents in a horizontal layer of fluid when the higher temperature is on the underside. *Phil. Mag.*, 32: 529–546.

Rampino, M.R. and Caldeira, K. (1993). Major episodes of geological change: correlations, time, structure and possible causes. *Earth Plan. Sci. Lett.*, 114: 215–227.

Rasskasov, S.V. (1994). Magmatism related to the Eastern Siberia Rift system and geodynamics. *Bull. Centres Rech. Explor. Proc. Elf Aquitaine*, 18: 437–452.

Rasskasov, S.V., Ivanov, A.V., Boven, A., Andre, L. and Brandt, S.V. (1996). Evolution of Late Cenozoic Udokan magmatism in the structures of the Baikal and Olekma-Stanovaya mobile systems. *Abstr. Workshop 'Continental Rift Tectonics and Evolution of Sedimentary Basins.'* Novosibirsk, pp. 36–38.

Renne, P.R., Glen, J.M., Milner, S.C. and Duncan, A.R. (1996). Age of Etendeca flood volcanism and associated instrusions in south-western Africa. *Geology*, 24: 659–662.

*Reverdatto, V., Sheplev, V.S. and Polyanskii, O.P. (1995). Metamorphism of subsidence and evolution of rift basins: ideal approximation. *Petrologiya*, 3(1): 35–48.

Ribe, N.M., Christensen, U.R. and Theiping, J. (1995). The dynamics of plume-ridge interaction. I: Ridge-centred plumes. *Earth Plan. Sci. Lett.*, 139: 155–168.

Ricard, Y. and Wuming, B. (1991). Inferring viscosity and the 3-D density structure of the mantle from geoid, topography and plate velocities. *Geophys. J. Intern.*, 105: 561–572.

Richards, M.A. and Engebretson, D.C. (1992). Large-scale mantle convection and the history of subduction. *Nature*, 355: 437–440.

Richards, M.A., Hager, B.H. and Sleep, N.H. (1988). Dynamically supported geoid highs over hot spots: observation and theory. *J. Geophys. Res.*, 93: 7690–7708.

Richardson, S.H., Harris, J.W. and Gurney, J.J. (1993). Three generations of diamonds from old continental mantle. *Nature*, 366: 256–258.

Richter, F.M. and Parsons, B. (1975). On the interaction of two scales of convection in the mantle. *J. Geophys. Res.*, 80(17): 2529–2541.

Richter, F.M. and McKenzie, D.P. (1981). On some consequences and possibilities of layered mantle convection. *J. Geophys. Res.*, 86: 6133–6142.

Rieget, H., Projahn, U. and Beer, H. (1982). Analysis of the heat transport mechanisms during melting around a horizontal circular cylinder. *Intern. J. Heat Mass Transfer*, 25(1): 137–147.

*Riffo, K. and Le Pichon, K. (1979). *FAMOUS Expedition.* Gidrometeoizdat, Leningrad, 222 pp.

Ring, U. and Brandon, M.T. (1994). Kinematic data for the Coast range fault and implications for exhumation of the Franciscan subduction complex. *Geology,* 22: 735–738.

Ringwood, A.A., Kerson, S.E., Hibberson, W. and Ware, N. (1992). Origin of kimberlites and related magmas. *Earth Plan. Sci. Lett.,* 113: 521–538.

Ringwood, A.E. (1966). The chemical composition and origin of the Earth. In: *Advances in Earth Science.* MIT Press, pp. 287–356.

Ringwood, A.E. (1974). *Origin of the Earth and the Moon.* Springer-Verlag, New York, 293 pp.

Ringwood, A.E. (1977). Composition of the core and implications for the origin of the Earth. *Geochem. J.,* 11: 111–135.

Ringwood, A.E. and Irifune, T. (1988). Nature of the 650-km seismic discontinuity: implications for mantle dynamics and differentiation. *Nature,* 331: 131–136.

Riter, J.C.A. and Smith, D. (1996). Xenolith constraints on the thermal history of the mantle below the Colorado plateau. *Geology,* 24: 267–270.

*Ronov, A.B. and Yaroshevskii, F.F. (1978). Chemical composition of the Earth's crust and its shells, In: *Tectonospheres of the Earth* (V.V. Belousov, ed.). Nauka, Moscow, pp. 379–402.

Ross, W.C., Halliwell, B.A., May, J.A., et al. (1994). Slope readjustment: a new model for the development of submarine fans and aprons. *Geology,* 22: 511–514.

Roy, R.F., Blackwell, D.D. and Birch, F. (1968). Heat generation of plutonic rocks and continental heat flow provinces. *Earth Plan. Sci. Lett.,* 5(1): 1–12.

Rubie, D.C. (1986). The catalysis of mineral reactions by water and restrictions on the presence of aqueous fluid during metamorphism. *Mineral. Mag.,* 50(3): 399–415.

Runcorn, S.K. (1992). Polar path in geomagnetic reversals. *Nature,* 356: 654–656.

Russo, R.M. and Silver, P.G. (1996). Cordillera formation, mantle dynamics and the Wilson cycle. *Geology,* 24: 511–514.

*Rykov, V.V. and Trubitsyn, V.P. (1995). Three-dimensional model of mantle convection with moving continents. *Vychislitel'naya Seismologiya,* vol. 27. *Teoreticheskie Problemy Geodynamiki i Seismologii,* pp. 21–41.

*Safronov, V.S. (1995). Problems of planet formation. In: *Origin of the Solar System.* Nauka, Moscow, pp. 5–23.

Salters, V.J.M. (1996). The generation of mid-oceanic ridge basalts from the Hf and Nd isotopic perspective. *Earth Plan. Sci. Lett.,* 141: 109–123.

Sandwell, D.T. and Shubert, G. (1992). Evidence of retrograde lithospheric subduction on Venus. *Science,* 257: 766–770.

Schilling, J.G. (1973). Iceland mantle plume: geochemical study of Reykjanes ridge. *Nature,* 242: 565–571.

Schmalze, J. and Hansen, U. (1993). Mixing properties of mantle convection. *Europ. Geol. Congress VII, Terra Abstr.,* Suppl. no. 1 to Terra Nova, no. 5, p. 56.

Schwarzacher, W. (1992). *Cyclostratigraphy and the Milankovitch Theory.* Elsevier, Amsterdam, 240 pp.

Sclater, J.G. and Francheteau, J. (1970). The implication of terrestrial heat flow observations on current tectonic and geochemical models of the crust and upper mantle of the Earth. *Geophys. J. Astrom. Soc.*, 20: 509–542.

Sclater, J.G. and Christie, P.A. (1980). Continental stretching: an explanation of the post-mid-Cretaceous subsidence of the central North Sea basin. *J. Geophys. Res.*, 85: 3711–3739.

Sclater, J.G., Helling, S. and Tapscott, C. (1977). The palaeobathymetry of the Atlantic Ocean from the Jurassic to the present. *J. Geol.*, 85: 509–552.

Scotese, C.R., Gahagan, L.M. and Larson, R.L. (1988). Plate tectonic reconstructions of the Cretaceous and Cenozoic ocean basins. *Tectonophysics*, 155: 27–48.

Sengor, A.M.Č. (1991). Timing of orogenic events: a persistent geological controversy. In: *Tectonics and Mountain Building*. Academic Press, pp. 405–473.

Sengör, A.M.Č., Natal'in B.A., and Burtman, V.S. (1993). Evolution of the Altaid tectonic collage and Paleozoic crustal growth in Eurasia. *Nature*, 364: 299–307.

*Sengör, A.M.Č., Natal'in, B.A. and Burtman, V.S. (1994). Tectonic evolution of the Altaide. *Geof. i Geol.*, 35(7–8): 41–58.

*Shadlun, T.N., Bortnikov, N.S., Bogdanov, Yu.A. et al. (1992). Mineral composition, texture and formation conditions of present-day sulphide ores in the rift zone of Magnus basin. *Geol. Rud. Mestorozhd.*, 34(5): 3–22.

*Sharapov, V.N. (1992). *Development of Endogenic Fluid Ore Formation Systems*. Nauka, Novosibirsk, 138 pp.

*Sharapov, V.N. (1994). Phenomenological description of magmatic facies and structural-dynamic systems. *Geol. i Geofiz.*, 35(9): 3–20.

*Sharapov, V.N., Kalinin, A.S. and Vasil'ev, E.N. (1977). Model of the fusion of the rocks of the Earth's crust in the presence of repeated convective thermal flows. *Geol. i Geofiz.*, 12: 3–10.

Sharaskin, A.Ya., Dobretsov, N.L. and Sobolev, N.N. (1980). Marianites—the clinoenstatite-bearing pillow lavas, associated with the ophiolite assemblages of Mariana trench. *Ophiolites*, Nicosia, pp. 473–479.

Shearer, P.M. (1991). Constraints on upper mantle discontinuities from observation on long-period reflected and convected phase. *J. Geophys. Res.*, 96: 18147–18182.

Shearer, P.M. and Masters, Y.G. (1992). Global mapping of topography on the 660-km discontinuity. *Nature*, 355: 791–796.

Shemenda, A.I. (1994). *Subduction Insights from Physical Modelling*. Kluwer Acad. Publ., 210 pp.

*Sheplev, V.S. and Reverdatto, V.V. (1994). A study of the model of rifting. *Dokl. RAN*, 334(1): 343–348.

Shimizu, N. and Sobolev, N.V. (1995). Young peridotitic diamonds from the Mir kimberlite pipe. *Nature*, 375: 394–397.

Shipley, T.H., Moore G.F., Bangs, N.L. et al. (1994). Seismically inferred dilatancy distribution, northern Barbados ridge decollement: implication for fluid migration and fault strength. *Geology*, 22: 411–414.

*Shkodzinskii, V.S. (1976). *Problems of Physicochemical Petrology and Genesis of Migmatites*. Nauka, Novosibirsk, 226 pp.

*Shlikhting, G. (1969). *Theory of Boundary Layer*. Nauka, Moscow, 742 pp.

Shreve, R.L. and Cloos, M. (1986). Dynamics of sediment subduction, melange formation and prism accretion. *J. Geophys. Res.*, 91: 10229–10245.

*Shvets, M.E. (1954). Solving parabolic equations. *Prikl. Mat. i Mekh.*, 18(2): 243–244.

*Silant'ev, S.A. (1995). Metamorphism of contemporary oceanic basins. *Petrology*, 3(1): 20–31.

Silver, P.G. and Russo, R.M. (1996). Flow-coupled plate interaction or how the Alps helped to make the Andes. *EOS AGU Spring Meeting, Abstr.*, p. 274.

Simon, G.W. and Weiss, N.O. (1991). Convective structures in the Sun. *Monat. Nat. Royal Astronom. Soc.*, 252: 1–5.

*Simonov, V.A., Dobretsov, N.L. and Buslov, M.M. (1994). Boninite series in the structures of the Palaeo-Asiatic Ocean. *Geol. i Geofiz.*, 7–8: 182–200.

Sinton, J.M. and Detrick, R.S. (1992). Mid-ocean ridge magma chambers. *J. Geophys. Res.*, 97: 197–216.

*Skinner, E.M., Clement, K.R., Gerin, D.D. et al. (1992). Spatial distribution and structural position of the kimberlites of southern Africa. *Geol. i Geofiz.*, 10: 33–40.

*Slezkin, N.A. (1955). *Dynamics of a Viscous Incompressible Fluid.* Gostekhteorizdat, Moscow, 520 pp.

Small, Ch. and Sandwell, D.T. (1994). Imaging mid-oceanic ridge transition with satellite gravity. *Geology*, vol. 22, no. 2, pp. 123–126.

Smith, J.R. and Kawamoto, T. (1997). Weldslayite II: A new high-pressure hydrous phase in the peridotite-H2O system. *Earth. Plan. Sci. Lett.*, 146: 9–16.

*Sobolev, A.V. (1996). Melt inclusions in minerals as a source of vital petrological information. *Petrologiya*, 4(3): 228–239.

Sobolev, A.V. and Chaussidon, M. (1996). H2O concentration in primary melt from suprasubduction zones and mid-oceanic ridges. *Earth Plan. Sci. Lett.*, 137: 45–55.

*Sobolev, N.V. (1974). *Deep Inclusions in Kimberlites and the Problem of the Composition of the Upper Mantle.* Nauka, Novosibirsk, 263 pp.

*Sobolev, N.V., Chepurov, A. I. and Malinovskii, I. Yu. (1983). Some aspects of the experimental study of diamond crystallisation. In: *Experimental Study of Diamond Crystallisation in Metallic Systems.* Inst. Geol. Geophys. SO AN SSSR, pp. 3–22.

*Sobolev, N.V., Pokhilenko, N.P. and Efimova, E.S. (1984). Xenoliths of diamond-bearing pyropic peridotites in kimberlites and the problem of origin of diamonds. *Geol. i Geofiz.*, 12: 63–80.

*Sobolev, S.V. (1979). Gravitational effect and the evolution of channels of low viscosity in the mantle. *Dokl. AN SSSR*, 246: 1070–1074.

*Sobolev, V.S. and Sobolev, A.V. (1977). Composition of deep-level pyroxenes and problems of 'eclogite barrier'. *Geol. i Geofiz.*, 12: 46–59.

*Sobotovich, E.V. (1984). Cosmochemical model of the origin of the Earth. *Geol. Zb.*, 2: 112–123.

*Somin, M.L. and Mil'yan, G. (1981). *Geology of Cuban Metamorphic Complexes.* Nauka, Moscow, 174 pp.

*Sorokhtin, O.G. (1974). *Global Evolution of the Earth.* Nauka, Moscow, 182 pp.

*Sorokhtin, O.G. and Ushakov, G.D. (1991). *Global Evolution of the Earth.* Izd. Moscow univ. Moscow, 466 pp.

Spakman, W. (1990). Topographic images of the upper mantle below central Europe and the Mediterranean. *Terra Nova*, 2(6): 542–554.

Spakman, W., Van der Lee, S. and Van der Hilst, R. (1993). Travel-time tomography of the European-Mediterranean mantle down to 1400 km. *Phys. Earth Planet Interior*, 79: 3–74.

Staltenpohl, M.G., Cymerman, Z., Krogh, E.J. and Kunk, M.J. (1993). Exhumation of eclogitised continental basement during Variscan lithospheric delamination and gravitational collapse, Sudety mountains, Poland. *Geology*, 21: 1111–1114.

Stein, C.A. and Stein, S. (1992). A model for the global variation in oceanic depth and heat flow with lithospheric age. *Nature*, 359: 123–129.

Storey, B.C. (1995). The role of mantle plumes in continental break-up: case histories from Gondwanaland. *Nature*, 3: 62–59.

Stothers, R.B. (1993). Hot spots and Sun spots: surface of deep mantle convection in the Earth and Sun. *Earth Plan. Sci. Lett.*, 116: 1–8.

Strangway, D.W. (1980). Magnetic fields in the early Solar system. *Precambr. Res.*, 10: 167–175.

Strehlau, J. and Meissner, R. (1987). Estimation of crustal viscosities and shear stress from an exploration of experimental steady-state flow data. In: *Composition, Structure and Dynamics of the Lithosphere-Asthenosphere System*. AGU, Washington, pp. 69–87.

Strong, H.M. and Hanneman, R.E. (1967). Crystallisation of diamond and graphite. *J. Chem. Phys.*, 46(9): 3668–3676.

Su, W.J., Woodward, R.L. and Dziewonski, A.M. (1994). Degree 12 model of shear velocity heterogeneity in the mantle. *J. Geophys. Res.*, 99: 6945–6980.

Suppe, J. (1979). Structural interpretation of the southern part of the northern Coast ranges and Sacramento valley, California: summary. *Geol. Soc. Amer. Bull.*, 90: 327–330.

*Surkov, V.S., Trofimuk, A.A. and Zhero, O.G. (1982). Triassic rift system of western Siberian platform and its effect on the structure and oil content of Meso-Cenozoic roof. *Geol. i Geofiz.*, 8: 3–15.

Swanberg, C.A. (1972). Vertical distribution of heat generation in the Idaho batholith. *J. Geophys. Res.*, 77(14): 2508–2513.

Tackley, P., Stevenson, D.J., Glatzmaier, G.A. and Schubert, G. (1993). Effects of an endothermic phase transition at 670 km depth in a spherical model of convection in the Earth's mantle. *Nature*, 361: 699–704.

Takahashi, E. (1986). Melting of a dry peridotite KLB-1 up to 14 GPa: implication for the origin of peridotitic upper mantle. *J. Geophys. Res.*, 91: 9367–9382.

Talbot, G.J., Ronnlund, P., Schmeling, H., Kayi, H. and Jackson, M.P.A. (1991). Diapiric spoke patterns. *Tectonophysics*, 188: 187–201.

Tatsumi, Y. (1989). Migration of fluid and genesis of basalt magmas in subduction zones. *J. Geophys. Res.*, 94: 4697–4707.

*Tauson, L.V. and Kuz'min, M.I. (1976). Geochemical characteristics and metallogeny of granitoids formed under different geodynamic conditions. In: *Geodynamics and Economic Minerals*. GKNT, Moscow, pp. 21–23.

Taylor, C.R. (1986). The origin of the Moon: geochemical considerations. *Proc. of Lunar and Planet. Inst.* (W.K. Hartman et al., eds.), pp. 453–487.

Triggvason, K., Husebye, E.S. and Stefansson, R. (1983). Seismic image of the hypothesised Iceland hot spot. *Tectonophysics*, 100: 97–118.

Tromp, J. (1995). Normal-mode-splitting observations from the Great 1994 Bolivia and Kuril Island earthquakes: constraints on the structure of the mantle and inner core. *GSA Today*, 5: 137–151.

*Trubitsyn, V.P. and Kharybin, E.V. (1988). Geodynamic model of differentiation of mantle material deep within the Earth. *Izv. AN SSSR, Fizika Zemli*, 4: 83–89.

*Trubitsyn, V.P. and Nikolaichik, V.V. (1991). Thermal convection regimes. *Izv. RAN, Fizika Zemli*, 6: 3–12.

Turcotte, D.I. and Schubert, G. (1982). *Geodynamics.* John Wiley and Sons, New York, vol. 1–2, 436 pp.

Turcotte, D.L. and Kay, S.M. (1992). On the coupling between plate tectonics and mantle convection. *EOS, AGU Spring Meeting, Abstracts*, S272.

Turner, G. (1989). The outgassing history of the Earth's atmosphere. *J. Geol. Soc. London*, 146(1): 147–154.

Ueda, S. (1982). Subduction zones: an introduction to comparative subductology. *Tectonophysics*, 81: 133–159.

*Unksov, V.A. (1981). *Plate Tectonics.* Nedra, Leningrad, 288 pp.

Unrug (1997). Rodinia to Gondwana: the Geodynamic map of Gondwana super-continent assembly. *GSA Today*, 7: 1–6.

Urabe, T., Yuasa, M., Nadao, S. et al. (1987). Hydrothermal sulphides from a submarine caldera in the Sichito-Jwajima ridge, NW Pacific. *Mar. Geol.*, 74(3–4): 295–299.

Vail, P.R., Mitchum, R.M. and Thompson, S. (1987). Seismic stratigraphy and global changes of sea level. In: *Seismic Stratigraphy. Applications to Hydrocarbon Exploration.* Amer. Assoc. Petrol. Geol. Mem., 26: 89–97.

*Valizer, P.M. and Lennykh, V.I. (1980). *Blueschist Amphiboles of the Urals.* Nauka, Moscow, 211 pp.

Van Bergen, M.J., Vroon, P.Z. and Hoogewerff, J.A. (1993). Geochemical and tectonic relationships in the east Indonesian arc-continent collision region: implications for the subduction of the Australian passive margin. *Tectonophysics*, 223: 97–116.

Van der Hilst, R. and Seno, T. (1993). Effect of relative plate motion on the deep structure and penetration depth of slabs below the Izu-Bonin and Mariana island arcs. *Earth Plan. Sci. Lett.*, 120: 395–407.

Van der Wahl, D. and Vissers, R.L.M. (1993). Uplift and emplacement of upper mantle rocks in the western Mediterranean. *Geology*, 21: 1119–1122.

*Vargaftik, N.B. (1972). *Handbook of Thermophysical Properties of Gases and Liquids.* Nauka, Moscow, 720 pp.

Vejbaek, O.V. (1989). Effect of asthenospheric heat flow in basin modelling exemplified by the Danish basin. *Earth Plan. Sci. Lett.*, 95: 97–114.

*Vernikovskii, V.A. (1995). Formation characteristics of metamorphic complexes of northern Taimyr in the Riphean and Palaeozoic. *Petrologiya*, 1: 64–83.

*Vernikovskii, V.A., Neimark, L.A., Ponomarchuk, V.A., Vernikovskaya, A.E., Kireev, A.D. and Kusmin, D.S. (1995). Geochemistry and age of collision granites and metamorphic rocks of the Kara microcontinent (North Taimyr). *Russ. Geol., Geophys.*, 12: 50–65.

Vine, F.J. and Matthews, D.H. (1963). Magnetic anomalies over oceanic ridges. *Nature*, 199: 947–949.

*Vityazev, A.V., Pechernikov, G.V. and Safronov, V.S. (1990). *Planets of the Earth Group. Origin and Early Evolution.* Nauka, Moscow, 296 pp.

*Vladimirov, A.G., Malykh, M.M., Dronov, V.I. et al. (1992). *Indosinian Magmatism and Geodynamics of Southern Pamir.* OIGGM SO RAN, 228 pp.

*Vysotskii, S.V. (1989). *Ophiolite Associations of Island Arc Systems of the Pacific Ocean.* Vladivostok, 196 pp.

Waissel, J.K., Hayes, D.E. and Herron, E.M. (1977). Plate tectonics synthesis: the displacements between Australia, New Zealand and Antarctica since the Late Cretaceous. *Mar. Geol.,* 25: 231–277.

Wang, Y.B., Weidner, D.J. and Guyot, F. (1996). Thermal equation of the state of $CaSiO_3$-perovskite. *J. Geophys. Res., Solid Earth,* 101: 661–662.

Warren, R.G. and Ellis, D.J. (1989). Mantle origin of fluid and genesis of basalt magmas in subduction zones. *J. Geophys. Res.,* 94: 4697–4707.

Weinberg, R.F. and Podladchikov, Y. (1994). Diapiric ascent through lower crust and mantle. *J. Geophys. Res.,* 99: 9543–9559.

Wentworth, C.W., Blake, M.C., Jones, D.L. et al. (1984). Tectonic wedging associated with emplacement of the Franciscan assemblage, California Coast Ranges. In: *Franciscan Geology of North California* (M.C. Blake, ed.). Field Trip Guidebook 43: 163–173.

Wernicke, B. (1985). Uniform-sense simple shear of the continental lithosphere. *Can. J. Earth Sci.,* 27: 108–125.

Wessel, P., Bercovici, D. and Kroenke, L.W. (1994). The possible reflection of mantle discontinuities in Pacific geoid and bathymetry. *Geophys. Res. Lett.,* 21: 1943–1946.

Whitehead, J.A. and Luther, D.S. (1975). Dynamics of laboratory diapir and plume model. *J. Geophys. Res.,* 80: 707–717.

Whittington, A.C. (1996). Exhumation overrated at Nanga Parbat, North Pakistan. *Tectonophysics,* 206: 216–226.

Wijbraun, J.R. and Van Hare (1974). Wave velocities in granulite facies rocks and eclogites to 10 kbar. *J. Geophys. Res.,* 79: 5427–5466.

Wijbraun, J.R., Van Wees, J.D., Stephenson, R.A. and Cloetingh, S. (1993). Pressure-temperature-time evolution of high-pressure metamorphism. *Geology,* 21: 443–446.

Willet, S., Beaumont, Ch. and Fukksack, Ph. (1993). Mechanical model for the tectonics of doubly vergent compressional orogens. *Geology,* 21: 371–374.

Willis, G.E. and Deardorff, J.W. (1967). Development of short-period temperature fluctuations in thermal convection. *Physics of Fluids,* 10(5): 931–937.

Wilson, J.T. (1965). Evidence from oceanic islands suggesting movement in the Earth. *Phil. Trans. Royal Soc. London,* Ser. A, 285: 145–146.

Wilson, J.T. (1923). Mantle plumes and plate motions. *Tectonophysics,* 19: 149–164.

Wilson, J.T. (1990). Continental drift and a theory of convection. *Terra Nova,* 2(6): 519–538.

Windley, B.F. and Allen, M.B. (1993). Mongolian plateau: evidence for a Late Cenozoic mantle plume under central Asia. *Geology,* 21: 295–298.

Wood, D.J., Stowell, H.H., Onstott, T.C. and Holister, L.S. (1992). ^{40}Ar-^{39}Ar constraints on the emplacement, uplift and cooling of the Coast Plutonic Complex Sill, SE Alaska. *Geol. Soc. Amer. Bull.,* 103: 849–860.

Woon-Shing Yeung (1989). Engineering analysis of heat transfer during melting in vertical rectangular enclosures. *Intern. J. Heat Mass Transfer,* 32(4): 689–696.

Wu, Y.K., Prud Homme and Hung Nguen (1989). Etude numerique de la fusion author d'un cylindre vertical soumis à deux tupes de conditions limites. *Intern. J. Heat Mass Transfer*, 32(10): 1927–1938.

Wyllie, D.J., Abelson, Ph.H., Adams, S.S. et al. (1993). *Solid-Earth Sciences and Society. Summary and Global Overview*. Nat. Acad. Press, Washington, 46 pp.

Wysession, M.E. (1995). Seismic images of the core-mantle boundary. *GSA Today*, 5: 238–257.

Wysession, M.E., Valenzuela, R.W., Bartko, I. and Zhu, A.N. (1995). Investigating the base of the mantle using differential travel times. *Phys. Earth Planet Interior*, vol. 39.

Yardley, B.W.D., Barber, J.P. and Gray, J.R. (1987). The metamorphism of the Dalradian rocks of western Ireland and its relation to tectonic setting. *Phil. Trans. Roy. Soc. London*, Ser. A, 321: 243–270.

*Yarmolyuk, V.V. and Kovalenko, V.I. (1995). Late Mesozoic-Cenozoic intraplate magmatism of central and eastern Asia. *Geol. i Geofiz.*, 36(8): 132–141.

Young, G.M. (1995). Are Neoproterozoic glacial deposits preserved on the margins of Laurentia related to the fragmentation of two supercontinents? *Geology*, 23: 153–156.

Yuen, D.A., Reuteler, D.M., Balachandar, S., Steinbach, V., Malevsky, A.V. and Smedsond, J.L. (1994). Various influences on three-dimensional mantle convection with phase transition. *Phys. Earth Planet Interior*, vol. 83.

*Zaikov V.V. and Zaikova, E.V. (1994). *Genesis of Minerals and Metallic Formations of Oceanic Rifts*. Inst. Mineralogy URO RAN, Miass, 188 pp.

Zen E-an (1988). Thermal modelling of stepwise anatexis in a thrust-thickened sialic crust. *Trans. Royal Soc. Edinburgh, Earth Sci.*, 79: 23–25.

Zerr, A. and Boehler, R. (1993). Melting of $MgSiO_3$-perovskite to 62.5 kbar: indication of a high melting temperature in the lower mantle. *Science*, 262: 535–555.

Zhang, S. and Yuen, D.A. (1995). The influences of lower mantle viscosity stratification on 3D spherical-shell mantle convection. *Earth Plan. Sci. Lett.*, 132: 157–166.

Zhao, W., Yuen, D.A. and Honda, S. (1992). Multiple phase transitions and the style of mantle convection. *Phys. Earth Planet Interior*, 72: 185–210.

*Zharkov, V.N. (1983). *Internal Structure of the Earth and Planets*. Nauka, Moscow. 415 pp.

Zoback, M.L. and Zoback, M.D. (1989). Global pattern of tectonic stress. *Nature*, 341: 291–298.

Zoetemeijer, R., Cloetingh, S., Sassi, W. and Roure, F. (1993). Modelling of stratigraphic sequences in piggy-back basins: record of tectonic evolution. *Tectonophysics*, 226: 253–269.

Zoetemeijer, R., Desegaulx, P., Cloetingh, S. et al. (1990). Lithospheric dynamics and the tectono-stratigraphic evolution of the Ebro basin. *J. Geophys. Res.*, 95: 2701–2711.

*Zonenshain, L.P. and Gorodnitskii, A.M. (1978). Palaeozoic and Mesozoic reconstructions of continents and oceans. *Geotektonika*, 2: 3–25.

*Zonenshain, L.P. and Savostin, L.L. (1979). *Introduction to Geodynamics*. Nedra, Moscow, 311 pp.

*Zonenshain, L.P. and Kuz'min, M.I. (1983). Intraplate magmatism and its importance for understanding the processes in the Earth's mantle. *Geotektonika*, 1: 28–45.

*Zonenshain, L.P. and Kuz'min, M.I. (1993a). *Palaeogeodynamics*. Nauka, Moscow, 192 pp.

*Zonenshain, L.P. and Kuz'min, M.I. (1993b). Deep-level geodynamics of the Earth. *Geol. i. Geofiz.*, 4: 3–13.

*Zonenshain, L.P., Kuz'min, M.I. and Moralev, V.M. (1976). *Global Tectonics, Magmatism and Metallogeny*. Nedra, Moscow, 231 pp.

Zonenshain, L.P., Kuz'min, M.I. and Kononov, M.V. (1985). Absolute reconstruction of the Palaeozoic oceans. *Earth Plan. Sci Lett.*, 74: 103–116.

*Zonenshain, L.P., Kuz'min, M.I. and Kononov, M.V. (1987). Absolute reconstructions of the position of continents in the Palaeozoic and Early Mesozoic. *Geotektonika*, 3: 16–27.

*Zonenshain, L.P., Kuz'min, M.I. and Natapov, L.N. (1990). *Tectonics of Lithospheric Plates of the USSR Territory*. Nedra, Moscow, 1: 327 pp; 2: 234 pp.

Zonenshain, L.P., Kuz'min, M.I. and Bocharova, N.Yu. (1991). Hot-field tectonics. *Tectonophysics*, 199: 165–192.

*Zonenshain, L.P., Kuz'min, M.I, Bogdanov, Yu.A. et al. (1989a). Geology of Juan de Fuca range in the region of Axial town (Pacific Ocean). *Tikhookean. Geol.*, 1: 11–23.

*Zonenshain, L.P., Kuz'min, M.I., Lisitsyn, A.P. et al. (1989b). Tectonics of Mid-Atlantic ridge rift valley between regions TAG and MARK (26–24°N. Lat.). *Geotektonika*, 4: 99–113.

Zonenshain, L.P., Kuz'min, M.I., Lisitsyn, A.P. et al. (1989c). Tectonics of the Mid-Atlantic rift valley between TAG and MARK areas: evidence of vertical tectonism. *Tectonophysics*, 159: 1–23.

Printed in India